EXPERIMENTAL HEMATOLOGY TODAY

EXPERIMENTAL HEMATOLOGY TODAY

edited by

SIEGMUND J. BAUM
G. DAVID LEDNEY

With 147 Illustrations

SPRINGER-VERLAG
NEW YORK HEIDELBERG BERLIN

Siegmund J. Baum
Chairman, Experimental Hematology Department
Armed Forces Radiobiology Research Institute
Defense Nuclear Agency
Bethesda, Maryland 20014

G. David Ledney
Head, Immunology Division
Experimental Hematology Department
Armed Forces Radiobiology Research Institute
Defense Nuclear Agency
Bethesda, Maryland 20014

This book contains selected papers presented at the Egon Lorenz Memorial
Symposia, Fifth Annual Meeting, International Society for Experimental
Hematology, August 17–20, 1976, Washington, D.C., U.S.A.

Designer: Robert Bull

Library of Congress Cataloging in Publication Data

Egon Lorenz Memorial Symposia, Washington, D.C., 1976.
    Experimental hematology today.

    Bibliography: p.
    Includes index.
    1. Hematology—Congresses. I. Baum, Siegmund J. II. Ledney, G.
David. III. International Society for Experimental Hematology.
IV. Title. [DNLM: 1. Hematology—Congresses. WH100 E96 1976]
QP91.E35 1976   612.11   77-8668

9 8 7 6 5 4 3 2 1

ISBN-13: 978-1-4612-9897-7   e-ISBN-13: 978-1-4612-9895-3
DOI: 10.1007/978-1-4612-9895-3

Preface

Experimental Hematology Today is divided into six sections which are related to the presentations given at the Egon Lorenz Memorial Symposia presented as part of the Fifth Annual Meeting of the International Society for Experimental Hematology held in Washington, D.C. in 1976. The first division deals with papers from Symposium I chaired by Dr. Dirk W. van Bekkum, Rijswijk, Netherlands. Its topic is "Characterization of the Multipotential Stem Cell (CFU-s)," and represents a report of the most recent attempts of separation and description of the multipotential hematopoietic stem cell. The papers presented in Symposium II entitled "Humoral and Cellular Control Agents" and chaired by Dr. Eugene P. Cronkite, Brookhaven, New York, discuss the latest findings of the influence of specific factors on hematopoietic cell production. The third division is comprised of presentations from Symposium III entitled "Physiology of Committed Stem Cells (CFU-e and CFU-m)." It was chaired by Dr. James E. Till, Toronto, Canada. The papers of this symposium discuss the function of the erythrocytic and megakaryocytic committed stem cells. The papers from Symposium IV entitled "Physiology of Committed Stem Cells (CFU-c)," chaired by Dr. Siegmund J. Baum, Bethesda, Maryland, are presented in the fourth division. This division deals with the latest research findings of the functional interrelationship of the leukocytic committed stem cells. The topic of Symposium V was "Bone Marrow Transplantation Immunology" chaired by Dr. John J. Trentin, Houston, Texas. The papers of this symposium comprise the fifth division. They relate the most recent research results of the immunologic problems in the field of bone marrow transplantation. The sixth division presents the papers of Symposium VI entitled "Experimental Models of Clinical Conditions in Hematology" which was chaired by Dr. Georges Mathé, Villejuif, France. It deals primarily with the utilization of proper animal models for the understanding and amelioration of human diseases of hematopoietic origin.

In general, the book is comprised of research papers describing the most recent discoveries in hematology from cellular to animal models and possible applications to clinical situations. Therefore, we believe that the material discussed in nearly all the papers presented will be of great interest to investigators, teachers, and certainly, as well, to clinicians.

SIEGMUND J. BAUM
G. DAVID LEDNEY

Acknowledgment

The members of the organizing committee of the Fifth Annual Meeting of the International Society for Experimental Hematology wish to acknowledge the generous assistance provided by the following agencies of the United States government:

Defense Nuclear Agency, Armed Forces Radiobiology Research Institute, Energy Research and Development Administration, National Cancer Institute;

And by the following commercial firms: Searle, Burroughs Wellcome Co., and Travenol.

Contents

Contents

xv

List of Contributors

Siegmund J. Baum, Armed Forces Radiobiology Research Institute, Bethesda, Maryland.

Simon Bol, Radiobiological Institute TNO, 150 Lange Kleiweg, Rijswijk (ZH), The Netherlands

Mortimer M. Bortin, Winter Research Laboratory, Mount Sinai Medical Center, New York

A. Brouwer, Radiobiology, Erasmus University, Rotterdam.

A. W. Burgess, Walter and Eliza Hall Institute of Medical Research, Victoria, Australia.

H. Burlington, Mount Sinai School of Medicine of the City University of New York

W. Byrt, Department of Biology, Boston University, Boston, Massachusetts.

W. Calvo, University of Ulm, Germany

A. D. Chanana, Brookhaven National Laboratory, Long Island, New York

D. G. Colley, Department of Biological Resources, Roswell Park Memorial Institute, New York State Department of Health, 666 Elm Street, Buffalo, New York.

L. Colly, Radiobiological Institute TNO, 151 Lange Kleiweg, Rijswijk (ZH), The Netherlands.

Eugene Cronkite, Brookhaven National Laboratory, Long Island, New York.

Surjit K. Datta, Baylor College of Medicine, Houston, Texas

J. B. DeMello, Department of Biology, Boston University, Boston, Massachusetts.

P. Q. Eichacker, Department of Biology, Boston University, Boston, Massachusetts.

T. M. Fliedner, University of Ulm, Germany

I. Florentin, Institut de Cancérologie et d'Immunogénétique (INSERM), Hôpital Paul-Brousse, 94800-Villejuif, France

R. P. Gale, UCLA School of Medicine, Los Angeles, California.

Michael T. Gallagher, Baylor College of Medicine, Houston, Texas

Robert C. Gallo, National Cancer Institute, Bethesda, Maryland

R. I. Garver, Department of Biology, Boston University, Boston, Massachusetts.

M. J. Gilio, Department of Biology, Boston University, Boston, Massachusetts.

David W. Golde, UCLA School of Medicine, Los Angeles, California

E. Goldwasser, Department of Biochemistry and Franklin McLean Memorial Research Institute, University of Chicago, Chicago, Illinois.

M. Y. Gordon, Institute of Cancer Research, Belmont, Sutton, Surrey, England.

C. J. Gregory, British Columbia Cancer Foundation, Vancouver, Canada.

Ton Hagenbeek, Radiobiological Institute TNO, 151 Lange Kleiweg, Rijswijk (ZH), The Netherlands.

R. M. Henkelman, British Columbia Cancer Foundation, Vancouver, Canada.

M. C. James, University of Washington

D. D. Joel, Brookhaven National Laboratory, Long Island, New York

Rolf Kiessling, Karolinska Institute, Stockholm, Sweden

M. Körbling, University of Ulm, Germany.

Sulabha S. Kulkarni, Baylor College of Medicine, Houston, Texas

J. Kurland, Department of Development Hematopoiesis, Sloan-Kettering Institute for Cancer Research, New York, New York.

William P. LeFeber, Winter Research Laboratory, Mount Sinai Medical Center, New York

Thomas J. MacVittie, Armed Forces Radiobiology Research Institute, Bethesda, Maryland.

Georges Mathé, Institut de Cancérlogie et d'Immunogénétique, Villejuif, France.

K. F. McCarthy, Armed Forces Radiobiology Research Institute, Defense Nuclear Agency, Bethesda, Maryland.

M. P. McGarry, Department of Biological Resources, Roswell Park Memorial Institute, New York State Department of Health, 666 Elm Street, Buffalo, New York.

D. Metcalf, Walter and Eliza Hall Institute of Medical Research, Victoria, Australia.

A. M. Miller, Department of Biological Resources, Roswell Park Memorial Institute, New York State Department of Health, 666 Elm Street, Buffalo, New York.

F. C. Monette, Department of Biology, Boston University, Boston, Massachusetts.

M. A. S. Moore, Department of Development Hematopoiesis, Sloan-Kettering Institute for Cancer Research, New York, New York.

Dries Mulder, Radiobiological Institute TNO, 151 Lange Kleiweg, Rijswijk (ZH), The Netherlands.

A. Nakeff, Section of Cancer Biology, Mallinckrodt Institute of Radiobiology, Washington University, School of Medicine, St. Louis, Missouri.

W. Nothdurft, University of Ulm, Germany.

V. Ober-Kieftenburg, Radiobiological Institute TNO, Rijswijk and Department of Radiobiology, 151 Lange Kleiweg, Erasmus University, Rotterdam.

L. Olsson, Institut de Cancérologie et d'Immunogénétique (INSERM), Höpital Paul-Brousse, 94800-Villejuif, France.

M. F. Peters-Slough, Radiobiological Institute TNO, Erasmus University, Rotterdam

Janet M. D. Plate, Transplantation Unit, Massachusetts General Hospital, Boston, Massachusetts.

U. Reincke, Leukemia Society of America

William C. Rose, Medical College of Wisconsin, Milwaukee, Wisconsin

W. M. Ross, University of Ulm, Germany.

Francis W. Ruscetti, Cell Biology, Litton Bionetics, Bethesda, Maryland

C. Rutkowsky, UCLA School of Medicine, Los Angeles, California.

George W. Santos, The Johns Hopkins Oncology Center, Carnegie 332, Baltimore, Maryland.

Rainer Schwerdtfeger, The Johns Hopkins Oncology Center, Carnegie 332, Baltimore, Maryland.

J. W. Singer, The Division of Oncology, The University of Washington, The Seattle Veterans Administration Hospital, and the Fred Hutchinson Cancer Research Center, Seattle, Washington.

Richard E. Slavin, The Johns Hopkins Oncology Center, Carnegie 332, Baltimore, Maryland.

J. Stevens, University of Minnesota, St. Paul, Minnesota

E. D. Thomas, The Division of Oncology, The University of Washington, The Seattle Veterans Administration Hospital, and the Fred Hutchinson Cancer Research Center, Seattle, Washington.

J. E. Till, Ontario Cancer Institute, Toronto, Ontario, Canada

John Trentin, Baylor College of Medicine, Houston, Texas.

Robert L. Truitt, Leukemia Society of America

Peter J. Tutschka, The Johns Hopkins Oncology Center, Carnegie 332, Baltimore, Maryland.

Dirk van Bekkum, Radiological Institute TNO, 151 Lange Kleiweg, Rijswijk (ZH), The Netherlands.

Gerrit J. van den Engh, Radiobiological Institute, 151 Lange Kleiweg, Rijswijk, The Netherlands.

G. van Zant, Department of Biochemistry and Franklin McLean Memorial Research Institute, University of Chicago, Chicago, Illinois.

Jan W. M. Visser, Radiobiological Institute, 151 Lange Kleiweg, Rijswijk, The Netherlands.

G. Wagemaker, Radiobiological Institute TNO, Rijswijk and Department of Radiobiology, 151 Lange Kleiweg, Erasmus University, Rotterdam

T. L. Weatherly, Armed Forces Radiobiology Research Institute, Bethesda, Maryland

Hans Wigzell, Upsala University, Sweden

Neil Williams, Radiobiological Institute, 151 Lange Kleiweg, Rijswijk (ZH), The Netherlands.

A. M. Wu, Department of Cell Biology, Litton Bionetics Inc., Bethesda, Maryland.

PART I

Characterization of the Multipotential Stem Cell (CFU-s)

D. W. van Bekkum, M.D.

INTRODUCTION

It is perhaps not too much of an exaggeration to claim that experimental hematology as it flourishes today originated largely from the pioneering attempts to protect lethally radiated animals (l) by shielding of hemopoietic tissues by L. O. Jacobson (9), and (2) by treatment with bone marrow suspensions by E. Lorenz and his collaborators (12). The site chosen for this annual meeting of the International Society for Experimental Hematology is given a special historic significance by the fact that it was 25 years ago that the first publication on this subject by Lorenz appeared from his laboratory at the National Institutes of Health. Lorenz's discovery marked the beginning of a period which lasted until 1956, during which the protection afforded by hemopoietic cell suspensions was confirmed by many. This soon led to an intensive scientific debate on the mechanism of this protective effect: was it due to a humoral factor produced and provided by the bone marrow—as Lorenz postulated—or to transplantation and subsequent proliferation of hemopoietic cells? This question was definitively answered in 1956 by evidence from three different laboratories (7, 15, 26), which demonstrated the origin of the cells in the repopulated tissues using a variety of cellular and immunologic markers. By the same token, these contributions marked the birth of radiation chimeras. Not only did the first clinical bone marrow transplantations spring from those revolutionary discoveries, but they also led to the gradual development of the notion of the existence of a multipotential hemopoietic stem cell (HSC) in the bone marrow. It took more than a decade—until 1970—until irrevocable proof of the existence and key role of the multipotential HSC was provided. Among the major contributors to this important evolution in hematology were Till and McCulloch, with their introduction of the spleen colony assay (19); Trentin et al., who demonstrated the development of a complete hematologic system out of the progeny of a single spleen colony (20); and Peter Nowell, who provided conclusive evidence for the inclusion of the lymphoid cells in that progeny (16). From then until the present day, many experimental hematologists have tried to identify the HSC and to describe its morphologic and physical characteristics. Table 1 lists the various cells which have been proposed as candidates to meet the requirements of multipotential stem cells. It cannot be by chance that most authors focus on mononuclear and lymphocyte-like cells, since such structures

The Appearance of the Multipotential Hemopoietic Stem Cell

1

D. W. van Bekkum

3

TABLE 1 Various Descriptions of Pluripotential Hematopoietic Stem Cells (HSC) in the Mouse

REF.	AUTHORS	DESCRIPTION	METHOD
(27)	Yoffey (1957)	Lymphocyte-transitional cell in bone marrow	Morphology, kinetics
(6)	Caffrey-Tyler and Everett (1966)	"Monocytoid" cell	Radiation parabiosis, labeling
(18)	Niewisch et al. (1967)	"B-cell" (from spleen)	Density gradient Ficoll
(5)	Bennett and Cudkowicz (1968)	Bone marrow "lymphocyte" 20% in bone marrow	Glass wool columns
(12)	Haas et al. (1971)	Bone marrow "lymphocyte"	In situ labeling
(17)	Murphy et al. (1971)	Large mononuclear cell	Density gradient Ficoll Electron microscopy of spleen colonies
(2)	Van Bekkum et al. (1971)	Primitive round cell distinct from lymphocyte	Electron microscopy of highly concentrated fractions (density gradient albumin)
(22)	Rubinstein and Trobaugh (1973)	Resembling (2)	Cryopreservation resistance
(1)	Barr et al. (1975)	Pluripotential lymphocytes (from human blood)	Combination of purifications

are considered to be characteristic of the most primitive cell types. Caffrey-Tyler and Everett (3), as well as Haas et al. (8), traced their HSC by way of kinetic studies of radioactively labeled cells regenerating in the bone marrow after radiation. All others employed concentration techniques for their attempts at identification, except Rubinstein and Trobaugh (18), who exploited the relative resistance of mouse HSC, as compared to most other cells in the bone marrow, to cryopreservation and thawing. Morphologic identification of a hitherto unknown cell type is feasible only if that cell type presents a uniform morphologic appearance and if it is present in a concentration of at least 5% to 10% of the cell preparations under study. This means that the number of HSC in those preparations has to be quantitatively established by functional tests, i.e., the spleen colony assay in the case of mouse HSC. "To judge a cell type on morphological characteristics alone without considering its function, origin and relationship to other cells, seems inadequate and adds confusion to an already complicated problem" (Caffrey-Tyler and Everett (3). This requirement was only met by Niewisch et al. (14) and by our own group (21). The starting material employed by Niewisch et al. was a spleen cell suspension which has a low HSC content (4 per 10,000). This was the reason why their concentration procedure did not yield a sufficiently high proportion of HSC for conclusive morphologic recognition. We started with mouse bone marrow (6 HSC per 1000 cells) and reached concentration factors of 10–30 by using discontinuous albumin density gradients combined with *in vivo* enrichment through pre-

treatment of the mice with vinblastine and mustine. The resulting cell populations had a calculated HSC content of 5% to 29% and were found to contain proportional numbers of relatively small mononuclear cells which could be clearly distinguished from lymphoid cells with the electron microscope. This candidate stem cell has been described in detail (21). Physical characterization of the HSC in mouse bone marrow by van den Engh (24) and Visser (25) has also revealed a homogenous cell population with a size range (7–9.2 μm diameter) which corresponds closely with that measured in electron microscopic preparations (7–10 μm in diameter).

By further density gradient centrifugation of monkey and human cells capable of producing colonies in soft agar culture system, cells closely resembling the murine candidate stem cells were identified (5). There is good evidence that this cell (the CFU-c) is a close descendant of the HSC, so that CFU-c determinations can serve to monitor concentrations of HSC. However, a direct estimate of HSC in primate bone marrow cannot be made, so that these experiments only support the likelihood of primate HSC having the same morphologic characteristics as mouse HSC. It was, therefore, of interest to pursue investigations along similar lines with bone marrow cells of the rat, which is the only other species for which a direct estimate of the HSC is, so far, technically feasible. The opportunity for identifying the HSC in rats arose from our studies on the fate of HSC in rat bone marrow in the course of the development of a transplantable myeloid leukemia (22). This study involved the separation of bone marrow cell populations on albumin density gradients, which led to an unexpectedly high concentration of normal HSC in the upper gradient factions using a one-step separation procedure.

THE SPLEEN COLONY ASSAY OF RAT HSC

The spleen colony assay of Till and McCulloch (19) has been modified for the rat by Comas and Byrd (4) by exposing recipients to lethal wholebody γ-radiation at a low dose rate of approximately 50 rad/h. Lahiri (10) subsequently compared this method to radiation of the recipients (WAG/Rij rats aged 5–7 weeks) with 550 rad of x-rays at a dose rate of 2500 rad/hr. and found similar CFU-s growth curves for the two groups, except for the 2 hr seeding efficiency factor (f) being 4.5% for the low dose rate γ-radiation group as compared to 13% for the x-ray-treated animals. Dunn (6) reported on a spleen colony assay in 6–8-week-old Wilkie Hooded rats conditioned with busulfan and aminochlorambucil, and calculated an f of 17% at 2 hr for bone marrow-derived CFU-s.

For the routine monitoring of rat HSC, the rat spleen colony assay is rather expensive. However, as shown by Rauchwerger et al. (17), rat bone marrow cells do produce spleen colonies in radiated mice, but the colony yield differs from one mouse strain to another. We have explored the possibilities of assaying rat bone marrow HSC in radiated rats and in mice of different strain combinations which were conditioned by high-dose, wholebody radiation. X-rays, 300 kV, 10 mA, 34 rad/min, total dose 850 rad, were used for the rats, and γ-radiation (^{137}Cs, 122 rad/min, total dose 1050 rad) for the recipient mice. Rats were 12–20 weeks old and weighed 250–350 g. Donors and recipients were always of the same sex. A linear relationship between number of bone marrow cells injected and number of spleen colonies scored was found previously between 2×10^5 and 10^6 cells (3). Mice were (CBA $\times$ C57BL) F_1 hybrids aged 11–15 weeks. The sex of these animals did not influence the results.

For the determination of f in rats, the primary recipients were injected i.v. with 2×10^8 or 4×10^7 bone marrow cells, and killed at intervals from 2 hr to 12 days later. For each point two cell doses were injected into groups of 12 rats each. The spleen colonies were counted 11–12 days after injection and fixation in Tellyesniczky fluid. Radiation mortality occurred in certain experiments before that time so that some estimates were based on only 4–6 spleens per cell dose.

To determine f of rat bone marrow in

mice, the primary mouse recipients received 4×10^7 rat bone marrow cells and were killed at similar intervals. The spleen cells of the primary recipients were assayed in the usual way for the total CFU-s content. Spleen colonies were counted at 9 days after cell injection. The seeding efficiency factor, f, for each was calculated according to the method of Lahiri et al. (11) by extrapolation of the linear part of the CFU-s growth curve to the interception with the ordinate (zero time). In Fig. 1 the bone marrow CFU-s recovery curves in Sprague-Dawley spleens are depicted. The extrapolated f values varied between 0.006 and 0.007, although the 2 hr values show a much wider variation. In order to avoid the uncertainties of extrapolating such recovery curves, an experiment was performed using primary recipients which had been radiated 3 days before being given the bone marrow cells. By this procedure the dip in the curve at 24–48 hr is completely avoided and the interception with the ordinate is more accurately obtained. Figure 2 shows that very good linearity (on semi-log scale) was obtained until day 10, and that an f of 0.008 agrees very well with that from the previous experiments. In the same experiment a gradient fraction with a 24-fold higher CFU-s

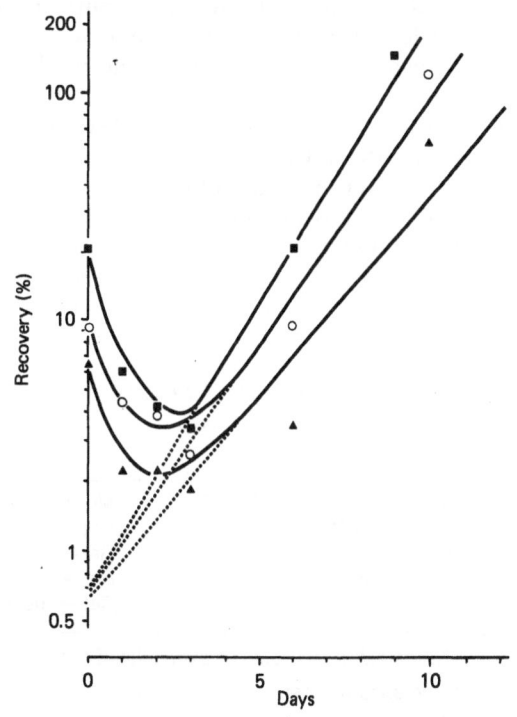

FIGURE 1. Recovery curves of Sprague-Dawley rat bone marrow CFU-s in rat spleen: ○ ♀♀ primary 2 x 10⁸ b.m.; ■ ♀♀ primary 1 x 10⁸ b.m.; ▲ ♂♂ primary 1 × 10⁸ b.m.

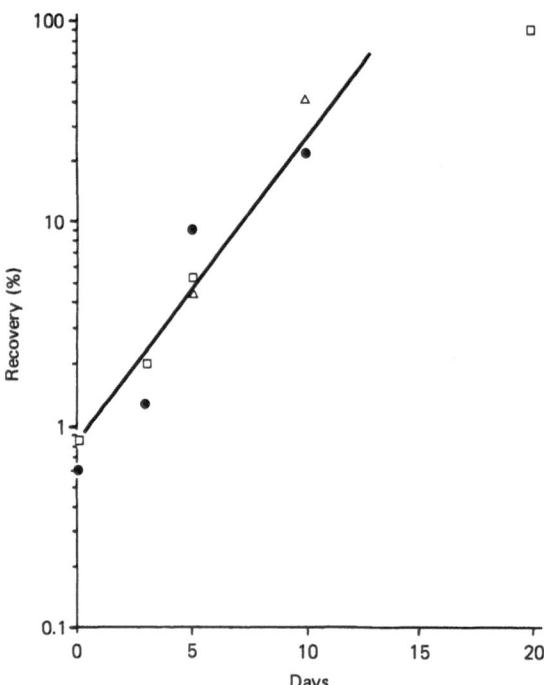

FIGURE 2. Recovery curves of Sprague-Dawley rat bone marrow CFU-s in rat spleen. Primary recipients were injected with bone marrow cells at 3 days after 850 rad X-radiation. □ primary 2×10^8 b.m.; ▲ primary 4×10^7 b.m.; ● primary 12×10^6 fr.3 ($\times$ 24).

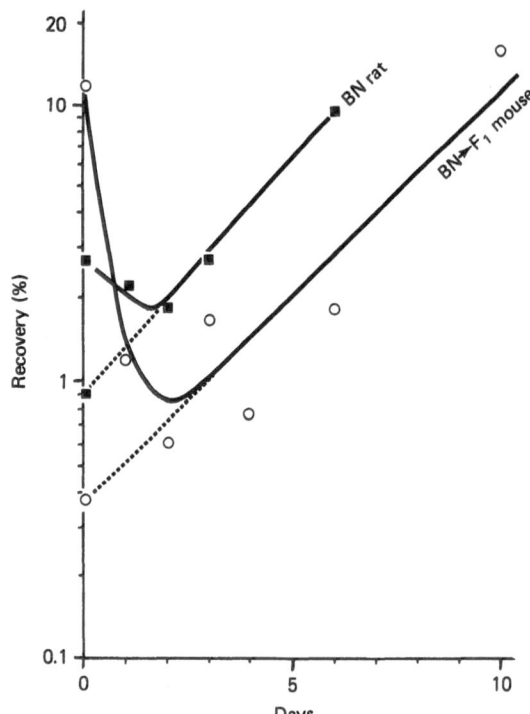

FIGURE 3. Recovery curves of rat bone marrow CFU-s in rat and mouse spleen. Primary recipients were injected with bone marrow cells on the day of radiation. ■ Rat, primary 2×10^8 b.m.; ○ mouse (F_1), primary 4×10^7 b.m.

content than normal rat bone marrow was studied, and the resulting recovery rate of these cells as well as f did not deviate significantly from the values for unfractionated bone marrow.

To enable us to interpret the spleen colony counts observed in *mice* following injection of rat bone marrow or of concentrated rat HSC, recovery curves of BN rat CFU-s in both rat spleen and mouse spleen were compared by transferring the spleen cells, at intervals, into secondary rat recipients and secondary mouse recipients, respectively. The results (Fig.3) show no differences of f (0.009) in BN rats as compared with that of the

Sprague-Dawley rats but f in the mouse (0.004) seemed to be somewhat lower. However, all the f's (seeding efficiency factors) for rat bone marrow determined so far in both rats and mice lie in the range 0.004 to 0.009, so that for practical purposes it seems justified to use an average f of 0.007. This factor is considerably *lower* than that determined for mouse bone marrow CFU-s (0.05). The present experiments show that this difference cannot be due to different properties of the recipients, but has to be attributed to differential homing characteristics between rat and mouse HSC. The lower factors for rat CFU-s can also be derived

TABLE 2 Estimates of f from CFU-s Assays

		DIP	CORRECTED[a]
Lahiri et al., 1970 (15)[b]	Mouse → mouse	0.09	0.05
Lahiri, 1969 (14)[b]	Rat → rat	0.008	0.002
Dunn, 1973 (9)[b]	Rat → rat	0.06	0.01
Present experiment	Rat → rat	0.02–0.04(4)[c]	0.006–0.009(6)[c]
	Rat → mouse	—	0.004

[a]Preirradiated recipients or extrapolated to 0 hr.

[b]Reference number.

[c]Number of experiments.

from the data of Lahiri (0.002) and from that of Dunn (0.010) (Table 2).

The number of HSC in rat bone marrow can now be calculated using the average of values obtained in rats and mice (0.007), and the average number of spleen colonies scored/10⁶ rat marrow cells (Table 3). This calculation yields 0.7% HSC for the rat as compared to 0.6% for the mouse bone marrow. The close similarity of HSC content in mouse and rat bone marrow implies that the number of isogeneic bone marrow cells needed for 50% protection of lethally radiated animals should be similar on a kg body weight basis for the two species. This has indeed been found to be so: 5×10^4 for a 20 g mouse and 10^6 for a 300–350 g rat. Previous reports (23) indicated that 10 times more bone marrow cells were needed to provide 100% 30-day survival of lethally radiated rats as compared to mice. The fact that protection of rats is presently being obtained with less bone marrow cells may be due to the use of SPF animals, which were not available at the time of the earlier experiments.

CONCENTRATION AND IDENTIFICATION OF RAT HSC

Discontinuous BSA gradients were prepared as described by Dicke et al. (5), but employing 1% steps between 19% and 24% albumin concentration. The osmolarity of the 35% BSA stock solution was 270 mOsm; tubes of 0.8 cm diameter were used and spun for 30 min at 2000 g after equilibration at 10°C. Samples of each layer were washed in Hanks's medium and appropriate numbers of nucleated cells were injected into

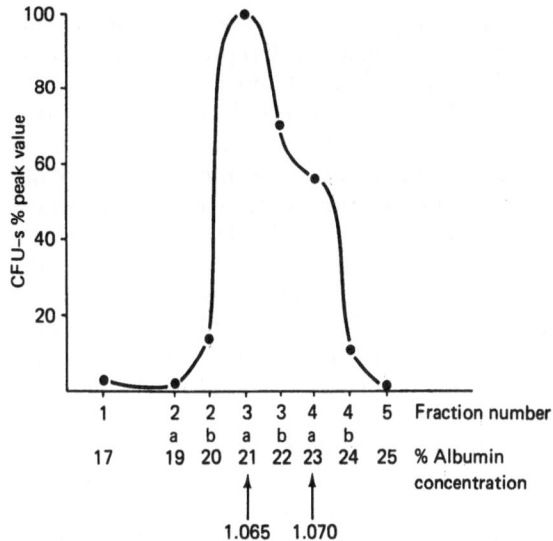

FIGURE 4. Distribution of rat bone marrow CFU-s on a discontinuous bovine serum albumin density gradient. Figures under arrows represent absolute densities in g/cm³. The rat marrow CFU-s assay was done with mice as was previously discussed.

γ-radiated mice for the determination of spleen colonies. Figure 4 shows that the CFU-s are predominantly found in fractions 2ᵇ, 3ᵃ, and 3ᶜ, that is, between densities 1.065 and 1.068. The great majority of nucleated cells settle at density 1.070 and over, which results in a high concentration of

FIGURE 5. Distribution of CFU-s and total nucleated cells of BN rat bone marrow on discontinuous density gradients. Average of 9 separations. Figures inside the graph indicate enrichment of CFU-s in each fraction.

TABLE 3 Calculated Hematopoietic Stem Cell (HSC) Incidence in Rat Bone Marrow and Restorative Capacity of Rat Bone Marrow Compared to Previously Established Values for Mouse Bone Marrow

F	SPLEEN COL./10⁶ B.M.
Rat → rat | 0.004–0.009	BN rat → rat | 40 / 58
Rat → mouse | av. 0.007	BN rat → mouse | av. 50

$$\text{HSC:} \quad \frac{1000}{7} \times 50 = 7000/10^6 b1 = b1\ 0.7\% b1$$

(mouse = 0.6%)

Number of isogeneic bone marrow cells for 50% radiation protection.

rat 3×10^6/kg
mouse 2.5×10^6/kg

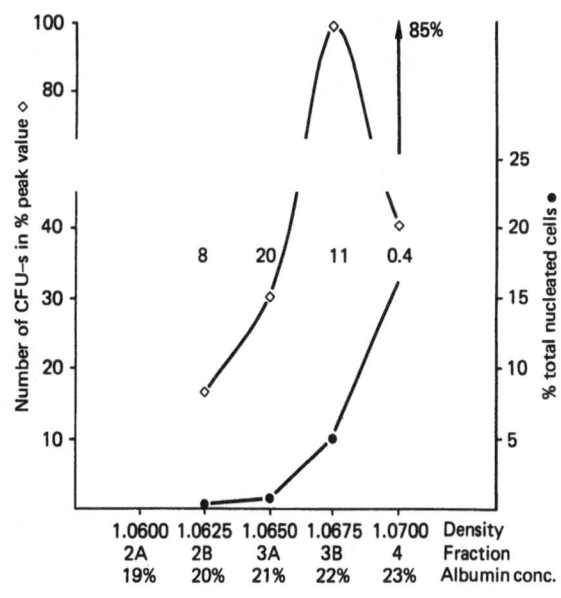

HSC in the lighter fractions of the gradient. This is depicted in Fig. 5, which represents the average distributions of nine different experiments. In fraction 3A an average HSC enrichment of 20 × was obtained, with a maximum of 28 ×. From a number of these HSC concentrates electron microscopic preparations were made and inspected for the presence of Cells Meeting Our Morphological Criteria for stem cells (CMOMC's), as described previously for the mouse. Such cells were indeed easily detected (Fig. 6), so that counts could be obtained from the electron microscopic preparations. In each sample at least 100 recognizable nucleated cells were inspected. The results (Table 4) indicate a very good agreement between the mean HSC values calculated and the mean numbers of CMOMC's scored. However, the individual results show some important deviations, in particular in the samples which were

TABLE 4 Electron Microscopic Analysis of Stem Cell Concentrates of BN Rat Bone Marrow

CONCENTRATION FACTOR	CALCULATED HSC (%)[a]	CMOMC COUNTED (%)
28	23	6
27	16	9
26	14	10
23	22	10
14	10	13
11	6	11
10	7	7
7	6	6
6	3	8
3	2	5
	$\overline{11}$	$\overline{8.5}$

[a]f, 0.007, rat → mouse spleen colony assay.

FIGURE 6. Electron microscopic picture of a candidate stem cell in fraction 3 of density albumin gradient of rat bone marrow. Magnification: 21,000.

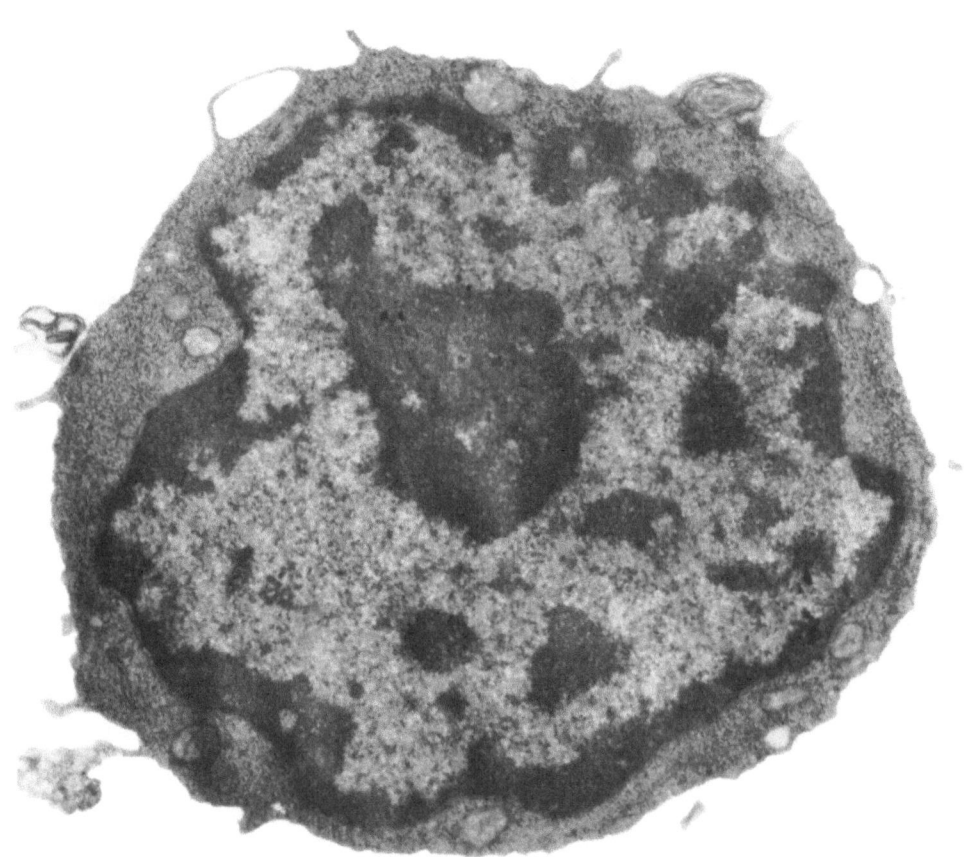

——— μm ———

calculated to contain over 10% HSC. Since all the HSC concentrates were collected from the bone marrow of normal rats of the same strain, it does not seem likely that we are dealing with different cell populations in the highly enriched fractions as compared to the less concentrated fractions. For the present, we speculate that some kind of systematic technical error is responsible for the underestimation of CMOMC counts in the highly enriched fractions. Actually, our scores of mouse CMOMC's also yielded a notable discrepancy between CMOMC counts and calculated CFU-S values in the highly concentrated stem cell fractions (5). This we thought to be due to the facts that only those cells were counted which were clearly recognized as belonging to a certain cell series, and that the proportion of poorly recognizable cells was relatively higher in the preparations containing high concentrations of stem cells. One other possibility is that our CMOMC's represent mainly G_0 cells, and that the HSC in cycle have a different appearance. Considering the slight differences in density of these two HSC forms (25), the highly concentrated fractions of CFU-s may have contained relatively more HSC in cycle, which were not scored with the electron microscopic technique. Further investigations have to be directed at determining the cause of the underestimations of CMOMC numbers in the highest concentrated CFU-s fractions of rat bone marrow.

SUMMARY

The development of the concept of the existence of the pluripotential HSC is largely based on the discoveries made 20–25 years ago to protect lethally radiated animals with hemopoietic cell suspensions. The functional description of the pluripotential HSC, and the discovery of a methodology to enumerate them, has permitted the development of techniques for concentrating HSC. This led to the identification of a morphologic entity in mouse bone marrow which is a realistic candidate for the HSC. Cells with similar morphologic characteristics were subsequently observed in enriched fractions of monkey and human bone marrow.

Experimental results are presented here for enumerations in rat bone marrow using spleen colony assays in rats and mice. For both assays the seeding efficiency factors, f, have been determined, so that the HSC content of various rat HSC concentrates could be calculated. The HSC content in rat bone marrow is estimated to be 0.7%. Electron microscopic inspections of HSC concentrates obtained by discontinuous albumin gradient centrifugation revealed cells with similar morphologic characteristics as previously described for mouse bone marrow HSC. The numbers of these cells counted agreed satisfactorily with the calculated numbers of HSC, except for the most enriched fractions, where underestimates were obtained with the morphologic method. This discrepancy may be due to a not-yet-identified systematic error, or to different morphologic characteristics of G_0 and cycling HSC.

ACKNOWLEDGMENTS

The expert technical assistance of Miss P.H.M. Scheffers and Mrs. S. Knaan-Shanzer is gratefully acknowledged. This research was funded by the Foundation for Medical Research (FUNGO) and the Foundation "Koningin Wilhelmina Fonds" (Dutch National Cancer League).

REFERENCES

(1) Barr, R. D., Whang-Peng, J., and Perry, S. Hemopoietic stem cells in human peripheral blood. *Science*, *190*:284, 1975.

(2) Bennett, M., and Cudkowicz, G. Hemopoietic progenitor cells with limited potential for differentiation: Erythropoietic function of mouse marrow "lymphocytes". *J. Cell. Physiol.*, *72*:129, 1968.

(3) Caffrey-Tyler, R. W., and Everett, N. B. A radioautographic study of hemopoietic repopu-

lation using irradiated parabiotic rats. Relation to the stem cell problem. *Blood, 28*:873, 1966.

(4) Comas, F. V., and Byrd, B. L. Hemopoietic colonies in the rat. *Radiat. Res., 32*:355, 1967.

(5) Dicke, K. A., van Noord, M. J., Maat, B., Schaefer, U. W., and van Bekkum, D. W. Attempts at morphological identification of the hemopoietic stem cell in primates and rodents. In Wolstenholme, G. E. W., and O'Connor, M., eds., *Haemopoietic Stem Cells: Ciba Foundation Symposium 13 (New Series)*. Amsterdam: Elsevier, p. 47, 1973.

(6) Dunn, C. D. R. The proliferative capacity of haemopoietic colony forming units in the rat. *Cell Tissue Kinet., 6*:55, 1973.

(7) Ford, C. E., Hamerton, J. L., Barnes, D. W. H., and Loutit, J. F. Cytological identification of radiation chimaeras. *Nature (London), 177*:452, 1956.

(8) Haas, R. J., Bohne, F., and Fliedner, T. M. Cytokinetic analysis of slowly proliferating bone marrow cells during recovery from radiation injury. *Cell Tissue Kinet., 4*:31, 1971.

(9) Jacobson, L. O., Marks, E. K., Robson, M. J., and Gaston, E. O. The role of the spleen in radiation injury. *Proc. Soc. Exp. Biol. Med., 70*:740, 1949.

(10) Lahiri, S. K. Personal communication.

(11) Lahiri, S. K., Keizer, H. J., and van Putten, L. M. The efficiency of the assay for haemopoietic colony forming cells. *Cell Tissue Kinet., 3*:355, 1970.

(12) Lorenz, E., Uphoff, D. E., Reid, T. R., and Shelton, E. Modification of irradiation injury in mice and guinea pigs by bone marrow injections. *J. Nat. Cancer Inst., 12*:197, 1951.

(13) Murphy, M. J., Jr., Bertles, J. F., and Gordon, A. S. Identifying characteristics of the haematopoietic precursor cell. *J. Cell Sci., 9*:23, 1971.

(14) Niewisch, H., Vogel, H., and Matioli, G. Concentration, quantitation, and identification of hemopoietic stem cells. *Proc. Nat. Acad. Sci. U.S.A., 58*:2261, 1967.

(15) Nowell, P. C., Cole, L. J., Habermeyer, J. G., and Roan, P. L. Growth and continued function of rat marrow cells in X-irradiated mice. *Cancer Res., 16*:258, 1956.

(16) Nowell, P. C., Hirsch, B. E., Fox, D. H., and Wilson, D. B. Evidence for the existence of multipoten-

tial lymphohematopoietic stem cells in the adult rat. *J. Cell. Physiol., 75*:151, 1970.

(17) Rauchwerger, J. M., Gallagher, M. T., and Trentin, J. J. "Xenogeneic Resistance" to rat bone marrow transplantation. I. The basic phenomenon. *Proc. Soc. Exp. Biol. Med., 143*:145, 1973.

(18) Rubinstein, A. S., and Trobaugh, F. E., Jr. Ultrastructure of presumptive hematopoietic stem cells. *Blood, 42*:61, 1973.

(19) Till, J. E., and McCulloch, E. A. Direct measurement of radiation sensitivity of normal bone marrow cells. *Radiat. Res., 14*:213, 1961.

(20) Trentin, J. J., and Fahlberg, W. J. An experimental model for studies of immunologic competence in irradiated mice repopulated with "clones" of spleen cells. In *Conceptual Advances in Immunology and Oncology: Sixteenth Annual Symposium, 1962*. New York: Harper & Row (Hoeber Medical Division), p. 66, 1963.

(21) van Bekkum, D. W., van Noord, M. J., Maat, B., and Dicke, K. A. Attempts at identification of hemopoietic stem cell in mouse. *Blood, 38*:547, 1971.

(22) van Bekkum, D. W., van Oosterom, P., and Dicke, K. A. *In vitro* colony formation of transplantable rat leukemias in comparison with human acute myeloid leukemia. *Cancer Res., 36*:941, 1976.

(23) van Bekkum, D. W., and de Vries, M. J. *Radiation Chimeras*. London: Logos Press/Academic Press, 1967.

(24) van den Engh, G. J. Early events in haemopoietic cell differentiation. Thesis, Rotterdam, 1976.

(25) Visser, J. W. M., van den Engh, G. J., and Williams, N. Physical separation of the cycling and noncycling compartments of murine haemopoietic stem cells. (this volume).

(26) Vos, O., Davids, J. A. G., Weyzen, W. W. H., and van Bekkum, D. W. Evidence for the cellular hypothesis in radiation protection by bone marrow cells. *Acta Physiol. Pharmacol. Neerl., 4*:482, 1956.

(27) Yoffey, J. M. Cellular equilibria in blood and blood forming tissues. In *Homeostatic Mechanisms* (Brookhaven Symp. Bibl. No. 10), pp. 1–25, 1958.

INTRODUCTION

The development of specific tools and methodologies to probe the nature of cell differentiation and proliferation in multicellular organisms is a crucial and essential step in the development of our capabilities to obtain definitive answers to these fundamental biologic problems. For the hematopoietic system, the colony assay technique introduced by Till and McCulloch (27) for pluripotent stem cells or CFU-s, the *in vitro* assays for unipotent myeloid [CFU-c (3)], erythroid [CFU-e (25)], and, recently, megakaryocytic [CFU-m (16, 21)] progenitors, as well as cell separation techniques (12, 17), provide us with invaluable tools for addressing ourselves to the nature of hematopoietic cell regulation.

Another fruitful methodology has been immunologic in nature. The finding by Reif and Allen (22) of a specific membrane-bound antigen on the surface of thymocytes (i.e., the theta or thy-1 antigen) leads one to speculate about the possible existence of other specific blood cell antigens. Since Reif and Allen noted that the theta antigen was also found in homogenates of adult brain, Golub (7) has prepared a potent anti-theta serum by immunizing animals with brain tissue rather than with thymus. This not only produced an antiserum with anti-theta activity, but one which has, since, also been shown to be selectively cytotoxic to other hemic cells, such as mature erythrocytes (8), the pluripotent stem cell (9), and perhaps B-cells as well (5). With selective tissue absorption it can be shown that the cytotoxic activity for each of these cells is due to a different antibody in the anti-brain serum. All four of the above cell-specific antigens must therefore be expressed by the brain itself.

For those interested in stem cell kinetics, the observation of a specific stem cell antigen is of obvious interest. This interest was intensified by the experiment reported by van den Engh and Golub (28) which showed that the stem cell antigen was lost during the step(s) in differentiation leading to the myeloid progenitor cell: although present on almost all CFU-s, the antigen could *not* be detected on CFU-c. With some modification in technique, this important observation appears to have been recently confirmed (24). If further confirmation of this apparent specificity is forthcoming, it would appear that the use of the anti-brain antiserum will offer a new dimension to cell kinetics studies. For example, there is one school of thought that envisions the CFU-s as a more closely related progenitor of the CFU-c (2, 11, 31) than of the erythroid progenitor, CFU-e (10). If this is the

An Immunologic Approach to Cell Cycle Analysis of the Stem Cell

2

Francis C. Monette,
Peter Q. Eichacker, William Byrt,
Robert I. Garver, Michael J. Gilio,
and John B. DeMello.

11

case and CFU-s do indeed have a parent-to-progeny relationship with CFU-c, then, extrapolating back to van den Engh and Golub's observation, it would appear that the stem cell antigen might be rapidly and dramatically lost as CFU-s become committed to the myeloid pathway. Whether this might be true for other early progenitor cells [such as the BFU of Axelrad (1)] remains to be seen.

The precise relationship of the cell cycle of stem cells to blood cell differentiation is not known. However, it is well-established that both the pluripotent stem cell and several unipotent progenitor cells are capable of extensive self-renewal (13). It is therefore highly probable that the step in differentiation between these two is in some way related to the cell cycle. The studies reported here addressed themselves to two questions which may have some bearing on this: (1) do all CFU-s express the stem cell antigen? and (2) do CFU-s in all phases of the cell cycle express the antigen? The demonstration of an antigen-free subpopulation of stem cells might provide a clue to the relationship between the cell cycle and the initial steps in stem cell differentiation.

MATERIALS AND METHODS
ANIMALS

Throughout, 9–12-week-old female swiss albino mice of the CF_1 or CD_1 strains (Charles River Labs, Wilmington, Mass.) were employed as donors and recipients. Female New Zealand White rabbits (Ancare Corporation, Long Island, N.Y.) initially weighing 1.5–2 kg were employed for immunization with mouse brain or served as a source of normal rabbit serum (NRS).

ANTISERA PREPARATION

The rabbit anti-mouse brain serum was prepared, by a procedure similar to that originally reported by Golub (7), as follows. Briefly, the brains of five adult mice were aseptically removed, homogenized in Eagle's MEM (Grand Island Biological Co., Grand Island, N.Y.), and emulsified with an equal volume of Freund's adjuvant (Difco Laboratories, Detroit, Mich.). A total of 2 ml was injected into rabbits at four subcutaneous sites. The procedure was repeated 7 days later and again at 4–6-week intervals for boosting. Blood was collected 7 days following the last boosting. All sera were heat-inactivated prior to use by incubating at 56°C for 1 hr. Normal rabbit sera (NRS) from uninjected rabbits served as

the source of the control sera. Prior to assay, both NRS and the experimental rabbit serum (ERS) were routinely absorbed with syngeneic mouse cells as follows. Cell suspensions of mouse bone marrow and RBC's were prepared, washed in MEM, and centrifuged. One milliliter of fresh MEM was added to the cell pellet, along with 0.5 ml of the heat-inactivated serum to be absorbed. Absorption of all sera occurred at 37°C for 1 hr followed by 4°C for 2 hr, with intermittent shaking. The tubes were then centrifuged at high speed to remove intact and disrupted cells, and the supernatant removed and stored at −70°C.

BONE MARROW CYTOTOXIC TEST AND STEM CELL ASSAY

To test for the anti-stem cell activity, pooled murine marrow was washed in MEM and resuspended to yield a cell concentration of 10^7 cells in 0.175 ml. There was then added to 10^7 marrow cells 0.3 ml of absorbed NRS or antiserum (ERS) (representing a 1:3 dilution of unabsorbed serum), and the mixture incubated at 4°C for 30 min, followed by incubation in a water bath at 37°C for 1 hr. In some experiments, when exogenous (tissue-absorbed) guinea pig serum complement was employed, it was added (1:5 dilution) prior to the 37°C incubation, as described by Golub (8, 9). Complement was not found to be necessary for cell killing in the *in vivo* assay, however, and it was subsequently omitted.

The stem cell assay was performed according to the method of Till and McCulloch (27). Briefly, syngeneic marrow cells were pooled and washed in MEM, centrifuged, and resuspended in fresh MEM prior to counting. Immediately following incubation, cells were centrifuged and the serum supernatant discarded. Fresh MEM was added and the cells washed and counted again prior to further dilution with MEM for injection into lethally radiated mice (950 rads). Nine-day macroscopic colonies were enumerated following fixation with Bouin's solution. Since the 37°C incubation itself resulted in some loss of CFU-s, the experimental data are expressed as a function of the NRS control.

REGENERATING BONE MARROW CFU-s

The method of Monette (18) was employed to induce CFU-s into cell cycle. Briefly, the cytotoxic drug hydroxyurea (HU) (Sigma Chemical Co., St. Louis, Mo.) was administered i.v. (0.9 mg/g) to normal mice 12 hr prior to sacrifice (for bone marrow collection. We have shown that this procedure induces 40–50% of marrow CFU-s into cell cycle (18).

VELOCITY SEDIMENTATION

The method of velocity sedimentation at unit gravity developed by Miller and Phillips (17) was employed, as reported by us elsewhere (18). Briefly, 170×10^6 pooled, washed, marrow cells were loaded into a sedimentation chamber after priming the system with MEM. This was followed by a 0.33% bovine serum albumin (BSA) "buffer" layer, which was in turn followed by a linear gradient of 1–2% BSA in MEM. Following a 4.5-hr sedimentation at 4°C, 15 ml fractions were collected, centrifuged, washed twice in MEM, and resuspended for enumeration of nucleated cells by a hemocytometer. Fractions were then pooled into groups by their sedimentation velocities as needed to ensure sufficient cell yields. The cell pools were then divided into two identical groups. One was incubated with NRS; the second with antiserum, as described above for unsedimented cells. The cells were then assayed for the fraction of CFU-s surviving the incubation with NRS and ERS, and graphed as a function of the modal sedimentation velocity for each cell pool.

RESULTS

Sera from rabbits immunized with mouse brain contain antibodies directed against marrow CFU-s (Table 1). When normal marrow was incubated in ERS (i.e., anti-brain serum) a 98% reduction in CFU-s relative to the NRS control was observed. Typically, the CFU-s killed in different experiments ranged between 92 and 98%, and was seldom 100%. This was true even though high cell numbers ($\sim 10^6$) were injected into radiated mice (Table 1) and extrapolated to the NRS control.

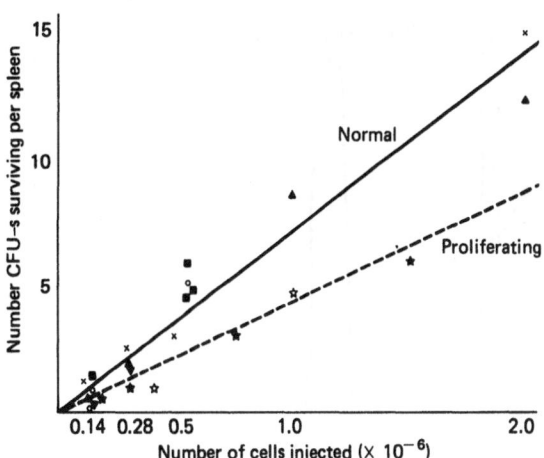

FIGURE 1. Survival of CFU-s in normal and proliferating bone marrow following incubation with anti-stem cell antiserum. Number of CFU-s surviving = average colony count per spleen ($N = 6$–8 mice). The different symbols refer to separate experiments. Lines fitted by the method of least squares.

This led us to consider the possibility that the ERS might not be cytotoxic for all CFU-s and that a small residual subpopulation of CFU-s might actually be resistant to the antiserum. To test this, varying numbers of marrow cells were injected into radiated recipient mice following incubation with ERS (Fig. 1). The number of CFU-s surviving the ERS incubation is clearly linearly related to the number of cells injected. In six separate experiments, the relative size of this resistant CFU-s subpopulation ranged between 3% and 5% of the unincubated marrow control (i.e., the total marrow CFU-s cell population).

To determine whether survival of some CFU-s following incubation was due to a high cell-to-serum ratio during the incubation procedure itself (and thus a subthreshold serum concentration for some cells), the following experiment was performed. A constant number of marrow cells (10^7) was incubated in varying amounts

TABLE 1 Reduction of CFU-s by Anti-Brain Antiserum[a]

EXPERIMENTAL GROUP	N	NUMBER CFU-s/SPLEEN	REDUCTION (%)
1. Unincubated bone marrow control[a]	8	14.1 ± 5.6	—
2. Normal serum control	7	16.0 ± 2.4	—
3. Experimental (anti-brain) serum	6	0.29 ± 0.4	98
4. Experimental (anti-brain) serum (7 × cell dose)	7	8.4 ± 2.0	93
5. Radiated controls	7	0.0	—

[a]Normal bone marrow; injected cell doses were 0.07×10^6, 0.14×10^6, 10^6, and no cells, for groups 1, 2 and 3, 4, and 5 respectively. N = number of mice injected.

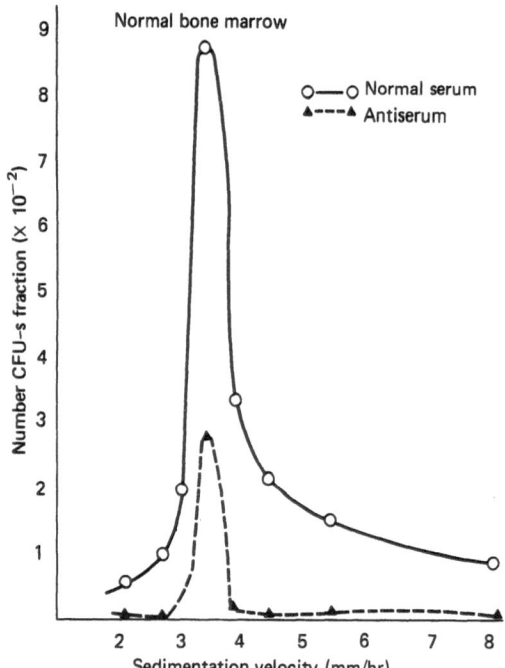

FIGURE 2. Velocity sedimentation profiles for normal bone marrow CFU-s incubated with normal rabbit serum or anti-stem cell serum following a 4.5 hr sedimentation at unit gravity. In some cases neighboring fractions were pooled. Six to 8 recipient mice per point. Representative of three separate experiments.

of NRS or ERS, ranging from the standard 0.1 ml up to 2.0 ml of serum. The results are presented in Table 2. Survival of a small number of CFU-s was observed regardless of the serum concentration.

Murine marrow CFU-s are normally relatively quiescent with respect to proliferation, having proportionally few cells positioned in phases of the cell cycle which are sensitive to cell cycle-specific cytotoxic agents, such as suicidal

doses of ^{3}H-TdR or hydroxyurea (14). It was therefore of interest to determine whether the resistant CFU-s subpopulation could be related in any way to its distribution in cell cycle. To test this, proliferating bone marrow (see the Materials and Methods section) was incubated with NRS or ERS and various cell numbers injected into radiated mice. It can be seen from the data in Fig. 1 for the proliferating marrow that the numbers of ERS-resistant CFU-s are also linearly related to the injected cell dose. In four separate experiments, the relative size of this resistant cell population also ranged between 3% and 5% of control. This was true even though the absolute number of CFU-s in HU-induced proliferating marrow was somewhat lower than in normal bone marrow (Fig. 1) It would therefore seem that the overall size of the CFU-s subpopulation resistant to the antiserum is not substantially affected by the proliferative state of these cells. However, the possibility remained that a substantial shift in their cell cycle kinetic properties may have occurred and that there was no actual correlation between the absence of the antigen and the cell cycle. It was therefore important to determine the cell cycle distribution of these resistant cells. That there were these two possibilities was apparent: either (1) the stem cell antigen would appear on CFU-s independently of the cell cycle, and resistant cells would also be randomly distributed throughout the cyle; or (2) the antigen would be nonrandomly expressed by CFU-s, i.e., in all but one subcycle phase. In an attempt to distinguish between these two alternatives, the distribution of resistant CFU-s throughout the stem cell cycle was observed by employing the technique of velocity sedimentation at unit gravity (17). MacDonald and Miller (15) demonstrated that this technique could be used to separate exponentially growing mouse L-cells, and we have shown (18)

TABLE 2 Survival of CFU-s Incubated in Varying Amounts of Antiserum

EXPERIMENTAL GROUP	NUMBER OF CELLS INJECTED	NUMBER CFU-s/SPLEEN	REDUCTION (%)
Unincubated bone marrow control	70,000	12.9 ± 1.9	—
2.0 ml NRS[a]	70,000	7.7 ± 4.2	(39)
2.0 ml ERS	70,000	0.44 ± 0.20	94
2.0 ml ERS	500,000	12.5 ± 1.8	86
1.0 ml ERS	70,000	0.34 ± 0.30	96
1.0 ml ERS	560,000	1.5 ± 0.5	98
0.1 ml ERS	70,000	0.32 ± 0.30	96
0.1 ml ERS	560,000	5.0 ± 2.3	95

[a]10^7 normal marrow cells incubated in different amounts of serum. Since these represent different serum batches, the quantity shown refers to the actual amount of raw serum used, regardless of the subsequent dilution as a consequence of absorption procedures.

that the technique is applicable to separating out proliferating from quiescent CFU-s of murine marrow. Both normal and proliferating marrow cells were separated by this method (see the Materials and Methods section) and the data are presented in Fig. 2 and 3, respectively. Following a 4.5-hr sedimentation, neighboring cell fractions were pooled, washed, and divided into two equal halves. One was incubated with NRS, the other with ERS, as noted in the Materials and Methods section. The data in Fig. 2 expresses the number of CFU-s surviving in each serum incubate as a function of increasing sedimentation velocity (and therefore size). In the normal serum groups, CFU-s were observed in all cell fractions ranging from 2 to 8 mm/hr in sedimentation velocity. On the contrary, CFU-s incubated in the antiserum were virtually eliminated in all fractions except for a distinct mode sedimenting between 3 and 4 mm/hr. This surviving population of CFU-s represents 5% to 10% of those incubated in NRS, and less than 5% of unincubated normal bone marrow controls (compare with unincubated marrow; for example, see Fig. 4). Thus, the antiserum-resistant CFU-s in normal marrow are homogeneously small cells with a modal sedimentation velocity of about 3.4 mm/hr.

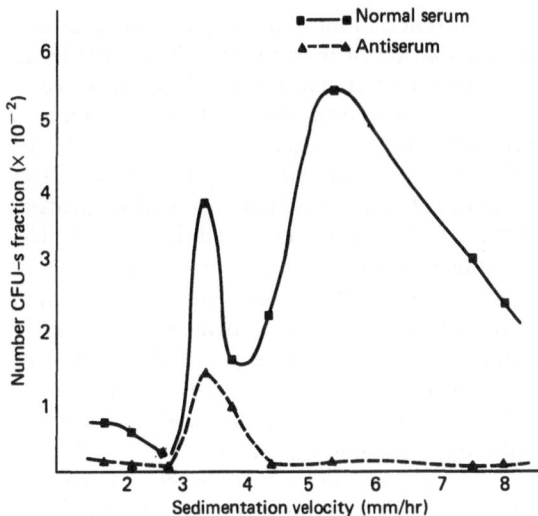

FIGURE 3. Velocity sedimentation profiles for proliferating bone marrow CFU-s incubated with normal rabbit serum or anti-stem cell serum following a 4.5-hr sedimentation at unit gravity. In some cases neighboring fractions were pooled. Six to 8 recipient mice per point. Representative of two separate experiments.

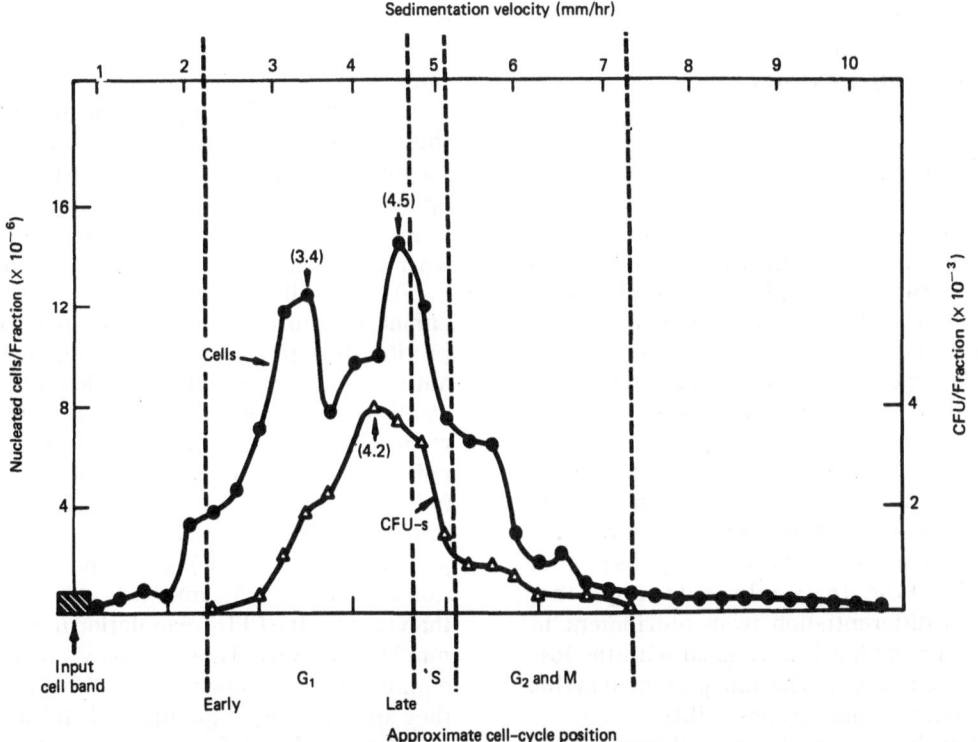

FIGURE 4. Composite velocity sedimentation profiles for unincubated normal mouse bone marrow. Shown are profiles for nucleated cells and 9 day CFU-s. On the lower abscissa is a depiction of the approximate cell-cycle position of the CFU-s. (Adapted from Monette et al. (18).)

15

As noted in the Materials and Methods section, incubation of marrow at 37°C results in an appreciable loss of cells (and CFU-s), thus necessitating the use of proper controls (i.e., incubation with NRS). It was generally assumed that this NRS incubation-induced cytolysis was nonselective as to cell cycle position. The cell separation experiments, however, indicated that large CFU-s (i.e., those sedimenting above 4 mm/hr) were more sensitive to the incubation procedure than small CFU-s (i.e., those sedimenting below 4 mm/hr). A distinct and reproducible peak of CFU-s was therefore observed sedimenting with a mode of ~ 3.4 mm/hr when normal marrow was incubated in the control serum.

We then addressed ourselves to the cell cycle distribution of antiserum-resistant cells in marrow which had been induced to enter cycle by HU-administration (Fig. 3). Again, although CFU-s were observed in all normal serum fractions (sedimentation velocity range: 1.7–8.0 mm/hr), the antiserum-incubated cells survived only in those cell fractions sedimenting between 3 and 4 mm/hr. This peak of resistant cells was similar in size to that observed for the normal marrow (Fig. 2). The substantial increase in the number of large cells observed in the cell fractions incubated with the NRS is consistent with the substantial recruitment of CFU-s into cycle which occurs as a consequence of HU-administration (see (18)). Clearly, antiserum-resistant CFU-s did not take part in this recruitment.

DISCUSSION

The existence of a stem cell antigen shared with brain tissue has been confirmed by these as well as other studies (6, 23). Although other hematopoietic cells cross-react with anti-brain serum, (e.g., T-cells, B-cells, and mature erythrocytes), this appears to be due to the presence of other antibodies in anti-brain serum preparations, and these can be absorbed out with the proper tissue (8, 9, 19, 23, 28). An indication of the possible specificity of the antiserum for the pluripotential stem cell is suggested by van den Engh and Golub's observations of the lack of an effect on the CFU-c. This would tend to imply that the step in differentiation from pluripotent to committed progenitor is associated with the loss of the stem cell antigen. The data presented in this report do not rule out this possibility.

That the process of stem cell inactivation by the serum is immunologic in nature was demonstrated by Golub (9) by absorbing the antisera with brain homogenates as well as other adult tissues. Only the brain removed the anti-stem cell activity, whereas the activity directed against other cells was removed by absorbing the anti-brain serum with other tissues. For example, anti-T-cell activity was lost with thymus absorption but not liver (9). In another report, we confirm and extend these observations on the immunologic nature of the anti-stem cell activity (19).

Not all CFU-s appear to be inactivated by the anti-stem cell serum, however. We report here the existence of a small subpopulation of CFU-s which appears to be very resistant to the cytotoxic effects of the antiserum. This was true even when large numbers of serum-incubated cells were injected into the assay animals (the assay efficiency is high with colony counts in the 8–12 range) (Fig. 1). Increasing the serum concentration did not appreciably alter the survival of these cells either (Table 2). This would eliminate the obvious artifactual possibility of insufficient antibody for complete cell killing. In addition, since the lots of serum used in these studies had very high titers (in excess of 1000 for 50% inactivation; see (19)), it was obvious that at the standard serum dilution employed in these studies the anti-stem cell activity was far in excess of that needed to inactivate all CFU-s in 10^7 marrow cells. This is borne out by the data in Table 2. It is worth noting that Golub (9) and others (6, 23), using other mouse strains, were also unable to show 100% CFU-s inactivation.

It should be emphasized that the size of this resistant cell population was small relative to the total CFU-s marrow population (3–5%). It also appeared to be largely unaffected by an increase in the rate of proliferation of stem cells (Fig. 1) (at least at the single time interval of 12 hr studied here). The cell separation studies also serve to emphasize this point. In both normal (Fig. 2) and proliferating (Fig. 3) marrow preparations, the antiserum-resistant cell population represented a small subpopulation of CFU-s (i.e., 3–5% of their respective controls). In addition, these studies point to the size homogeneity of resistant cells (with a range in sedimentation velocity of ~ 3–4 mm/hr, compared to ~ 2–8 mm/hr for the CFU-s population as a whole). Their modal sedimentation rate of ~ 3.3–3.4 mm/hr is also clearly lower than that for the CFU-s population in toto (3.9–4.2 mm/hr, (18, 30)). Thus not only are these cells quantitatively a minor stem cell population, but they appear to differ qualitatively from the CFU-s population when taken as a whole. One possible explanation for this can be attempted by relating these cells to their cell cycle position. Fig. 4 represents a summary of the results of a series of exper-

16

iments, some of which have been presented, in part, elsewhere (18). We have shown that the wide distribution in size of CFU-s can be attributed in part to their position within the cell cycle. It has long been established that mammalian cells double in size as they make their progress through cycle (15, 26). Our previous study showed that a subpopulation of only large CFU-s was eliminated by a suicidal dose of an S-phase-specific agent (also, see (31)). This has led us to propose that the bulk of CFU-s with sedimentation velocities between ~ 4.7 and 5.1 mm/hr are in the S-phase of cycle (Fig. 4). Presumably, larger cells are in succeeding phases of cycle (i.e., G_2 and M). Small CFU-s, which sediment from 2 to over 4.0 mm/hr, are in a majority in normal bone marrow. Their small size suggests a cell cycle position in G_1. The range of cell sizes for the G_1 cells (2.0 to > 4.0 mm/hr) is greater than that for any other phase (see Fig. 4). The antiserum-resistant cells reported here for both normal and proliferating marrow CFU-s fall wholly within this size range, and thus may represent a G_1 cell subpopulation.

It is tempting to speculate on the possible nature of these G_1 antiserum-resistant CFU-s. There are at least two alternative theories which might be tested with such a model (Fig. 5). One simple explanation for the existence of small numbers of antigen-free CFU-s is that these cells might represent the small fraction of stem cells in transition from pluripotential to unipotential states. This possibility receives support from the study of van den Engh and Golub (28). The loss of the stem cell antigen would, however, have to precede the loss of pluripotential capabilities in order to explain the presence of antigen-free CFU-s. Perhaps differentiation to the committed state is a multi-event step. Secondly, it is also intriguing to consider an alternative explanation for these resistant CFU-s. Could they represent a stem cell reserve, i.e., the true G_0 stem cell (Fig. 5)? One can immediately argue effectively against this possibility from the abundant data showing that normal marrow stem cells show little (4, 14) or modest (20) rates of cell renewal, and thus must be comprised largely of "quiescent" or "G_0" cells.

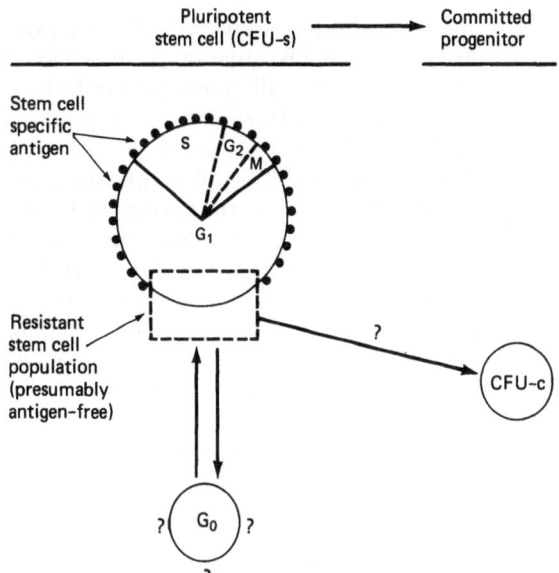

FIGURE 5. A graphic representation for a possible relationship between antigen-free pluripotent stem cells (CFU-s), the cell cycle, and stem cell differentiation into the committed progenitor (CFU-c) compartment.

The numbers of resistant CFU-s observed here would not be consistent with a major G_0 cell population. However, the fact that these antigen-free CFU-s otherwise behave like typical pluripotent stem cells (i.e., they produce macroscopic-spleen colonies) invites a certain amount of caution before excluding the G_0 notion as a possibility. A potentially fruitful approach in characterizing resistant CFU-s as G_0 cells might be to study the growth kinetics of these cells in various systems. These studies are currently underway.

It is clear that we need to learn a great deal more about the properties of this subpopulation of stem cells before any direct application to models of stem cell differentiation can be earnestly attempted. It does appear, however, that the presence or absence of antigens on the membranes of stem cells may provide us with an important marker for analyzing the relationship between the stem cell and its immediate progeny.

SUMMARY

The observations of Golub (8) showing that mouse brain shares an antigenic component with pluripotent hematopoietic stem cells (CFU-s) has led to the production of an antiserum directed against stem cells. A brief incubation with this antiserum, obtained from immunized rabbits, will therefore inactivate 90–98% of marrow CFU-s relative to normal rabbit serum controls. The specificity of this antibody for CFU-s was also noted by Golub, since the antibody appeared to have no cytotoxic effect on marrow progenitor cells committed to granulopoiesis (CFU-c). Since CFU-c are considered by many investigators to be close progeny of CFU-s, it would appear that an early step in the differentiation of CFU-s to CFU-c may involve the loss of this antigenic component. However, the precise relationship of differentiation of CFU-s to the cell cycle still remains obscure. Since passage through cell cycle may serve as an initial step in CFU-s differentiation, we addressed ourselves to the question of whether CFU-s retain the antigen *throughout* their cell cycle. To test this, both normal (largely G_0) and regenerating marrow CFU-s were separated according to size by velocity sedimentation at unit gravity. We have previously shown this to be an effective method for obtaining CFU-s in different phases of the cell cycle (18). Fractions of cells were split into two groups, one incubated with normal rabbit serum, the second with the anti-stem cell antiserum. Following incubation at 4°C and 37°C cells were injected into lethally irradiated syngeneic mice to determine CFU-s numbers.

The results of five separate experiments show that CFU-s in most phases of the cell cycle (i.e., with sedimentation velocities ranging from 2.5 to 9.0 mm/hr) are largely inactivated (average reduction was 95%) by incubation with the antiserum. A persistent exception was observed in both normal as well as proliferating CFU-s for a small segment of cells with a modal sedimentation velocity of 3.6 mm/hr. Since these cells which lack the antigen and are thus resistant to the antiserum appear to correspond in size to cells in early to mid-G_1, this may represent the position in cell cycle which is associated with the transition from CFU-s to CFU-c. In any case, the majority of CFU-s were observed to retain the antigen throughout the remainder of the cell cycle (mid-G_1 through M).

ACKNOWLEDGMENTS

The authors wish to express their gratitude for the use of the ^{137}Cesium radiation facilities generously provided by Dr. Shirley Ebbe of St. Elizabeth's Hospital, Brighton, Massachusetts. This work was supported by grant AMDD-17735 from the N.I.H., and by an N.I.H. Biomedical Support Grant to Boston University (No. GRS 449-BI).

REFERENCES

(1) Axelrad, A. A., McLeod, D. L., Shreeve, M. M., and Heath, D. S. Properties of cells that produce erythrocytic colonies *in vitro*. In Robinson, W. A., ed., *Proceedings of the Second International Workshop on Hemopoiesis in Culture, Airlie House, Virginia*. New York: Grune and Stratton, 1973, p. 226.

(2) Bennett, M., Cudkowicz, G., Foster, Jr., R. S., and Metcalf, D. Hemopoietic progenitor cells of W anemic mice studied *in vivo* and *in vitro*. J. Cell. Physiol., *71*: 211, 1968.

(3) Bradley, T. R., and Metcalf, D. The growth of mouse bone marrow cells *in vitro*. Aust. J. Exp. Biol. Med. Sci., *44*: 287, 1966.

(4) Becker, A. J., McCulloch, E. A., Siminovitch, L., and Till, J. E. The effect of differing demands for blood cell production on DNA synthesis by hemopoietic colony-forming cells of mice. Blood, *26*: 296, 1965.

(5) Filppi, J. A., Rheins, M. S., and St. Pierre, R. L. *In vivo* effects of rabbit antimouse brain serum on

B-bearing lymphocytes of AKR mice. *Immunology, 28*: 659, 1975.

(6) Filppi, J. A., Rheins, M. S., and Nyerges, C. A. Antigenic cross-reactivity among rodent brain tissues and stem cells. *Transplantation, 21*: 124, 1976.

(7) Golub, E. S. Brain-associated θ antigen: Reactivity of rabbit anti-mouse brain with mouse lymphoid cells. *Cell. Immunol., 2*: 353, 1971.

(8) Golub, E. S. Brain-associated stem cell antigen: An antigen shared by brain and hemopoietic stem cells. *J. Exp. Med., 136*: 369, 1975.

(9) Golub, E. S. Brain associated erythrocyte antigen: An antigen shared by brain and erythrocytes. *Exp. Hematol., 1*: 105, 1973.

(10) Gregory, C. J., McCulloch, E. A., and Till, J. E. Erythropoietic progenitors capable of colony formation in culture: State of differentiation. *J. Cell. Physiol., 81*: 411, 1973.

(11) Haskill, J. S., McNeill, T. A., and Moore, M. A. S. Density distribution analysis of *in vivo* and *in*

18

vitro colony forming cells in bone marrow. *J. Cell. Physiol.*, *75*: 167, 1970.

(12) Hulett, H. R., Bonner, W. A., Sweet, R. G., and Herzenberg, L. A. Development and application of a rapid cell sorter. *Clin. Chem.*, *19*: 813, 1973.

(13) Lajtha, L. G. Stem cell kinetics. In Gordon, A. S., ed., *Regulation of Hematopoiesis*. New York: Appleton-Century-Crofts, 1970, vol. 1, p. 111.

(14) Lord, B. I., Lajtha, L. G., and Gidali, J. Measurement of the kinetic status of bone marrow precursor cells: Three cautionary tales. *Cell Tissue Kinet.*, *7*: 507, 1974.

(15) MacDonald, H. R., and Miller, R. G. Synchronization of mouse L-cells by a velocity sedimentation technique. *Biophys. J.*, *10*: 834, 1970.

(16) Metcalf, D., MacDonald, H. R., Odartchenko, N., and Sordat, B. Growth of mouse megakaryocyte colonies *in vitro*. *Proc. Nat. Acad. Sci. U.S.A. 72*: 1744, 1975.

(17) Miller, R. G., and Phillips, R. A. Separation of cells by velocity sedimentation. *J. Cell. Physiol.*, *73*: 191, 1969.

(18) Monette, F. C., Gilio, M. J., and Chalifoux, P. Separation of proliferating CFU from G_0 cells of murine bone marrow. *Cell Tissue Kinet.*, *7*: 443, 1974.

(19) Monette, F. C., Eichacker, P. Q., Garver, R. I., Byrt, W., and Gilio, M. J. Characterization of the antistem cell activity of antimouse brain antiserum. *Exp. Hematol.* (Submitted.)

(20) Morse, B. S., Rencricca, N. J., and Stohlman, Jr., F. The mechanism of action of erythropoietin in relationship to cell cycle kinetics. In Stohlman, F., Jr., ed., *Hemopoietic Cellular Proliferation*. New York: Grune and Stratton, 1970, p. 160.

(21) Nakeff, A., and Daniels-McQueen, S. *In vitro* colony assay for a new class of megakaryocyte precursor: Colony-forming unit megakaryocyte (CFU-M). *Proc. Soc. Exp. Biol. Med.*, *151*: 587, 1976.

(22) Reif, A. E., and Allen, J. M. V. The AKR thymic antigen and its distribution in leukemias and nervous tissues. *J. Exp. Med.*, *120*: 413, 1964.

(23) Rodt, H., Thierfelder, S., and Eulitz, M. Antilymphocytic antibodies and marrow transplantation. III. Effect of heterologous anti-brain antibodies on acute secondary disease in mice. *Eur. J. Immunol.*, *4*: 25, 1974.

(24) Schachner, M., Wortham, K. A., and Kincade, P. W. Detection of nervous-system specific cell surface antigen(s) by heterologous anti-mouse brain antiserum. *Cell. Immunol.*, *22*: 369, 1976.

(25) Stephenson, J. R., Axelrad, A. A., McLeod, D. L., and Shreeve, M. M. Induction of colonies of hemoglobin-synthesizing cells by erythropoietin *in vitro*. *Proc. Nat. Acad. Sci. U.S.A. 68*: 1542, 1971.

(26) Terasima, T., and Tolmach, L. J. Growth and nucleic acid synthesis in synchronously dividing populations of HeLa cells. *Exp. Cell Res.*, *30*: 344, 1963.

(27) Till, J. E., and McCulloch, E. A. A direct measurement of the radiation sensitivity of normal mouse bone marrow cells. *Radiat. Res.*, *14*: 213, 1961.

(28) van den Engh, G. J., and Golub, E. S. Antigenic differences between hemopoietic stem cells and myeloid progenitors. *J. Exp. Med.*, *139*: 1621, 1974.

(29) Visser, J. W. M., and van den Engh, G. J. Physical separation of the cycling and non-cycling compartments of murine hemopoietic stem cells. This volume.

(30) Worton, R. G., McCulloch, E. A., and Till, J. E. Physical separation of hemopoietic stem cells from cells forming colonies in culture. *J. Cell. Physiol.*, *74*: 171, 1969.

(31) Wu, A. M., Siminovitch, L., Till, J. E., and McCulloch, E. A. Evidence for a relationship between mouse hemopoietic stem cells and cells forming colonies in culture. *Proc. Nat. Acad. Sci. U.S.A.*, *59*: 1209, 1968.

INTRODUCTION

Till and McCulloch (22) have demonstrated that mouse hemopoietic tissue contains a class of cells capable of giving rise to macroscopic colonies in the spleens of radiated mice. These colonies have been shown to be derived from single cells (3), the "colony-forming cells," which are capable of extensive proliferation (3, 22), of differentiation (12, 30), and of self-renewal (23, 24). These findings support the hypothesis that blood-forming tissue contains a pluripotent, hemopoietic stem cell which has the capacity to give rise to all blood cells, including copies of itself. Pluripotent stem cells then give rise to colonies in the spleens of radiated mice. Since only a portion of the injected stem cells reach the spleen and form colonies there, the cells that form colonies have been termed colony-forming units in the spleen, or CFU-s. The number of CFU-s is then directly proportional to the number of stem cells. The possibility of quantifying the stem cell has led to the initiation of many experiments with the aim of further characterizing it.

Several investigators have demonstrated the existence of subpopulations of CFU-s which differ with respect to physical characteristics and function. Worton et al. (29) showed that subpopulations of CFU-s which were separated by velocity sedimentation may differ in self-renewal capacity. Haskill and Moore (7) reported differences in buoyant densities and sedimentation rates between CFU-s from embryonic liver and CFU-s from adult bone marrow. Metcalf et al. (15) found that the CFU-s population of bone marrow is heterogenous with respect to adherence to glass beads. Micklem et al. (16) demonstrated that CFU-s from peripheral blood differ from marrow CFU-s with respect to self-renewal capacity or homing site. Monette et al. (19) observed that subpopulations of CFU-s which were separated by velocity sedimentation have different sensitivities to hydroxyurea. They reported that CFU-s in DNA synthesis which were destroyed by hydroxyurea were separated from other more slowly sedimentating CFU-s which were insensitive to this cytotoxic drug. They concluded that the latter CFU-s are probably in the G_0-phase of the cell cycle.

The existence of a G_0 compartment for the pluripotent hemopoietic stem cell has been suggested by the results of several investigators. It has been reported that only 5% to 20% of CFU-s is destroyed by treatment of bone marrow cells with highly labeled tritiated thymidine or with hydroxyurea (4, 9, 19, 26). With bone marrow cells

Physical Separation of the Cycling and Noncycling Compartments of Murine Hemopoietic Stem Cells

3

Jan Visser, Ger van den Engh, Neil Williams, and Dries Mulder

obtained from mice recovering from radiation damage, 40% to 60% of CFU-s was found to be destroyed by these treatments (4, 9, 26). These data suggest that 60% to 90% of the stem cells is quiescent (in G_0 or in a prolonged G_1) in unradiated mice.

We have studied the separation of CFU-s in different phases of the cell cycle in more detail by equilibrium density centrifugation (20) and by velocity sedimentation techniques (17), using mouse bone marrow containing different concentrations of cycling CFU-s, and by *in vitro* treatment of the separated subpopulations with highly labeled tritiated thymidine (8). The results indicate that quiescent and proliferating stem cells are physically distinguishable.

THEORY

As reported by McDonald and Miller (11), differential velocity sedimentation at unit gravity can be used to separate cells on the basis of their phase in the cell cycle. Briefly, since cells before and after mitosis (in G_2 and G_1, respectively) differ in volume by a factor of two (1, 21), their sedimentation rates will differ by a factor of 1.59. Furthermore, it has been shown that the resolution of the velocity sedimentation technique depends on the sedimentation rate. Experimentally, the half-width of the volume distribution obtained after a velocity sedimentation separation with cells of equal volume was determined to be 20% of the sedimentation rate (18, 27). Therefore, both half-width and extent of a band in the volume distribution are linearly proportional to the sedimentation rate.

Consider an exponentially growing population of cells in which the G_1 and G_2 phases account for an equal portion of the cell cycle. At any time point, there are twice as many cells in early G_1 as there are in G_2. According to the above consideration, the volume distribution of such a cell suspension will be as shown in Fig. 1. Although the number of cells in early G_1 and that in late G_2 differ by a factor of two, the height of the peak of the G_2 population is only 0.31 times that of the G_1 peak.

MATERIALS AND METHODS
PREPARATION OF CELL SUSPENSIONS

The cells used for the experiments described were obtained from the femurs of young female F_1 hybrids of CBA and C57BL/Rij mice (8–12 weeks of age), as described by van den Engh (5). Cells were suspended in Hanks's balanced salt solution (HBSS), buffered with Hepes buffer, and kept at 0°C unless stated otherwise. Cell concentrations were determined by use of a hemocytometer.

For several experiments, bone marrow cells of mice recovering from radiation damage were used (9). Such cells were obtained from the femurs of mice which were reconstituted with a graft of 2.5×10^6 marrow cells (0.25 femur), after having received a lethal dose of γ-rays (1250 rad). The mice were sacrificed 8 days later.

SPLEEN COLONY ASSAY

CFU-s were measured by injecting adequate cell numbers intraveneously into lethally radiated mice (1025 rad γ-radiation, total body) (22). After 8 to 10 days, the mice were sacrificed and their spleens were fixed in Tellyesniczky's solution.

VELOCITY SEDIMENTATION PROCEDURE

A detailed description of the method of differential velocity sedimentation at unit gravity has been given by Miller (18). Streaming was prevented by loading less than 2×10^8 cells and by

FIGURE 1. Hypothetical sedimentation rate profile of an exponentially growing cell population of a single cell type. It is assumed that there are twice as many cells in G_1-phase as in G_2. The width of distribution is due to volume changes during the cell cycle. G_2 cells sediment 1.59 times as fast as G_1 cells.

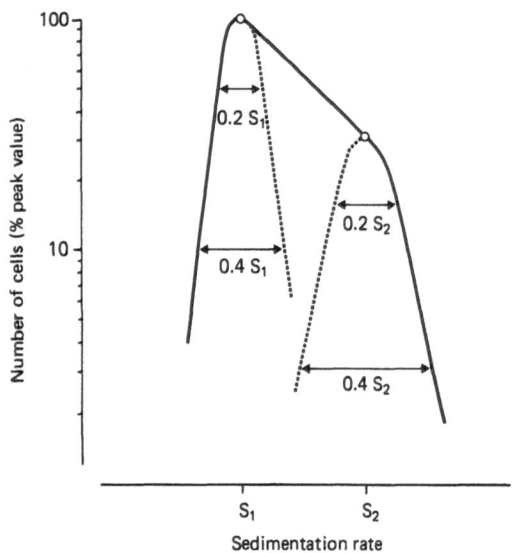

application of a "buffered-step" gradient from 0.2% to 2% bovine serum albumin (BSA, Sigma, St. Louis, Mo.) in phosphate-buffered saline (PBS). Cells (generally 10^8) suspended in 0.2% BSA (30 ml) were loaded into a cylindrical sedimentation chamber with a diameter of 19.5 cm. After sedimentation for 3.5 to 4.5 hr at 5°C, typically 25 fractions (50 ml each) were collected. The cells in each fraction were recovered by centrifugation for 7 min at 800 g; the pellet was resuspended in HBSS with Hepes buffer and 10% calf serum. Cell counts on each fraction were made with a hemocytometer. The mean yields of cells were 80% of CFU-s and of all nucleated cells.

EQUILIBRIUM DENSITY CENTRIFUGATION PROCEDURE

Separation of cells on the basis of their buoyant densities was performed as described by Shortman (21) and Williams and van den Engh (28). The mean yields of cells were 60% to 70% of CFU-s and 80% to 90% of all nucleated cells.

THYMIDINE CYTOCIDE

The fraction of CFU-s in the S-phase of the cell cycle was determined by treatment with tritiated thymidine (^{3}H-TdR) as described by Iscove (8). Samples of cell suspension (1 ml each) were incubated for 30 min at 37°C with 15 μCi ^{3}H-TdR/ml. Cell concentrations did not exceed 5 × 10^6 cells/ml. The specific activity of the ^{3}H-TdR was 22 Ci/mM.

RESULTS AND DISCUSSION
SEDIMENTATION RATES

The sedimentation rate distribution of nucleated bone marrow cells from normal and from radiated mice were found to be different (Fig. 2). Cells from radiated mice were found to sediment more rapidly than those from normal mice. Similar distribution profiles obtained with unradiated mice have been reported by Worton (19) and Monette (29). These authors observed that the peak of slowly sedimenting cells consisted largely of nucleated erythroid cells and lymphocytes, and that the second peak of more rapidly sedimenting cells was mainly granulocytes.

Figure 3 shows the sedimentation rate distribution of CFU-s with normal bone marrow and this distribution after exposure of all fractions

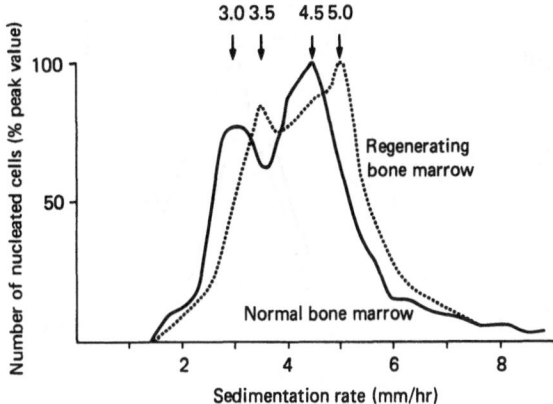

FIGURE 2. Sedimentation rate distribution of nucleated marrow cells. Solid line: bone marrow from normal mice; dashed line: bone marrow from mice which had been lethally radiated and reconstituted with 0.25 femur 8 days previously. Both curves represent the average of four separate experiments. The total number of cells per fraction is expressed as a percentage of the maximum value.

from the sedimentation experiment to highly labeled tritiated thymidine. With untreated bone marrow, the majority of CFU-s were found in a band at 4.0 mm/hr which spreads out at higher sedimentation rates (up to 7.5 mm/hr). These characteristics are in good agreement with those

FIGURE 3. Sedimentation rate distribution of CFU-s from mouse bone marrow before and after thymidine cytociding. The profiles show the average of four independent experiments. Solid line: CFU-s content of untreated fractions; dashed line: CFU-s content of fractions after exposure to 15 μCi/ml of ^{3}H-TdR. The control fractions were handled in the same manner as the ^{3}H-TdR-treated fractions. The values are expressed as a percentage of the maximum of the control fractions. The area under the graph of the ^{3}H-TdR-treated fractions is 81% of the area under the graph of the untreated controls.

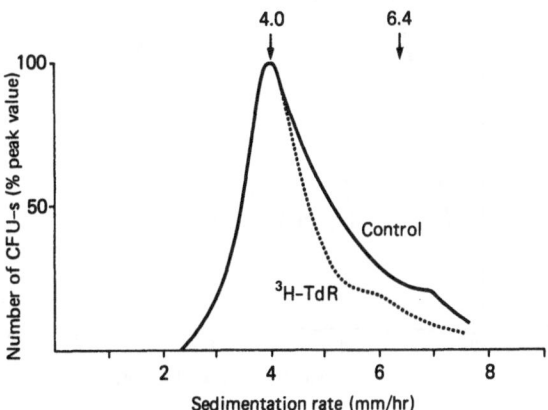

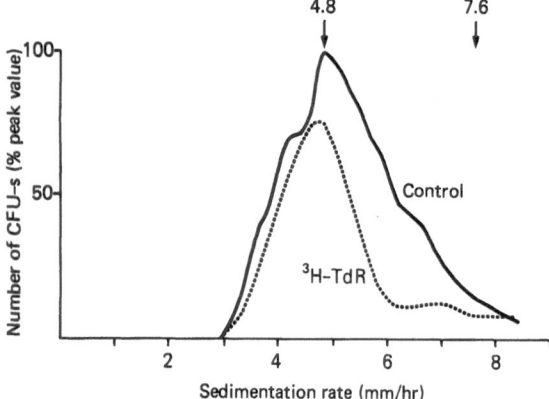

FIGURE 4. Sedimentation rate distribution of a rapidly proliferating population of CFU-s before and after thymidine cytociding. The bone marrow cell suspension was obtained from mice which had been lethally radiated and reconstituted with 0.25 femur 8 days previously. The curves show an average profile from two experiments. The values are expressed as a percentage of the maximum value of the untreated controls.

FIGURE 5. The construction of the sedimentation rate profile of cycling CFU-s from normal marrow. Line A represents the average profile of CFU-s from normal mouse marrow (Fig. 3) in a semi-log fashion. The theoretical distribution of the noncycling cells (line B) was obtained by the construction of a mirror image of the rising slope at the half-height of the curve such that the band width at that point is 0.2 times the sedimentation rate. The profile of the cycling cells (line C) is obtained by subtraction of the noncycling cells from the total population. The area under line C is approximately 50% of the total area under line A.

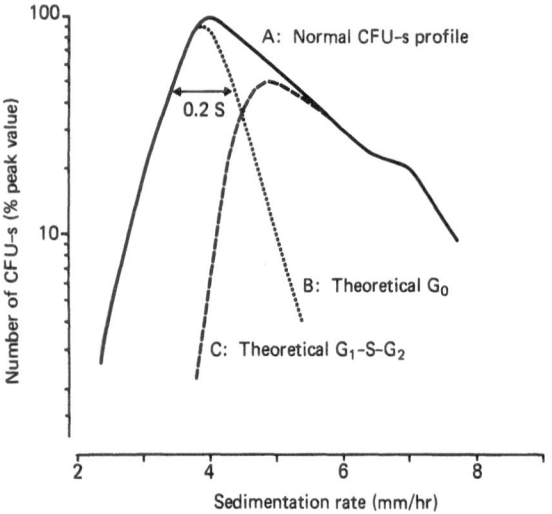

reported by Worton (19) and Monette (29). Exposure to ^{3}H-TdR (Fig. 3) reduced the total number of colonies by 10% to 20%; the treatment did not affect the number of CFU-s which sedimented at a rate of 4.0 mm/hr. This suggests that CFU-s sedimenting at a rate of 4.0 mm/hr are quiescent cells (2, 11, 19, 27). This hypothesis was tested by velocity sedimentation with bone marrow cells of radiated mice.

With mice recovering from radiation damage (Fig. 4), the peak of the CFU-s sedimentation distribution profile is found at 4.8 mm/hr. In fractions containing cells with higher sedimentation rates, the number of CFU-s gradually decreases. A shoulder can be seen at 4.0 mm/hr. In agreement with the results of other investigators (5, 9), treatment with ^{3}H-TdR killed 50% to 60% of CFU-s of reconstituted mice; this indicates that all CFU-s were proliferating. Therefore, a comparison of the distribution profiles shown in Figs. 3 and 4 strongly suggests that CFU-s in the G_0 phase of the cell cycle sediment at a rate of 4.0 mm/hr. The distribution profile shown in Fig. 4 indicates that G_1 cells sediment at a rate of 4.8 mm/hr.

An analysis of the sedimentation rate distribution of CFU-s from normal bone marrow (Fig. 3) as described above (see the Theory section) indicates that these CFU-s do not follow the theoretical distribution of a homogeneous cell population. Figures 5 and 6 compare the measured distribution of cycling CFU-s with a profile of normal marrow CFU-s from which a hypothetical G_0 population has been subtracted. At low sedimentation rates, the profile of normal marrow CFU-s forms a slope leading to the major peak, which deviates little from that of a band caused by cells of equal volume (half-width: 20% of the sedimentation rate (18, 27)). Furthermore, it is shown (Fig. 3) that the major peak consists of CFU-s which are insensitive to high doses of ^{3}H-TdR. This indicates that this peak represents a homogeneous population of noncycling cells. The theoretical distribution of these cells was subtracted from the measured profile of normal marrow CFU-s (Fig. 5). The resultant distribution was found to almost completely coincide with the measured profile of CFU-s from radiated mice (Fig. 6). This supports the view that the CFU-s under normal conditions consist of a relatively homogeneous population of G_0 cells and more rapidly sedimentating cycling cells. Therefore, G_0 cells are physiologically different from G_1 cells. The difference in sedimentation rates between these cells may be due to either size or density differences between cycling and noncycling stem cells.

BUOYANT DENSITIES

The buoyant density profiles of CFU-s of bone marrow from normal and from radiated mice are shown in Fig. 7. The distribution profile of normal marrow CFU-s shows a broad peak around 1.070 g/cm³ and a shoulder at approximately 1.075 g/cm³. The range of CFU-s activity (1.067 to 1.077 g/cm³) is much wider than the reported resolution of the method (0.004 g/cm³) (28). This also indicates that normal marrow CFU-s form a heterogeneous cell population. When these values are corrected for the differences between the osmolarity of the separation medium used (308 mOsm) and the osmolarity employed by Metcalf et al. (15) and Messner et al. (13), there is a good agreement among the results. Figure 7 also shows the density distribution of CFU-s from mice which are recovering from radiation damage. About 80% of CFU-s were found in a density band peaking at 1.075 g/cm³. These data indicate that resting and cycling stem cells have modal densities of 1.070 and 1.075 g/cm³, respectively.

SIZES AND CELL CYCLE PARAMETERS

By combining the density and the sedimentation rate measurements, it is possible to calculate the size heterogeneity of the CFU-s population (11). It was assumed that the medium in the sedimentation chamber had a viscosity of 1.567 centipoise (water at 4°C) and a density of 1.075 g/cm³ (1.5% BSA). Using these values and those of the measurements described above, the modal diameters of G_0, G_1, and G_2 cells were calculated to be 7.0, 7.3, and 9.2 μm, respectively. Assuming spherical cells, these values correspond to volumes for G_0, G_1, and G_2 cells of 180, 204, and 408 μm³, respectively. The accuracy of these values was estimated from the accuracy of the methods to be within 5%. Therefore, the limits of the range of CFU-s diameters are given by the values 6.6 to 9.7 μm. The calculated variation in the size of CFU-s is in good agreement with the size distribution of the candidate stem cell described by van Bekkum et al. (25). These authors consider the hemopoietic stem cell to measure 7 to 10 μm in diameter. Furthermore, the inaccuracy of the calculations is such that the difference in sedimentation rate between G_0 and G_1 cells may be completely explained by differences in densities between these cells. Their volumes may be considered to be equal. The difference in volume between these cells and G_2 cells, however, is evident.

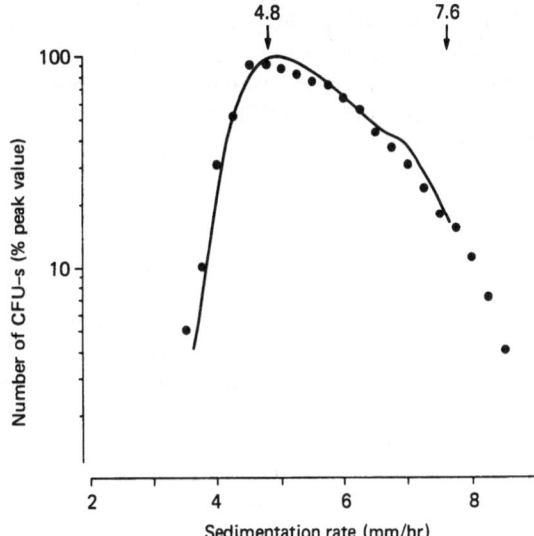

FIGURE 6. Sedimentation rate distribution of cycling CFU-s. Drawn line is equal to line C from Fig. 5, normalized as a percentage of its peak value. Solid circles represent the sedimentation rate distribution of cycling CFU-s from the marrow of reconstituted mice measured as for Fig. 4, except for thymidine cytociding. The measured distribution represents the average of two independent experiments.

Analysis of the data in terms of the cell cycle gives the following results. Figure 5 shows that 60% of normal marrow CFU-s is in the G_0-phase of the cycle. Experiments with ³H-TdR indicated that 20% of normal marrow CFU-s and

FIGURE 7. The buoyant density distribution of CFU-s from the bone marrow of normal mice (open circles), and of mice which had been lethally radiated and reconstituted with 0.25 femur 8 days previously (solid circles). The profiles are expressed as cells per fraction per density increment, with each point normalized to the maximum value of the experiment.

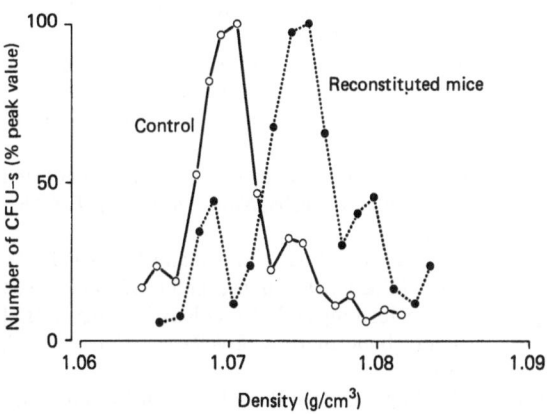

50% of proliferating CFU-s are in the S-phase of the cycle. The data of Fig. 4 indicate that the G_1 and the ($G_2 + M$) compartments contain 25% and 15% of the proliferating CFU-s, respectively, and, thus, 13% and 7% of normal marrow CFU-s, respectively. These values are in good agreement with cell cycle parameters of CFU-s reported by Vassort et al. (26). These authors derived the following values for the various stages of the cell cycle: G_1, 2 hr; S, 4 hr; $G_2 + M$, 2 hr.

SUMMARY

Physical properties of the pluripotent hemopoietic stem cell of the mouse were studied by use of a combination of methods: spleen colony assay (CFU-s), equilibrium density centrifugation, velocity sedimentation at unit gravity, and thymidine cytocide. Normal bone marrow CFU-s from adult mice were found to have sedimentation rates and density distributions which did not correspond to distributions of a single cell type. It appeared that bone marrow CFU-s from mice recovering from radiation damage had a uniform density and sedimentation rate distribution. The heterogeneity of normal marrow CFU-s can be attributed to the existence of two subpopulations: a slowly sedimentating population of quiescent cells, and a more rapidly sedimenting population of proliferating cells. The difference in sedimentation rates can be explained by a slight density increase (from 1.070 to 1.075 g/cm³) at the transition from the resting into the cycling state. The noncycling CFU-s are considered to be a homogeneous population with a diameter of $7.0 \pm 0.4\ \mu$m. The subpopulation of cycling CFU-s is calculated to vary in diameter from $7.3 \pm 0.4\ \mu$m to $9.2 \pm 0.5\ \mu$m depending on their phase in the cell cycle.

ACKNOWLEDGMENTS

This investigation is part of a study on the regulation of hemopoiesis which is supported by a program grant of The Netherlands Foundation for Medical Research (FUNGO), which is subsidized by The Netherlands Organization for the Advancement of Pure Research (ZWO).

We would like to thank Mrs. M. G. C. Platenburg and Mrs. H. Jackson for their skillful assistance.

REFERENCES

(1) Anderson, E. C., Bell, G. I., Peterson, D. F., and Tobey, R. A. Cell growth and division. IV. Determination of volume growth rate and division probability. *Biophys. J.*, 9:246, 1969.

(2) Baserga, R. Biochemistry of the cell cycle: A review. *Cell Tissue Kinet.*, 1:167, 1968.

(3) Becker, A. J., McCulloch, E. A., and Till, J. E. Cytological demonstration of the clonal nature of spleen colonies derived from transplantated mouse marrow cells. *Nature (London)*, 197:452, 1963.

(4) Blackett, N. M., Milliard, R. E., and Belcher, H. M. Thymidine suicide *in vivo* and *in vitro* of spleen colony forming and agar colony forming cells of mouse bone marrow. *Cell Tissue Kinet.*, 7:309, 1974.

(5) van den Engh, G. J. In *Early Events in Haemopoietic Cell Differentiation*. Thesis, Leiden, 1976.

(6) Haskill, J. S., McNeill, T. A., and Moore, M. A. S. Density distribution analysis of *in vivo* and *in vitro* colony forming cells in bone marrow. *J. Cell. Physiol.*, 75:167, 1970.

(7) Haskill, J. S., and Moore, M. A. S. Two dimensional cell separation: Comparison of embryonic and adult haemopoietic stem cells. *Nature (London)*, 226:853, 1970.

(8) Iscove, N. N., Till, J. E., and McCulloch, E. A. The proliferative states of mouse granulopoietic progenitor cells. *Proc. Soc. Exp. Biol. Med.*, 134:33, 1970.

(9) Lahiri, S. K., and van Putten, L. M. Distribution and multiplication of colony forming units from bone marrow and spleen after injection in irradiated mice. *Cell Tissue Kinet.*, 2:21, 1969.

(10) Lahiri, S. K., and van Putten, L. M. Location of the G_0-phase in the cell cycle of the mouse

haemopoietic spleen colony forming cells. *Cell Tissue Kinet.*, *5*:365, 1972.

(11) McDonald, H. R., and Miller, R. G. Synchronization of mouse L-cells by a velocity sedimentation technique. *Biophys. J.*, *10*:834, 1970.

(12) McCulloch, E. A. Les clones de cellules hématopoïétiques "in vivo." *Rev. Fr. Etud. Clin. Biol. 8*:15, 1963.

(13) Messner, H., Till, J. E., and McCulloch, E. A. Density distributions of marrow cells from mouse and man. *Ser. Haematol.*, *5*:22, 1972.

(14) Metcalf, D., and Moore, M. A. S. In: Neuberger, A., and Tatum, E. L., eds., *Haemopoietic Cells.* Amsterdam: North-Holland Publishing Company, 1971.

(15) Metcalf, D., Moore, M. A. S., and Shortman, K. Adherence column and buoyant density separation of bone marrow stem cells and more differentiated cells. *J. Cell. Physiol.*, *78*:441, 1971.

(16) Micklem, H. S., Anderson, N., and Ross, E. Limited potential of circulating haemopoietic stem cells. *Nature (London)*, *256*:41, 1975.

(17) Miller, R. G., and Phillips, R. A. Separation of cells by velocity sedimentation. *J. Cell. Physiol.*, *73*:191, 1969.

(18) Miller, R. G., Separation of cells by velocity sedimentation. In Pain, R. H., and Smith, B. J., eds., *New techniques in Biophysics and Cell Biology.* London: John Wiley and Sons, vol. I, 1973.

(19) Monette, F. C., Gilio, M. J., and Chalifoux, P. Separation of proliferating CFU from G_0 cells of murine bone marrow. *Cell Tissue Kinet.*, *7*: 443, 1974.

(20) Shortman, K. The separation of different cell classes from lymphoid organs. II. The purification and analysis of lymphocyte populations by equilibrium density gradient centrifugation. *Aust. J. Exp. Biol. Med. Sci.*, *46*:375, 1968.

(21) Terasima, T., and Tolmach, L. J. Growth and nucleic acid synthesis in synchronously dividing populations of Hela cells. *Exp. Cell Res.*, *30*:344, 1963.

(22) Till, J. E., and McCulloch, E. A. A direct measurement of the radiation sensitivity of normal mouse bone marrow cells. *Radiat. Res.*, *14*:213, 1961.

(23) Till, J. E., McCulloch, E. A., and Siminovitch, L. A stochastic model of stem cell proliferation based on the growth of spleen colony-forming cells. *Proc. Natl. Acad. Sci. U.S.A.*, *51*:29, 1964.

(24) Trentin, J. J., and Fahlberg, W. J. An experimental model for studies of immunologic competence in irradiated mice repopulated with "clones" of spleen cells. In *Conceptual Advances in Immunology and Oncology, 16th Annual Symposium, Houston.* New York: Harper and Row, 1963, p. 66.

(25) van Bekkum, D. W., van Noord, M. J., Maat, B., and Dicke, K. A. Attempts at identification of hemopoietic stem cell in mouse. *Blood, 38*:547, 1971.

(26) Vassort, F., Winterholer, M., Frindel, E., and Tubiana, M. Kinetic parameters of bone marrow stem cells using *in vivo* suicide by tritiated thymidine or by hydroxyurea. *Blood, 41*:789, 1973.

(27) Williams, N., and Moore, M. A. S. Sedimentation velocity characterization of the cell cycle of granulocytic progenitor cells in monkey haemopoietic tissue. *J. Cell. Physiol.*, *82*:81, 1973.

(28) Williams, N., and van den Engh, G. J. Separation of subpopulations of *in vitro* colony forming cells from mouse marrow by equilibrium density centrifugation. *J. Cell. Physiol.*, *86*:237, 1975.

(29) Worton, R. G., McCulloch, E. A., and Till, J. E. Physical separation of hemopoietic stem cells differing in their capacity for self-renewal. *J. Exp. Med.*, *130*:91, 1969.

(30) Wu, A. M., Till, J. E., Siminovitch, L., and McCulloch, E. A. Cytological evidence for a relationship between normal hemapoietic colony-forming cells and cells of the lymphoid system. *J. Exp. Med.*, *127*:455, 1968.

INTRODUCTION

These investigations deal with an attempt to identify and characterize the multipotential stem cell, present in peripheral blood of dogs. Various morphologic and functional tests have been employed in an endeavor to accomplish this task. Multipotential stem cells are known to be present in blood because of the complete and lasting regeneration of hemopoiesis induced by transfusion of leukocyte suspensions collected from blood by leukocytopheresis (6).

However, in the allogeneic situation, a graft-versus-host reaction (GvHR) occurs which presents a severe threat to the duration of the take and to survival (13). Previous studies by other groups have dealt with the feasibility of purification of bone marrow cell suspensions in order to increase the proportion of hemopoietic stem cells and to eliminate the immune-competent cells responsible for GvHR (3–5).

With this in mind we have set out to attain the following goals:

1. The collection of *blood* mononuclear cells (including stem cells) by leukocytopheresis.

2. An increase in the yield of collected CFU-c (a monitor for the presence of multipotential stem cells in the dog model) by means of administration of mobilizing agents, such as dextran sulfate.

3. The characterization (morphologic and functional) of blood mononuclear cells endowed with stem cell potential.

4. The segregation of cells responsible for restoration of the various hemopoietic systems.

5. The allogeneic (as well as autologous) transplantation of sufficient numbers of segregated stem cells to restore specific cell lineages and to allow survival.

EXPERIMENTAL APPROACH

The techniques used will be described in detail elsewhere (8). They allow the collection and purification of a relatively concentrated stem cell suspension from the peripheral blood of dogs (1–3-year-old beagles). The procedure is comprised of three steps: the collection of a leukocyte-rich suspension, containing stem cells, from the peripheral blood; the elimination of most "contaminating" erythrocytes from the suspension; and the segregation of a stem cell-rich suspension by density gradient centrifugation.

Figure 1 shows the overall approach schematically: the donor dog, having a surgically

Characterization of Bone Marrow and Lymph Node Repopulating Cells by Transplanting Mononuclear Cells into Radiated Dogs

W. M. Ross, M. Körbling,
W. Nothdurft, W. Calvo,
and T. M. Fliedner

4

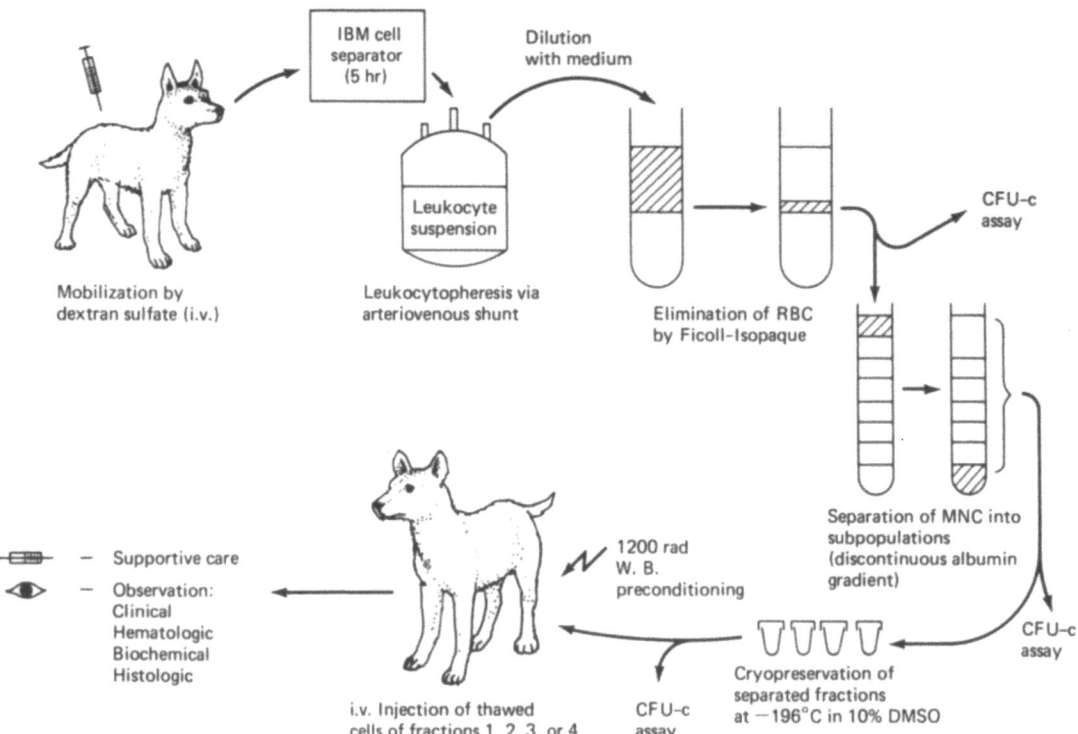

FIGURE 1. Schematic presentation of experimental approach for characterization and allogeneic transfusion of hemopoietic stem cells from blood.

inserted arteriovenous shunt, was given 15 mg/kg body weight of a stem cell-mobilizing agent, dextran sulfate, intravenously; stem cell mobilization is based on the estimation of CFU-c content. Thirty minutes later connections were established between the shunt and an IBM Continuous Flow Cell Separator for a 5 hr leukocytopheresis. Leukocytes were collected in a Fenwal Transfer Pack, and afterward diluted 1:3 with Hanks's medium, containing heparin and antibiotics.

Next, most erythrocytes (and some of the leukocytes) were eliminated by placing the suspension over Ficoll-Isopaque and centrifuging; the "ring" suspension was removed, washed in Hanks's, suspended in 17% albumin, and analysed for CFU-c content in soft agar culture, based on the method described by Bradley and Metcalf (1).

Finally, the leukocyte-rich suspension was placed atop a discontinuous albumin gradient, similar to that used by Dicke (3), and centrifuged to segregate the leukocytes on the basis of cell density. The gradient osmolality and other factors had to be adapted for use with dog leukocytes; the optimal conditions were determined in a pilot study. Each fraction was removed separately and cryopreserved in 10% DMSO at −196°C.

CFU-c assays were performed before and after cryopreservation, and it appears that over 90% of the CFU-c were recoverable on thawing. CFU-c, considered to be committed stem cells, are assumed to be indicative of the presence of multipotential stem cells, although there is no evidence, as yet, to indicate the kind of relationship between the two. A 10-ml aliquot of fraction 1, 2, 3, or 4 was injected intravenously into a DLA-identical, MLC-negative, recipient dog, preconditioned with 1200 rad wholebody x-radiation just prior to transfusion. Supportive care was necessary to bring the dog through the gastrointestinal phase of the acute radiation syndrome. The animals were placed under observation for their entire time of survival, with regular monitoring of the hematologic, biochemical, and clinical states for signs of recovery and/or GvH reaction.

RESULTS

MOBILIZATION

Dextran sulfate, when dissolved in saline to a concentration of 15 mg/ml, can be administered intravenously at a dosage of 15 mg/kg body weight. This will cause a mobilization of stem cells (CFU-c) into the peripheral blood. A maximum mobilization of CFU-c is observed about 3 hr after injection (Table 1). Thereafter, a gradual decline occurs over the next few hr until normal values are again reached after 24 hr. As a control, the same dogs were given the same injection volume of saline 1–2 weeks prior to the dextran sulfate experiments; no mobilization was seen. Several other dogs have received 15 mg/kg and have also displayed a mobilization of CFU-c from less than 100 to about 2000 CFU-c per ml blood within 3 to 5 hr after dextran sulfate administation. Leukocytic mononuclear cells (MNC) were mobilized by a factor of 2 to 3 within 3 hr.

LEUKOCYTOPHERESIS

Table 2 summarizes the results of blood leukocyte collection from several leukocytophereses. All collections were performed using the IBM Cell Separator on different dogs (with the exception of dog 5, which was used twice). The first group (dogs 1–4) received no dextran sulfate, while the second group (dogs 5–7) did. The differences in yield between the two groups are quite evident. In the 5 hr of leukocytopheresis, about 400 ml of leukocyte-rich cell suspension were collected; *without* mobilization, about 1×10^9 polymorphonuclear cells (PMN) and a like number of MNC were collected, whereas *with* mobilization there were about 6 times as many PMN and 9 times as many MNC.

TABLE 1 Influence of a Mobilizing Agent (Dextran Sulfate) on the Concentration of CFU-c in Peripheral Blood of Dogs[a]

HOURS AFTER INJECTION	SALINE	15 MG/KG
0	26.7 ± 6.3	203 ± 58
1	45.7 ± 5.7	708 ± 125
3	89.6 ± 11.7	1986 ± 614
5	74.2 ± 12.0	1060 ± 94
7	72.3 ± 10.4	824 ± 80

[a]Mean ± 1 S.E.; $N = 4$.

Under conditions of mobilization, it was usual to find many more MNC than PMN in the collected suspension. Most striking of all was the increase in the number of CFU-c collected, about 60 times greater than when no dextran sulfate was given. Erythrocyte numbers were recorded only to have a rough estimate of the position in the centrifuge bowl from which the cells were collected.

PURIFICATION OF A CFU-c SUSPENSION

The collected cell suspension was diluted and placed on Ficoll-Isopaque to remove most erythrocytes. The loss of mononuclear cells was variable from experiment to experiment (30–50%). The resultant cell suspension was washed, and placed on a discontinuous albumin gradient. The resultant distribution of leukocytes is shown diagramatically in Fig. 2. Most of the polymorphonuclear cells (and erythrocytes) went to the most dense fractions, 5 and 6. Lymphocytes were seen primarily in fractions 3 and 4. CFU-c were mainly to be found in fraction 2 (sometimes 2 and

TABLE 2 Total Cell Yield of Various Blood Cell Types from Leukapheresis of Peripheral Blood (Dogs) Using the IBM Continuous Flow Cell Separator, with or without a Mobilizing Agent[a]

DOG NO.	MOBILIZING AGENT	PMN ($\times 10^9$)	MNC ($\times 10^9$)	CFU-c ($\times 10^5$)	RBC ($\times 10^9$)
1	—	1.08	1.09	0.37	51.00
2	—	1.11	1.07	0.50	82.00
3	—	1.07	1.05	0.33	73.00
4	—	0.48	0.84	0.71	121.00
5	+	2.30	6.50	50.00	69.00
5	+	4.20	6.30	34.00	105.00
6	+	7.60	12.00	14.00	95.00
7	+	10.04	10.02	27.00	98.00

[a]Dextran Sulfate; 15 mg/kg body weight.

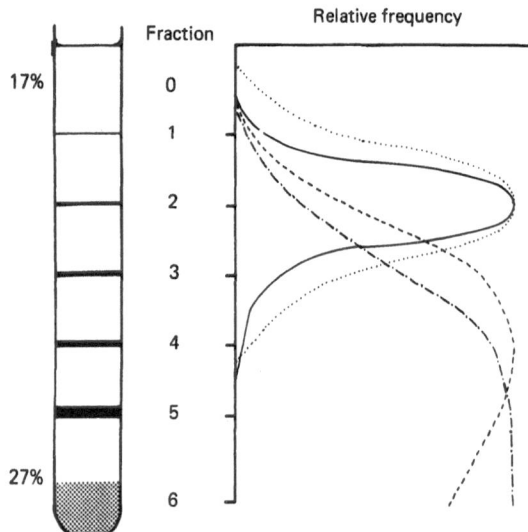

FIGURE 2. Representative distribution frequency of blood leukocytes along a discontinuous albumin gradient (diagrammatic): Monocytes ....; CFU-c ——————— Lymphocytes ---------; PMN -.-.-.-. .

3), as were monocytes. The cell suspension loaded on the gradient usually contained 1 CFU-c per 500 MNC. In one experiment, fraction 2 had 1 CFU-c per 13 MNC, a remarkable purification. Fractions 3 and 4 had about 1 CFU-c per 500 MNC and 1 CFU-c per 7000 MNC, respectively.

EM MORPHOLOGY

In an attempt to characterize multipotential stem cells morphologically, an electronmicroscopic study was performed which uncovered two possible candidates. Figure 3a shows an undifferentiated element, not yet displaying morphologic features characteristic of any particular

cell line; it was found in fractions 2 and 3. Another candidate, also undifferentiated, is shown in Fig. 3b; it was found in fractions 2, 3, and 4.

HEMOPOIETIC REGENERATION AND SECONDARY DISEASE

The data from two sets of experiments (7 dogs) are summarized in Table 3. All dogs received a frozen and thawed MNC suspension within 6 hr after radiation. Only those dogs receiving fraction 2 cells were able to survive and to show a permanent restoration of hemopoiesis; these two dogs are still alive (to date, 7 and 3 months, respectively). All other dogs died within 1 month.

At the time of autopsy, the dog that had received cells of fraction 1 and that died 11 days later showed very poor lymph node cellularity. On the other hand, the dogs receiving cells from fraction 3 showed, at death on days 25 and 32, a partial repopulation of the paracortical area of the lymph nodes. To our surprise, the dogs receiving fraction 4 cells exhibited a remarkable plasma cell hyperplasia of the lymph nodes. There was a massive infiltration of all areas of the lymph node by plasma cells (Fig. 4).

All dogs demonstrated a "take" in the bone marrow on day 10, characterized by the appearance of myelopoietic, erythropoietic, and/or megakaryopoietic cells in bone marrow aspirates. However, only those dogs transfused with cells of fraction 2 had, on day 10, a well-advanced hemopoietic recovery. Hemopoietic regeneration appeared to be transitory in the others.

On autopsy, the dog receiving fraction 1 cells displayed some hemopoietic regeneration in the marrow. The dogs receiving fraction 3 or 4 showed less regeneration, but there had apparently been a decline in the degree of hemopoietic

TABLE 3 Transfusion of Cells from Different Gradient Fractions to Restore Hemopoietic Systems: Cellular Effects and Clinical Outcome[a]

GRADIENT FRACTION	TRANSFUSED CELLS (× 10⁶)		SURVIVAL (DAYS)	LYMPH NODE CELLULARITY[a]	BONE MARROW HEMOPOIESIS[b]	GvH SYMPTOMS		
	MNC	CFU-c				A	B	C
1	0.7	0.02	11	− −	+++	0	0	0
2	30.0	3.23	Alive				0	0
	24.0	0.75	Alive				0	0
3	400.0	0.33	25	±	++	+	+	++
	290.0	1.01	32	±	++	+	+	++
4	900.0	0.12	21	+++	+	+	+	+
	765.0	0.12	16	+++	+	0	0	0

[a]A = Autopsy findings; B = biochemical results; C = clinical observations.
[b]At time of death.

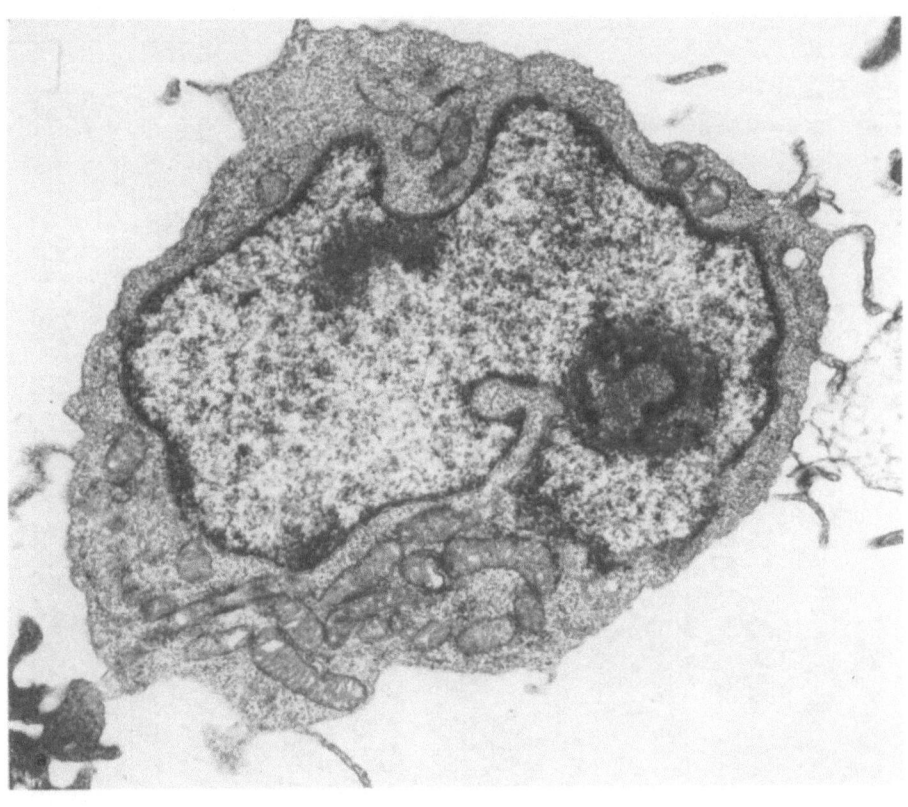

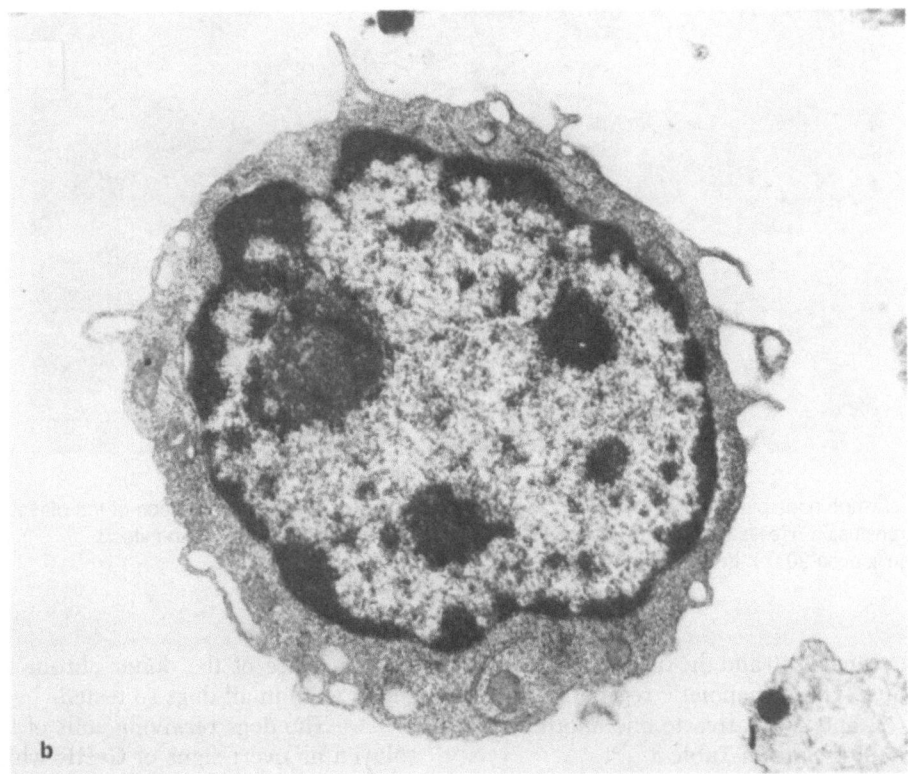

FIGURES 3a and 3b.   Candidate hemopoietic stem cells found among fraction 2 cells from the discon-tinuous albumin gradient (EM magnification: 22,800×).

33

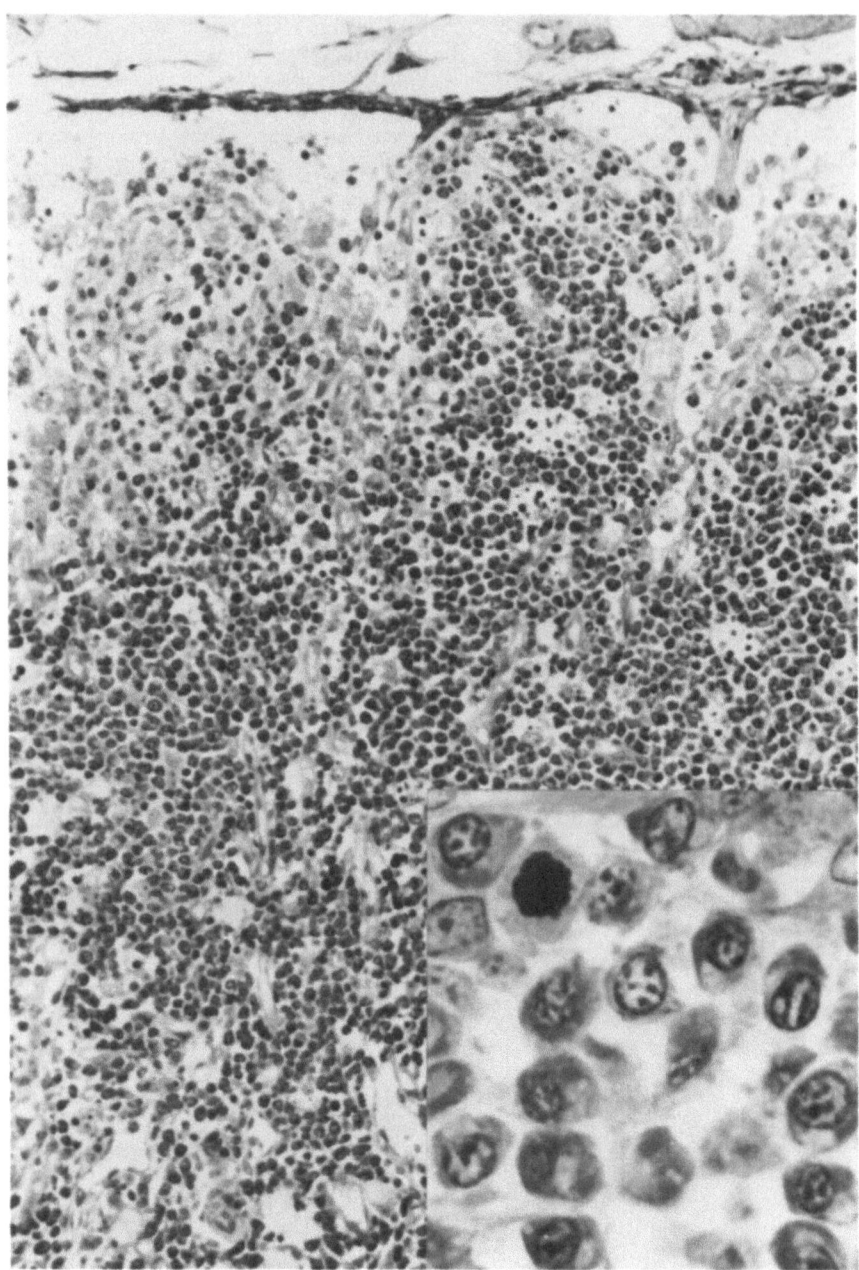

FIGURE 4. Lymph node plasma cell hyperplasia following transfusion of cells from gradient fraction 4 (magnification 200×). Insert (lower left) gives greater magnification of the plasma cells (1250×) and shows one in mitosis.

activity between day 10 and the time of death several days later. The hemopoietic regeneration of fractions 1, 3, and 4 (relative to one another) is indicated by plus signs in Table 3.

In some cases, the hemopoietic cells, those in mitosis, could be shown to be of donor origin only. Due to a sex difference between donor (XX) and recipient (XY), cytogenetic analysis revealed the presence of the donor chromosome complement (XX) in all dogs so tested.

The dogs receiving cells of fraction 2 displayed no overt signs of GvHR while those with fraction 3 or 4 had obvious signs and symptoms of GvH disease, mainly alterations in the mucous membranes. Exceptions were the dog with fraction 1 cells and one dog with fraction 4 cells, both of which died rather early, the latter from bronchial pneumonia. A range of biochemical tests was performed regularly on serum, the most indicative being a marked increase in the serum glutamic oxaloacetic transaminase (GOT) level shortly before death. On autopsy, tissue changes characteristic of GvH reaction were noted, especially in the intestinal tract, liver, and skin. In the skin, there were small areas of basal cell vacuolization and infiltration by mononuclear cells.

Recovery of myelopoiesis and of the stem cell pool was indicated (Table 4) by the level of CFU-c and granulocytes in peripheral blood. Following radiation and transfusion, the CFU-c population fell immediately to zero. In dogs that received fraction 2 cells, samples from blood and bone marrow on day 5 were negative with respect to CFU-c; however, on day 10, the marrow samples contained 55 CFU-c per 0.2×10^6 MNC. On the other hand, CFU-c reappearance in blood was later, between days 15 and 20. A normal concentration was not attained until day 50. When cells of fraction 3 or 4 were given, CFU-c did not usually reappear in the marrow; if they did, it was in minimally low numbers, and usually on only a single occasion between days 10 and 17. CFU-c did not reappear in the blood before death.

Granulocytes, after a rapid fall to zero on day 5 or 6 in peripheral blood following radiation, reappeared quite soon in dogs receiving fraction 2 cells. A fairly respectable level was attained by day 27. The second dog to receive fraction 2 cells did not attain as high a level for several weeks; he had been transfused with fewer CFU-c (only 0.75×10^6, as opposed to 3.23×10^6, CFU-c).

Table 4 shows the data for one experiment with respect to granulocyte recovery and number of CFU-c transfused. It is clear that a correlation exists not only with respect to the rate of recovery of the granulocyte level in blood, but also to the level attainable. Reappearance was slightly earlier, and the plateau in the recovery phase was higher and more quickly reached, when a greater number of CFU-c were transfused.

DISCUSSION

Leukocytopheresis performed on peripheral blood of dogs to obtain stem cells for hemopoietic restoration was shown to be effective in both autologous and allogeneic situations (2, 6). With the establishment of suitable cryopreservation methods, any desirable number of mononuclear blood cells can be accumulated for transfusion purposes. The separation and purification of the hemopoietic stem cells from immune-competent, GvH-reactive cells appears feasible for routine use. Dicke and others (5, 7) have attempted this with bone marrow cell suspensions and found that a very pure stem cell suspension is possible, but with few cells remaining that are capable of transfusion and effective hemopoietic restoration. With the procedure used in our present study it appears possible to prepare any desired number of blood-derived stem cells.

STEM CELL MOBILIZATION

Certain polyanions have been shown capable of mobilizing mononuclear cells into the peripheral blood of rats and other mammals (11, 12). Dextran sulfate of low molecular weight has proven to be able to mobilize not only MNC, but also CFU-c, and seems to be nontoxic (10, 16). In preliminary experiments for the present study, this heparinoid polyanion was observed to increase greatly the yield of CFU-c from blood leukapheresis. Although the mechanism and site of action have not yet been elucidated, it is likely that multipotential stem cells are also mobilizable by dextran sulfate. This must remain as an assumption, based on the correlation between number of CFU-c transfused and the degree of hemopoietic regeneration; only the multipotential stem cells would be capable of restoring hemopoiesis and maintaining it over an extended period of time. A mobilization of hemopoietic

TABLE 4 Recovery of Granulocyte Cellularity in Peripheral Blood following Lethal Radiation and Transfusion of Different Gradient-Separated Leukocyte Suspensions

FRACTION	CFU-c TRANSFUSED ($\times 10^6$)	CELL NUMBER PER MM³ BLOOD							
		Preradiation	Day 5	Day 6	Day 10	Day 20	Day 64	At death	(day)
1	0.02	8812	0	12.5	282.0	—	—	387.5	(11)
2	3.23	5355	0	14.0	312.0	2054.0	7744.0	—	
3	0.33	6262	2.5	0	109.0	240.0	—	244.0	(25)
4	0.12	4237	0	0	12.5	42.0	—	42.0	(21)

stem cells has also been observed soon after endotoxin or about 21 days after specific chemotherapy (9, 15).

The relatively large number of stem cells so made available encouraged us to attempt the purification of a stem cell suspension by density gradient centrifugation.

TRANSFUSION OF SEPARATED CELL SUSPENSIONS

When a leukocyte suspension derived from a single 5 hr leukocytopheresis was purified, the resultant gradient fraction 2 still had more than enough CFU-c (and multipotential stem cells also, as evidenced by the rapid hemopoietic regeneration) for transfusion into radiated hosts. The loss of important cells seemed to be relatively small during the separation procedures, and the segregation of immune-competent cells quite effective, as judged from the results obtained by transfusing fraction 2 cells.

The CFU-c assays revealed that there was no functional impairment of CFU-c due to cryopreservation. In addition, the capacity for hemopoietic regeneration was, also, not noticeably impaired. These results imply the feasibility of a "cyrobank" of purified stem cell suspensions ready for immediate thawing and transfusion in the event of an emergency, such as a radiation accident.

While fraction 2 had the most stem cells of the type with complete hemopoietic restoration potential, fraction 4 apparently contained quite a different stem-cell subpopulation. These cells had proliferative potential which was directed towards plasma cell differentiation in the lymph nodes, resulting in a marked hyperplasia. This has not, to date, been observed in dogs that had received whole, unseparated leukocyte suspensions (6). This might be indicative of cell interaction and of the influence exerted by different cell types—or different proportions of them—on the differentiation of stem cells when a given segregated leukocyte suspension is transfused. On the other hand, the previous leukocytophereses were not accompanied by mobilization. Therefore, it could also be possible that lymphopoietic-determined stem cells are also mobilized by dextran sulfate, traversing the gradient to settle in fraction 4 with a small number of CFU-c and multipotential stem cells. In this case, these lym-

phopoietic stem cells would not be capable of forming colonies in the so-called CFU-c assay and would not, therefore, be detectable.

STEM CELL MORPHOLOGY

Electronmicroscopic studies have revealed a multipotential stem cell candidate (Fig. 3a) which has the cellular structure that would be expected of a multipotential stem cell. It is an undifferentiated element, not belonging as yet to any particular cell line, and its presence in fraction 2 and absence in fraction 4 correlates well with the capability for complete hemopoietic restoration. Figure 3b is an electron micrograph of a second candidate, found in both fractions 2 and 4. Its presence in fraction 4 may lessen the likelihood that it is a multipotential stem cell. This second candidate also displays the primary characteristics of the presumptive stem cell suggested by van Bekkum in 1971 (14) for mice. However, it may not be possible to define a multipotential stem cell by a single definite set of morphologic parameters. Such a cell may only be identifiable by its general morphologic appearance: that is an undifferentiated cell without any morphologic characteristics of a specific hemopoietic lineage.

The yield of stem cells from leukocytopheresis of blood can be significantly increased by mobilizing agents, such as dextran sulfate. The blood MNC population can be segregated into subpopulations by using density gradient centrifugation (a discontinuous albumin gradient. In fraction 2 of such a gradient, the presence of the bulk of CFU-c signals the presence of stem cells capable of restoring primarily the bone marrow cell systems. Furthermore, the separation procedure appears to allow the segregation of cells that are specifically endowed with the potential to repopulate the lymph nodes with plasma cells. The biologic and clinical significance of this phenomenon remains to be investigated. This purification also leads to a minimization or elimination of the cells responsible for instigation of GvH reaction, thus increasing greatly the chances for survival.

Finally, the findings suggest the use of blood stem cells in the treatment of aplastic anemia and combined immune deficiency disease, and following severe radio- or chemotherapeutic regimens. Studies are in progress to transfer this preclinical model to the clinical level.

SUMMARY

The present investigations deal with an attempt to identify and characterize the multipotential stem cell present in mononuclear cell (MNC) suspensions collected from the peripheral blood of dogs by leukocytopheresis; various morphologic and functional tests have been employed in an endeavor to accomplish this task. Blood MNC suspensions contain such stem cells since they have been found capable of repopulating, on a long-term basis, the hemopoietic system of lethally radiated dogs. Leukocytes were collected from the peripheral blood by an IBM Experimental Blood Cell Separator. In an attempt to increase the yield of MNC and, in particular, of stem cells, the presence of which was assumed to be indicated by CFU-c (colony forming units in agar), dextran sulfate (DS) was administered i.v. (15 mg/kg) 30 min before the beginning of leukocytopheresis. DS has been found to be an effective CFU-c mobilizing agent, capable of increasing the number of CFU-c in peripheral blood by 7 to 10 times within 3 hr. During a 4 hr leukocytopheresis, about $6.5–14 \times 10^9$ MNC were collected. To eliminate erythrocytes, a Ficoll-Isopaque gradient was employed. A discontinuous albumin gradient was prepared with 6 fractions (17–27% albumin, 350 mOsm) in an attempt to obtain a cell suspension with an improved ratio of CFU-c to PHA-reactive lymphocytes. Lymphocytes accumulated predominantly in fractions 4 to 6. CFU-c were found primarily in fraction 2; one cell out of 13 MNC was a CFU-c and 82% of the CFU-c was found here. In contrast, the majority of PHA-responsive cells was found in fractions 3 and 4. Morphologic examination of the cells in fraction 2 by light and electron microscopy revealed a fairly good correlation between a particular type of MNC and the number of CFU-c determined to be present therein. Various fractions (1, 2, 3, and 4) were cryopreserved and stored at $-196°C$ until ready for transplantation into lethally radiated (1200 rad, wholebody) dogs of the opposite sex, to allow cytogenetic detection of chimerism. It will be shown that, in four recipients closely related to the donor, each having received a different fraction, a "take" was obtained in all instances despite the relatively low numbers of MNC transfused. The rate of regeneration showed a correlation with the number of CFU-c transfused, as measured by peripheral blood and bone marrow cellularity and, ultimately, by histologic examination of hemopoietic tissues at the time of death. There was a transient regeneration of the marrow with fraction 4, and very strong lymph node repopulation (plasma cell hyperplasia) and severe GvH disease attendant. This would seem to indicate that this fraction had a large number of lymphopoietic cells and few multipotential stem cells. On the other hand, fraction 2 produced a strong, permanent take in the marrow, as well as regeneration of lymphopoietic tissues. This can be interpreted as evidence that this fraction had a much larger number of multipotential stem cells than fraction 4. It is to be emphasized that in the dog receiving fraction 2 there was a complete absence of GvH symptoms, in contrast to the dog receiving fraction 4. The repopulation of blood cell-forming organs was accomplished by donor cells, as evidenced by chromosomal studies. These studies support our contention that transfusion of a MNC suspension rich in CFU-c (mobilized by DS and "purified" on a discontinuous albumin gradient) is a valuable tool in the treatment of hemopoietic failure and in the attempt to identify the elusive multipotential stem cell.

ACKNOWLEDGMENTS

We would like to thank the following, among others, for their assistance in the performance of this project: Miss E. Rüber, Miss I. Fache, Mrs. I. Schöntag, and Mr. S. Binder. Special thanks go to Dr. M. Dakkouri for preparing the dextran sulfate, Dr. H. P. Schnappauf for his surgical skills, and Dr. I. Steinbach for her skillful animal care.

REFERENCES

(1) Bradley, T. R., and Metcalf, D. The growth of mouse bone marrow cells in vitro. *Aust. J. Exp. Biol. Med. Sci., 44*: 287, 1966

(2) Calvo, W., Fliedner, T. M., Herbst, E. Hügl, E., and Bruch, C. Regeneration of blood-forming organs after autologous leukocyte transfusion in lethally-irradiated dogs. II. Distribution of cellularity of the marrow in irradiated and transfused animals. *Blood, 47*: 593, 1976

(3) Dicke, K. A., and van Bekkum, D. W. Avoidance of acute secondary disease by purification of haematopoietic stem-cells with density gradient centrifugation. *Exp. Hematol., 20*: 126, 1970

(4) Dicke, K. A., and van Bekkum, D. W. Allogeneic bone marrow transplantation after elimination of immunocompetent cells by means of density gradient centrifugation. *Transplant. Proc., 3*: 666, 1971

(5) Dicke, K. A., and van Bekkum, D. W. Preparation and use of stem-cell concentrates for restoration of immune deficiency diseases and bone marrow aplasia. *Rev. Eur. Etud. Clin. Biol., 17*: 645, 1972

(6) Fliedner, T. M., Flad, H. D., Bruch, C., Calvo, W., Goldmann, S. F., Herbst, E., Hügl, E., Huget, R., Körbling, M., Krumbacher, K., Nothdurft, W., Ross, W. M., Schnappauf, H. P., and Steinbach, I. Treatment of aplastic anemia by blood stem cell transfusion: A canine model. *Haematologica, 61*(2):141, 1976

(7) Good, R. A., and Bach, F. H. Bone marrow and thymus transplants: Cellular engineering to correct primary immunodeficiency. *In* Bach, F. H., and Good, R. A., eds., *Clinical Immunology.* New York: Academic Press, vol. 2, p. 63.

(8) Körbling, M., Nothdurft, W., Ross, W. M., Goldmann, S. F., Calvo, W., Fache, I., Flad, H. D., and Fliedner, T. M. In-vitro and in-vivo properties of canine blood mononuclear leukocytes separated by discontinuous albumin density gradient centrifugation. Proceedings of 7th Workshop on Leukocyte Cultures, Ulm, April, 1976. Submitted to *Biomedicine.*

(9) Richman, C. M., Weiner, R. S., and Yankee, R. A. Increase in circulating stem-cells following chemotherapy in man. *Blood, 47*: 1031, 1976

(10) Ricketts, C. R., Walton, K. W., van Leuven, B. D., Birbeck, A., Brown, A., Kennedy, A. C., and Burt, C. C. Therapeutic trial of the synthetic heparin analogue dextran sulphate. *Lancet, II*: 1004, 1953

(11) Ross, W. M., Martens, A. C., Robev, S., and van Bekkum, D. W. Lymphocytosis induced by polyanions in rats. (Submitted to *Chem. Biol. Interact.*)

(12) Ross, W. M., Martens, A. C., and van Bekkum, D. W. Polymethacrylic acid: Induction of lymphocytosis and tissue distribution. *Cell Tissue Kinet., 8*: 467, 1975

(13) Storb, R., Epstein, R. B., Ragde, H., Bryant, J., and Thomas, E. D.: Marrow engraftment by allogeneic leukocytes in lethally irradiated dogs. *Blood, 30*: 80, 1967

(14) van Bekkum, D. W., van Noord, M. J., Maat, B., and Dicke, K. A.: Attempts at identification of hemopoietic stem-cell in mouse. *Blood, 38*: 547, 1971

(15) Vos, O., Buurman, W. A., and Ploemacher, R. E. Mobilization of haemopoietic stem-cells (CFU) into the peripheral blood of the mouse: Effects of endotoxin and other compounds. *Cell Tissue Kinet., 5*: 467, 1972

(16) Walton, K. W.: Investigation of the toxicity of a series of dextran sulphates of varying molecular weight. *Br. J. Pharmacol., 9*: 1, 1954

PART II

Humoral and Cellular Control Agents

E. P. Cronkite, M.D.

INTRODUCTION

In dogs the neutrophilic granulocyte count in the blood responds quickly to infection. Marsh et al. (22) showed that neutrophils were two to three times normal within 6 hr after induction of bacterial pneumonia in dogs. Furthermore, the ratio of nonsegmented to segmented neutrophils increased progressively, showing a rapid release from the bone marrow of less mature cells. Under normal steady-state conditions there are modest fluctuations in the blood granulocyte count, presumably the result of an interplay among factors regulating the sojourn time in the blood, rate of release of granulocytes from the bone marrow, their rate of production in the bone marrow, and the factors that inhibit marrow granulocyte production. In fact, Morley et al. (26) believe there is a periodic cycling of normal granulocyte count, in man, as a result of this interplay.

The average sojourn time of granulocytes in the blood of man is very variable, with a large standard error (5). After induction of infection, marked changes in the granulocytic specific activity curve (GSAC) are observed, following autotransfusion of labeled cells. Early in infection the GSAC shows a rapid decrease, implying a faster input of unlabeled cells from the bone marrow. Later, the GSAC approaches normal and then, during recovery from infection, becomes longer than normal, implying that input of unlabeled granulocytes from the marrow is drastically reduced (22). There may, however, be other explanations for this phenomenon. The mechanism of release of granulocytes from the marrow has been reviewed by Cronkite and Vincent (6).

It appears clear that leukopenia, endotoxin, infection, and sterile inflammation result in production of increased serum levels of leukocytosis-inducing factor (LIF) (17). This factor, when administered to animals, produces a prompt granulocytic leukocytosis, a prompt decrease in number of mature granulocytes in the marrow, and a shorter transit time through the nonproliferating bone marrow pool (13). The relationship of this factor to colony-stimulating factor (CSF) is not clear. However, in ascribing a biologic importance to a factor isolated from blood plasma one must be certain that it is not contaminated with endotoxin—a caveat not always heeded.

Low molecular weight-inhibitors of granulopoiesis have been shown to be present in sera of animals and in media conditioned by mature granulocytes. The granulocytic chalone was ap-

Concepts and Observations on the Regulation of Granulocyte Production

E. P. Cronkite, H. Burlington,
A. D. Chanana, D. D. Joel,
U. Reincke, and J. Stevens

5

41

parently first described well by Rytomaa and Kiviniemi (31, 32). Subsequent studies are reviewed by MacVittie and McCarthy (21), who have also shown, in an *in vivo* diffusion chamber (DC) culture system, that the granulocytic chalone has no detectable effect upon the pluripotent stem cells (CFU-s), and that a major locus of its action is on the committed granulocytic stem cell (CFU-c), thus decreasing the amplification at this level. Whether it also suppressed production in the differentiated granulocytic pool is not clear. However, the net result is a reduction in yield of granulocytes in the DC.

The extensively studied CSF has major effects in *in vitro* culture systems and is the leading candidate for a granulopoietic hormone (granulopoietin which may be comparable to erythropoietin). An unquestioned *in vivo* effect has not yet been demonstrated. The *in vivo* studies of Metcalf and Stanley (24) with human urinary CSF produced granulo- and monocytosis in neonatal mice, and less striking effects in adult mice. Their material was not checked for presence of endotoxin. The CSF acts presumably upon a precursor of the granulocyte and macrophage series, depending upon time in culture and other factors. With longer times in culture, macrophage colonies tend to predominate. CSF, its action, and its properties have recently been reviewed (35).

From the experience in studying the regulation of erythropoiesis and the assignment of roles for erythropoietin one gains insight into problems that confront similar studies on granulopoiesis. Studies on erythropoiesis were facilitated by the availability of reliable and accurate techniques for the measurement of red cell life span and of the mass of circulating red cells, led to the estimation of the red cell death rate in the blood. Similar techniques for study of the

granulocyte, although available in part, are very time-consuming and poorly reproducible, and true granulocyte life span is in question (14). These, coupled with the lability of the granulocyte count (shifts between marginal and circulating pools of granulocytes and diurnal variation), its life in hours compared to months for the red cell, along with uncertainties about the existence of a tissue pool of granulocytes in addition to blood and marrow pools, complicates the study of granulopoiesis and its regulation. Ideally, one should know the entire quantitative structure (number/kg), from the earliest stem cell through all proliferating, differentiated cytological stages, the mature nondividing pool in the bone marrow and in the blood, and whether there is a tissue pool that exchanges with the blood and marrow. The preceding data, combined with relative production rates derived from DNA or mitotic time and fraction in DNA or mitosis (convenient markers), allow an estimation of absolute production rates. A DNA marker allows measurement of the replacement rate at the first nondividing stage in the production line, and also of the minimum time from the last myelocyte division to the appearance of the first cells in the peripheral blood. DNA labeling is complemented by, and must be consistent with, measurement of the DNA content by microspectroscopy of successive, cytologically defined, proliferating compartments of granulopoiesis.

A DETERMINISTIC MODEL OF GRANULOCYTOPOIESIS IN MAN

Cronkite and Vincent (6) have developed a model of granulocytopoiesis in man, shown in Fig. 1, which is based on: the tritiated thymidine labeling index (7), the mitotic index (18–20), the time for DNA synthesis of 12 hr (37), the proportional distribution of granulocytic cell types (6), the absolute number in different phases of granulocytic cell types (6), the absolute number of different phases of granulocytopoiesis (12), the flow from the proliferating to the nonproliferating pool (6, 7), the time from labeling at the myelocyte level until release into the peripheral blood (16), and the peripheral granulocyte turnover rate determined by DFP-32 studies, primarily from the Salt Lake City group (5). Full details of the logic and data that go into this model are presented by Cronkite and Vincent (6). In summary, there is an influx of approximately 0.4 × 10⁷/kg of stem cells per hr into granulocytopoiesis. At the myeloblast level there is one mitosis with a cycle time of

FIGURE 1. Deterministic model for human granulopoiesis. See text for details. The area under the curve (flux × time) represents the size of each phase in number per kilogram.

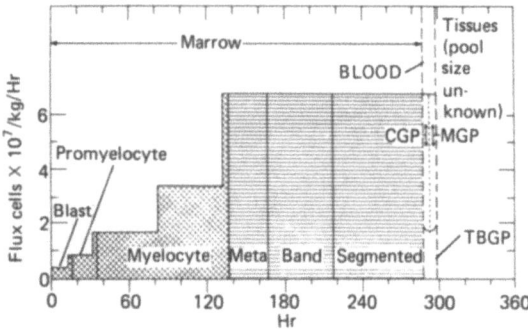

about 18 hr. At the promyelocyte stage there is another mitosis with a cycle time of approximately 24 hr. The myelocyte undergoes two successive myelocytic divisions, with cycle times of about 54 hr each. The average time spent in the metamyelocyte compartment has been determined by the replacement of unlabeled metamyelocytes by tritiated thymidine-labeled metamyelocytes, and is shown to be approximately 3.3 per hr, for an average replacement time of 33 hr (7). From the proportional distribution of cells, nearly 60 hr is spent on an average at the band-segmented level, and nearly 70 hr at the segmented neutrophil level in the bone marrow (7). The average time spent in the blood of segmented granulocytes is about 9.8 hr (27). Since DNA synthesis time is about 12 hr, the G_2 about 1–3 hr, and mitosis less than 1 hr, one can estimate G_1 to be about 39 hr. Hence, the fraction of diploid myelocytes should be equal to the ratio of the time in G_1 to the total cycle time, or approximately 72%.

Our studies summarized herein are aimed at elucidating the mechanisms by which the mammal produces and sustains a granulocytosis, using granulokinetic methodology after ^{3}H-thymidine-labeling of DNA by intravenous injection combined with sterile inflammation and measurement of DNA content of cells. In addition, we have taken advantage of a granulocytosis-producing and CSF-secreting tumor to study possible mechanisms of action of CSF *in vivo*.

MATERIALS AND METHODS

Sterile inflammation was induced in female mongrel dogs by injecting 2 ml of turpentine into the uterus under general anesthesia (36). For cell labeling ^{3}H-thymidine (SA 1.9 Ci/mM) was injected intravenously in a dosage of 0.2 μCi/g body weight in a single injection.

Bone marrow and blood samples were taken at regular intervals to observe changes in proportion of cell types in the bone marrow, development of granulocytosis, flow of labeled cells through the bone marrow, diminution in grain count over labeled cells, ratio of mitotic figures (myeloid-erythroid), and appearance of labeled cells in the peripheral blood. Bone marrow aspirations were done after local anesthesia.

DNA measurements techniques will be described in detail elsewhere (1). The procedures for Wright-Giemsa staining of blood and bone marrow, and for radioautography have been published (15).

RESULTS
DNA CONTENT OF HUMAN MYELOCYTES

In Fig. 2, the DNA content of myelocytes is plotted against the projected surface area of the nuclei of the myelocytes. The DNA content is on the ordinate and the nuclear area on the abscissa. This study was made with the hope that nuclear area, a simpler measurement, would substitute for the more time-consuming DNA measurement. In fact, for myelocytes of man this is satisfactory. The number of diploid DNA myelocytes is approximately 70%, as required by the deterministic model, thus fulfilling this requirement. This model allows one to suggest that a potential mechanism of increasing granulocyte production may reside at the myelocyte level. For example, if there were a mechanism by which the G_1 phase of nearly 39 hr could be substantially reduced, one could then have more mitoses during the same time interval and give a prompt and greater amplification of the stem cell-input into granulocytopoiesis.

INFLUENCE OF STERILE INFLAMMATION ON GRANULOPOIESIS

Peripheral Blood Granulocytes Granulocyte changes after the injection of turpentine are seen in Fig. 3. The band and segmented neutrophils decrease from an average of 4400/mm^3 to 2500/mm^3 at 1 hr after injection of turpentine due to margination from anesthesia and migration into inflammatory area. This was followed by a very rapid initial linear increase in the granulocyte count at a rate of 2500/mm^3/hr for the ensuing 5 hr. The rate of increase thereafter decreased and the maximum granulocyte count was reached 13 hr after induction of inflammation. Thereafter, it declined nearly linearly up to 56 hr, and then more slowly. During the linear decrease the concentration was decreasing at about 250/mm^3/hr.

One can estimate the apparent entrance rate of granulocytes into the blood by correcting for the loss of granulocytes analogous to Deubelbeiss et al. (10, 11), who studied serial changes in GSAC after injection of ^{3}H-thymidine in dogs. In Table 1 the crude estimates of rate of entry of granulocytes into the blood are tabulated. In making these estimates certain assumptions had to be made in respect to total body granulocyte pool (TBGP) and granulocyte half-time in the blood. On the basis of the studies of Deubelbeiss et al., it was assumed that the circulating granulocytic pool (CGP) and the marginal granulocytic pool (MGP) were about equal in the control period, and

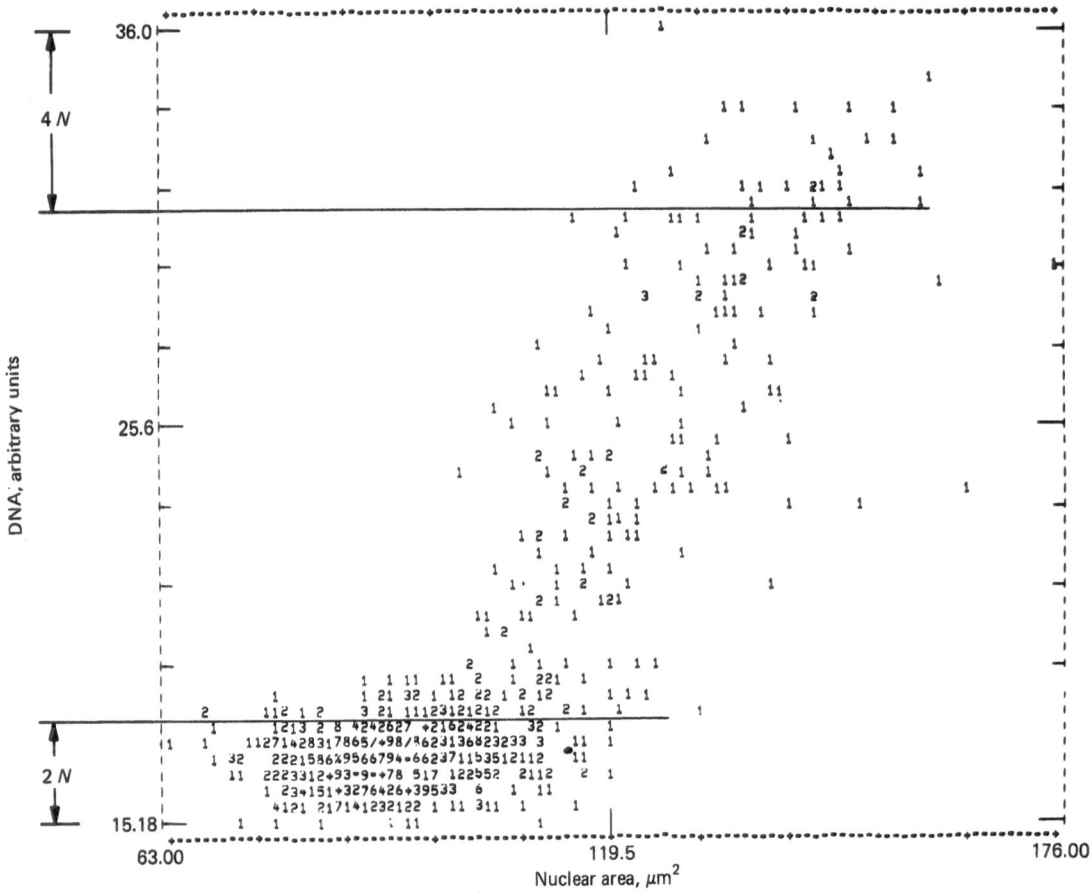

FIGURE 2. DNA content of myelocytes plotted against projected nuclear area.

FIGURE 3. Granulocytosis produced by injection of turpentine in dogs and by repeated bone marrow sampling in dogs.

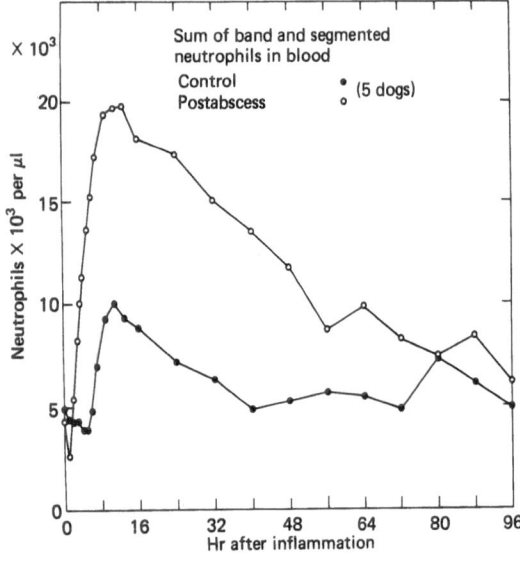

TABLE 1 Apparent Rate of Entry of Granulocytes into the Peripheral Blood of Dogs After Sterile Inflammation Induced by Intrauterine Turpentine[a,b]

TIME AFTER INDUCTION OF INFLAMMATION (HOURS)	RATE OF ENTRY OF GRANULOCYTES	
	Number of ccm³/hr × 10⁻³	Number per kg/hr × 10⁹
0	0.82	0.063
1[c]	3.01	0.22
4	4.2	0.31
7	5.56	0.41
8	4.38	0.32
10	4.32	0.31
12	3.35	0.25
14	1.50	0.11
16	1.46	0.10
24	1.28	0.09
32	1.10	0.08
40	0.97	0.07
48	0.77	0.06
56	0.61	0.05

[a] Assuming CGP = MGP.

[b] CGP = circulating granulocyte pool. MGP = marginal granulocyte pool. TBGP = total body granulocyte pool.

[c] 1 through 56 hr it is assumed that CGP approximates the TBGP.

we used the average half-time obtained by these authors of 7.6 hr. Since Marsh et al. (22) showed that the average half-time in the blood decreased promptly after infection was induced, we used their half-time of 2.8 hr, corrected to 4.09 hr for elution of DF^{32}P (10, 11). This half-time was used for all calculations made after induction of inflammation, even though Marsh et al. (22) suggested that entrance into the blood decreased and almost ceased in later stages after acute infection. With these caveats, one notes that the apparent input into the blood is 3.7 × the control level 1 hr after inflammation, increases to 6.8 × the control value by 7 hr after inflammation, and, thereafter, declines, and by 48 to 56 hr is below normal "steady-state" levels. It is also to be noted that, during the control period, repeated blood sampling and bone marrow aspirations under local anesthesia, and the resulting subcutaneous hemorrhage and trauma to bone, resulted in a lesser degree of granulocytosis.

FRACTION OF MYELOCYTES IN DNA SYNTHESIS AND INTENSITY OF LABELING

These studies in progress show that the labeling index of myelocytes during the control period rose to a maximum of 16%. After the induction of sterile inflammation, the fraction of myelocytes labeled rose more rapidly and attained nearly 40%, remaining above the control level throughout the 96 hr of the study. Such observations have been made in three dogs and suggest a shortened cycle time at the expense of the G_1 phase of the cell cycle, and also imply that there are more mitoses at the myelocytic level.

Further insight is also obtained from an analysis of the mean grain count overlying the myelocytes. After inflammation the mean grain count rises more rapidly and to a much higher level than in the control period, and declines to control values within about 25 hr, suggesting either that the DNA synthesis time is shortened or that tritiated thymidine is more efficiently incorporated into DNA during an inflammatory process. In addition, the rapid decrease to control levels implies a much greater rate of proliferation at the myelocyte level with a shorter cycle time and more successive mitoses to increase the amplification of the stem cell input.

RATIO OF MYELOID TO ERYTHROID MITOSIS

Another line of evidence suggesting a greater production rate at the myelocyte level is obtained from a study of the ratio of myeloid to erythroid mitotic figures. This is illustrated in Fig. 4. The ratio of myeloid to erythroid mitoses is shown on the ordinate and on the abscissa the time after induction of inflammation is indicated. Note the nearly constant ratio of myeloid to erythroid mitotic figures during the control period. After the induction of inflammation there is a detectable increase in the myeloid erythroid mitotic ratio by 4 hr after inflammation. This ratio progressively rises to a maximum 24 hr after inflammation, and then falls; but at 96 hr after inflammation it is still higher than the control levels. These studies imply that there are more mitoses within the proliferative granulocytic compartments unless mitotic time has become longer after inflammation. The latter appears unlikely at a time when the granulocyte production rate is markedly increased.

TURNOVER RATE OF METAMYELOCYTES

The replacement of unlabeled metamyelocytes by labeled metamyelocytes is faster after the induction of sterile inflammation. For example, in our studies and also in those of Ohkita (27), the normal replacement rate of metamyelocytes is 2% to 3% per hr. After inflammation the replacement rate varied from 4% to 8% per hour. The flow through the storage compartment from metamyelocytes through band neutrophils and segmented neutrophils into the peripheral blood is normally of the order of 60–70 hr in the dog. This is shortened to as little as 24 hr after induction of inflammation.

These studies in the dog, albeit incomplete, strongly suggest that peripheral inflammation is sensed in the marrow within 1 hr, and a sequence of events is initiated in the differen-

FIGURE 4. The ratio of myeloid to erythroid mitotic figures in bone marrow (BM) during control period and after sterile inflammation.

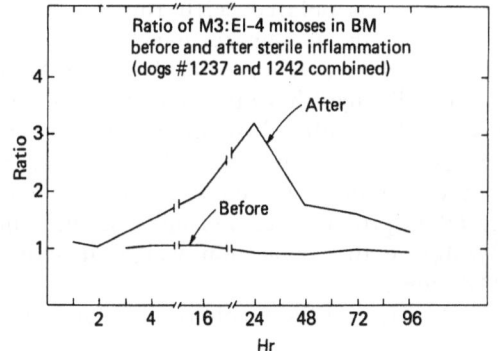

tiated pools of the bone marrow. First, stored marrow band and segmented neutrophils are mobilized. Whether mobilization of blood marginal neutrophils precedes or is concomitant with that of marrow neutrophils is not clear. This is followed by a much shorter transit time through the nonproliferating granulocytic pool in the bone marrow with more mitoses in the myelocytic compartment to give greater amplification of the stem cell input. Of course, this does not preclude, in addition, a greater input of stem cells into granulopoiesis.

STUDIES ON GRANULOPOIESIS AFTER INOCULATION OF A GRANULOCYTOSIS-PRODUCING TUMOR IN THE MOUSE

During the studies on ^{55}Fe-erythrocytocide (29) in the mouse, an adenocarcinoma of the mammary glands developed which was associated with a striking granulocytosis of up to and exceeding 100,000/mm^3. This transplantable tumor is very malignant and kills mice within 3 weeks. Upon transplantation, within 3 to 4 days a granulocytosis develops, increasing exponentially for about 2 weeks. Extirpation of the tumor results in a temporary decrease in the granulocytosis. This tumor has been placed in tissue culture and grows either as a monolayer or in suspension culture without difficulty. Burlington et al. (3) have shown that in monolayer and suspension cultures it consistently produces colony-stimulating activity (CSA) assayable in the culture medium. CSA is present in successive media collections made at 3–8-day intervals over a period extending up to 75 days. Heat abolishes the CSA activity and unmasks an inhibitor of CSA. The media conditioned by primary cultures of mouse lung and kidney, and the cell line derived from rat prostatic tumor associated with leukemia development, had negligible CSA when assayed simultaneously as controls for the mammary tumor culture media. Ultrafiltration studies on the CSA associated with the adenoma-carcinoma of the mammary gland shows it to have a molecular weight greater than 30,000 daltons. This distinguishes this factor from that of the material isolated from another granulocytosis-producing tumor by Delmonte et al. (9), in which they showed the granulocytosis material is protein-free, thermally stable, and dialyzable, with a molecular weight less than 2000 daltons.

To equate CSA to colony-stimulating fac-

tor (CSF), one needs many lines of evidence. One strong bit of evidence that the material is close to CSF comes as a result of studies with antibody (supplied by Dr. Richard Shadduck, Montefiore Hospital, Pittsburgh, Pa.) to L-cell CSF (34). Antibody to L-cell CSF completely neutralized the activity of tumor-conditioned medium at dilutions up to 1:64. Antibody dilution at 1:128 neutralized only 20% of CSF of L-cell-conditioned medium and still abolished 72% to 88% of the activity of the tumor culture-conditioned medium. This suggests CSA is immunologically closely related to CSF.

IN VIVO EFFECTS OF INOCULATION OF GRANULOCYTOSIS-PRODUCING TUMOR

A reproducible sequence of events is observed in the hematopoietic tissues after inoculation. The first observable change is an increase in the band neutrophils in the blood, observed 3 days after inoculation. This is followed by rapid increases in the segmented neutrophils and bands in the peripheral blood. These increase exponentially to a maximum around 10–15 days after inoculation. The total cellularity, the granulocytic, and the myelocytic populations of the spleen decrease during the first 3 days, apparently by mobilization into the blood. This is then followed by striking increases in the granulocytic and myelocytic populations of the spleen, peaking at 15 days after inoculation. The bone marrow myelocytic and granulocytic cells increased modestly 10–15 days after inoculation.

There are also striking serial changes in the CFU-s and the CFU-c populations of the spleen and bone marrow. In the spleen, the CFU-s and CFU-c decrease during the first 3 days, presumably due to differentiation into the granulopoietic lines. This is then followed by striking increases, peaking 14 days after inoculation at levels many times the control level.

In the bone marrow CFU-s changes seen were surprising. They are below the control levels at most points, suggesting there is a very rapid differentiation of these cells. Unfortunately, the studies on the fraction of CFU-s in DNA synthesis did not satisfactorily answer the question of whether there is a greater number of cells in DNA synthesis with greater production rate over that of the control animals. The bone marrow CFU-c follows the same trend as the splenic CFU-c, with a lesser magnitude of change. It appeared as if the spleen were the primary site of action in producing granulocytosis. Accordingly, Carsten et al. (4)

splenectomized mice and inoculated tumors. The splenectomized animals developed a comparable leukocytosis.

The effect of inoculating the CSA-producing tumor is successively seen in:

1. Mobilization of stored cells in the spleen and bone marrow

2. An increased production rate at the myelocyte level, as observed in the dog

3. An increased differentiation from stem cells in granulopoiesis, as indicated by the initial drop in the number of CFU-s and CFU-c, particularly in the spleen

4. Increased production and number of CFU-s and CFU-c, particularly in the spleen, and the latter to a lesser extent in the bone marrow

These studies clearly show that a tumor which produces CSA in culture has a striking effect at all levels of the production pathway for granulocytopoiesis. Its activity is inhibited *in vitro* by antibody against L-cell-conditioned medium containing CSF. It is logical to ask the question of whether one should conclude that these observations prove that CSF is an *in vivo* hormone acting at the CFU-c and/or CFU-s level. The evidence is not conclusive. The locus of effect in the dog and in the mouse in these studies appears to move successively from the storage pool to the myelocytic level to the CFU-c to the CFU-s. However, to conclusively prove this point will require repetitive serial studies with more frequent sampling intervals.

WORKING HYPOTHESIS ON REGULATION OF GRANULOPOIESIS

In our studies and many other studies the first event is the release of cells from the storage pool. This is followed by an increased amplification of the input of stem cells into granulopoiesis, primarily at the myelocytic level. The release of granulocytes from the bone marrow in several studies published in the literature appears to be humorally mediated by leukocytosis-inducing factor. The mediation of increased myelocyte production is unclear. Is it from a humoral stimulation, or from an intramedullary short-range humoral feedback loop (hemopoietic inductive microenvironment—HIM) that is sensed when the storage pool decreases in size? Does a reduction in the segmented cells in the marrow remove an inhibitory (chalone) effect on myelocyte production? The same applies to the increased proliferation in the CFU-c and CFU-s. Is it mediated by a long-range humoral effect from the periphery, or is there some intramedullary sensing HIM by these more primitive cells of changes in the size of the differentiated, more mature pools of cells in the bone marrow?

There is growing evidence from the work of Vassort et al. (39) on *in vivo* tritiated thymidine suicide and Reincke et al. (30) on iron-55 erythrocytocide and others suggesting strongly that there are, in fact, *in vivo* fast-acting intramedullary feedback loops that are mediated by obscure mechanisms resulting in rapid reduction in the CFU-s stem cell population, suggesting an intramedullary direction of differentiation of CFU-s into the differentiated cell lines. A reduced suppressive effect of granulocytes in the bone marrow may be largely responsible for the effects observed. *In vivo* regulation of granulocytopoiesis is not clearly established. However, humoral factors effective *in vitro*, chalone suppression, and intramedullary sensing of changes in population density with quick responses to these changes, are emerging as probable determinants in the regulation of granulocytopoiesis.

SUMMARY

Granulopoiesis is regulated within physiologic ranges by a series of feedback loops, faster and slower, that determine rates of release and production rates to meet the variable needs. Population size thus fluctuates within modest limits. The granulopoietic system also responds quickly to life threatening events such as bacterial infection. The initial response is release of mature cells from the marrow by leukocytosis inducing factor (LIF) thus depleting the storage pool. Whether the reduction in size of the storage pool that is closely intermingled with the proliferating and stem cell pools is sensed by these more immature precursors through a cell-to-cell interaction is a matter of conjecture at this point. In principle, the production rate may be accelerated by: (1) increased differentiation of stem cells down the granulopoietic pathway, (2) increased amplification of stem cell input, or (3) a combination of both. Sterile inflammation produces: (1) leukocytosis, (2) decreased transit time through bone marrow, (3) decreased fraction of myelocytes with 2n DNA content, (4) decrease in DNA synthesis time, (5) increase in ratio of myelocytic to erythroid mitoses by factor of 3–4. These data show in the dog that granulocyte production rate is in part regulated by an increase in the amplification of the stem cell input by increasing the number of mitoses in the differentiated granulopoietic pool with shorter mean cycle time. Granulocytosis producing tumors (GPT) have been described in mice by others. The serendipitous discovery of one in our mice permitted exploitation of its mechanism of action. It grows in culture and releases colony stimulating activity (CSA) into the culture medium. Studies with the CSA producing tumor in mice show: (1) band and segmented neutrophils increase in blood by mobilization; (2) absolute splenic cellularity, number of granulocytic stem cells in spleen, number of CFU-s in spleen increase strikingly, commencing 3 days after inoculation of tumor; (3) the bone marrow CFU-s remains constant as does cellularity but the CFU-c content increases some—all of these sustain the mobilization granulocytosis; (4) splenomegaly becomes evident but splenectomy does not prevent granulocytosis. In splenectomized tumor inoculated mice there is little change in the bone marrow CFU-s but a major expansion of the CFU-c in bone marrow. The marrow is sufficient itself to produce the granulocytosis.

From these results in dogs and mice it is concluded that granulocytosis and increased granulocytes production rates is accomplished by humoral (LIF) mobilization of stores and *in vivo* CSA stimulated expansion of granulocytic committed and pluripotent committed stem cell pools with concomitant increase in amplification in the differentiated granulocytic poliferative pool. LIF and CSA may be the same or related entities.

ACKNOWLEDGMENT

This work was supported by the U.S. Energy Research and Development Administration and NIH Grant # HL 15685. The research described in this report involved animals maintained in animal care facilities fully accredited by the American Association for Accreditation of Laboratory Animal Care.

REFERENCES

(1) Boecker, W. R., Ernst, P., Cronkite, E. P., Killmann, S., and Robertson, J. S. Correlation of human myelocyte size, DNA content and synthesis: Implications for granulocytopoiesis. To be published.

(2) Bradley, T. R., and Metcalf, M.D. The growth of mouse bone marrow cells *in vitro*. *Aust. J. Exp. Biol. Med. Sci., 44*: 287, 1966.

(3) Burlington, H., Cronkite, E. P., Laissue, J. A., Reincke, U., and Shadduck, R. K. Colony stimulating activity in cultures of granulocytosis inducing tumor. Submitted to *Proc. Soc. Exp. Biol. Med.*

(4) Carsten, A. L., Burlington, H., Cronkite, E. P., and Reincke, U. To be published.

(5) Cartwright, G. E., Athens, J. W., and Wintrobe, M. M. The kinetics of granulopoiesis in normal man. *Blood, 24*: 780, 1964.

(6) Cronkite, E. P., and Vincent, P. C. Granulocytopoiesis. *Ser. Haematol., 3*: 43, 1969.

(7) Cronkite, E. P., Bond, V. P., Fliedner, T. M., and Killmann, S. A. The use of tritiated thymidine in the study of haemopoietic cell proliferation. In Wolstenholme, G. E. W., and O'Connor, M., eds., *Ciba Foundation Symposium on Haemopoiesis*. London: J. and A. Churchill Ltd., 1960.

(8) Cronkite, E. P. To be published.

(9) Delmonte, L., and Liebelt, R. A. Species and strain dependence of the response to a granulocytosis-promoting factor (GPF) extracted from a mouse tumor. *Proc. Soc. Exp. Biol. Med., 121*: 1231, 1966.

(10) Deubelbeiss, K. A., Dancey, J. T., Harker, L. A.,B., and Finch, C. A. Marrow erythroid and neutrophil cellularity in the dog. *J. Clin. Invest., 55*: 825, 1975.

(11) Deubelbeiss, K. A., Dancey, J. T., Harker, L. A., and Finch, C. A. Neutrophil kinetics in the dog. *J. Clin. Invest., 55*: 833, 1975.

(12) Donohue, D. M., Reiff, R. H., Hansen, M. L., Betson,

Y., and Finch, C. A. Quantitative measurements of the erythrocytic and granulocytic cells of the marrow and blood. *J. Clin. Invest., 37*: 1571, 1958.

(13) Dornfest, B. S., LoBue, J., Handler, E. S., Gordon, A. S., and Quastler, H. Mechanisms of leukocyte production and release. II. Factors influencing leukocyte release from isolated perfused rat legs. *J. Lab. Clin. Med., 60*: 777, 1962.

(14) Dresch, C., Najean, Y., and Bauchet, J. Kinetic studies of ^{51}Cr and DF^{32}P labeled granulocytes. *Br. J. Haematol., 29*: 67, 1975.

(15) Fliedner, T. M., Cronkite, E. P., and Robertson, J. S. Granulocytopoiesis. I. Senescence and random loss of neutrophilic granulocytes in human beings. *Blood, 24*: 402, 1964.

(16) Fliedner, T. M., Cronkite, E. P., Killmann, S. A., and Bond, V. P. Granulocytopoiesis. II. Emergence and pattern of labeling of neutrophilic granulocytes in human beings. *Blood, 24*: 683, 1964.

(17) Gordon, A. S., Neri, R. O., Siegl, C. D., Dornest, B. S., Handler, E. S., LoBue, J., and Eisler, M. Evidence for a leukocytosis inducing factor. *Acta Haematol., 23*: 323, 1960.

(18) Killmann, S. A., Cronkite, E. P., Fliedner, T. M., and Bond, V. P. Mitotic indices of human bone marrow cells. III. Duration of some phases of erythrocytic and granulocytic proliferation computed from mitotic indices. *Blood, 24*: 267, 1964.

(19) Killmann, S. A., Cronkite, E. P., Fliedner, T. M., Bond, V. P., and Brecher, G. Mitotic indices of human bone marrow cells. II. The use of mitotic indices for estimation of time parameters of proliferation in serially connected multiplicative cellular compartments. *Blood, 21*: 141, 1962.

(20) Killmann, S. A., Cronkite, E. P., Fliedner, T. M., and Bond, V. P. Mitotic indices of human bone marrow cells. I. Number and cytologic distribution of mitoses. *Blood, 19*: 743, 1962.

(21) MacVittie, T. J., and McCarthy, K. F. The influence of a granulocytic inhibitor(s) on hemopoiesis in an *in vivo* culture system. *Cell Tissue Kinet., 8*: 553, 1975.

(22) Marsh, J. C., Boggs, D. R., Cartwright, G. E., and Wintrobe, M. M. Neutrophil kinetics in acute infection. *J. Clin. Invest., 46*: 1943, 1967.

(23) Maximow, A. A., and Bloom, W., eds. *Textbook of Histology.* Philadelphia: Saunders, 1935.

(24) Metcalf, D., and Stanley, E. R. Hematological effects in mice of partially purified colony stimulating factor (CSF) prepared from human urine. *Br. J. Haematol., 21*: 481, 1971.

(25) Moore, M. A. S., Williams, N., and Metcalf, D. Clarification and characterization of the *in vitro* colony forming cell in monkey hemopoietic tissue. *J. Cell. Physiol., 79*: 283, 1972.

(26) Morley, A. A., and Melb, M.D. A neutrophil cycle in healthy individuals. *The Lancet*, II: p. 1220, Dec., 1966.

(27) Ohkita, T. The kinetics of granulocytopoiesis. *Acta Haemat. Jap., 30*: 507, 1967.

(28) Pluznik, D. H., and Sachs, L. The cloning of normal blood cells in tissue culture. *J. Cell. Comp. Physiol., 66*: 319, 1965.

(29) Reincke, U., Burlington, H., Cronkite, E. P., Hillman, M., and Laissue, J. Selective damage to erythroblasts by ^{55}Fe. *Blood, 45*: 801, 1975.

(30) Reincke, U., Burlington, H., Cronkite, E. P., and Pappas, N. Iron incorporation into hematopoietic stem cells, assayed by ^{55}Fe cytocide of CFU-s after storage at $-196°$C. To be published.

(31) Rytomaa, T., and Kiviniemi, K. Control of granulocyte production: I. chalone and antichalone to specific humoral regulators. *Cell Tissue Kinet., 1*: 329, 1968.

(32) Rytomaa, T. Granulocytic chalone and antichalone. *In Vitro, 4*: 47, 1968.

(33) Seki, M. Hematopoietic colony formation in a macrophage layer provided by intraperitoneal insertion of cellulose acetate membrane. *Transplantation, 16*: 544, 1973.

(34) Shadduck, R. K., and Metcalf, D. Preparation and neutralization characteristics of an anti-CSF antibody. *J. Cell. Physiol., 86*: 247, 1975.

(35) Stanley, E. R., Hansen, G., Woodcock, J., and Metcalf, D. Colony stimulating factor and the regulation of granulopoiesis and macrophage production. *Fed. Proc., 34*: 2272, 1975.

(36) Stevens, J. To be published.

(37) Stryckmans, P., Cronkite, E. P., Fliedner, T. M., and Ramos, J. DNA synthesis time of erythropoietic and granulopoietic cells in human beings. *Nature (London), 211*: 717, 1966.

(38) Till, J. E., and McCullock, E. A. A direct measurement of the radiation sensitivity in normal mouse bone marrow cells. *Radiat. Res., 14*: 213, 1961.

(39) Vassort, F., Winterholer, M., Frindell, E., and Tubiana, M. Kinetic parameters of bone marrow stem cells using *in vivo* suicide by tritiated thymidine or by hydroxyurea. *Blood, 41*: 789, 1973.

(40) Wickraminsinghe, S. N., and Moffatt, B. Observations on cell proliferation in human myelocytes. *Acta Haematol., 46*: 193, 1971.

(41) Yoshida, K., and Seki, M. *In vitro* colony formation wth layers of peritoneal macrophages and fibroblasts. *Exp. Hematol.* (in press).

The Regulatory Role of the Macrophage in Normal and Neoplastic Hemopoiesis

J. Kurland and M. A. S. Moore

6

INTRODUCTION

The granulocyte-macrophage progenitor cell (colony-forming unit culture, CFU-c) can be detected by its ability to undergo clonal proliferation in semisolid medium when provided with stimulatory macromolecules operationally termed colony-stimulating factor (CSF) (3, 26, 35). A similar if not identical class of molecules also stimulates the proliferation of a precursor cell solely committed to macrophage differentiation, the colony-forming unit-peritoneal macrophage (CFU-pm) (21, 23). The principal cells capable of elaborating CSF have been identified as the monocyte (7, 29) or tissue macrophage (11, 14, 29), and therefore the control of granulopoiesis and macrophage production has been postulated to occur through a positive feedback involving the action of CSF on the two hemopoietic progenitor cells, CFU-c and CFU-pm (21, 27). The requirement for CSF of both the CFU-c and CFU-pm, and the inability of CSF to influence the clonal proliferation of two other hemopoietic cells capable of growth in soft agar (the B-lymphocyte colony-forming cell and the murine myelomonocytic leukemic cell line, WEHI-3 (38)), suggest cell-type specificity of this regulator of *in vitro* granulopoiesis (21). However, it is becoming increasingly apparent that the control of steady-state granulopoiesis and the response of the system to perturbation cannot be accounted for solely on the basis of variations in CSF levels, and that possible inhibitory influences may play an important role in counteracting the proliferative stimulus of CSF.

The questions of the origin and identity of these possible inhibitory factors have been difficult to elucidate since they may represent a diverse class of molecules with a range of activities—for example, those capable of directly inhibiting the stem cell or progenitor cell, or those influencing the production or secretion of CSF. We have recently reported that the E-series prostaglandins (PGE) represent a somewhat unique inhibitor of hemopoietic proliferation, because of their ability to make both the CFU-c and CFU-pm less responsive to the actions of CSF (21). The capacity of increased concentrations of available CSF to counteract the inhibitory effects of PGE suggests a possible physiologic antagonism, not operative in the regulation of B-lymphocyte or WEHI-3 (CSF-independent) colony formation (21, 27).

Though synthetic PGE can suppress CFU-c proliferation, it is important to determine whether its synthesis and release by cells of the hemopoietic system constitute a regulatory influence effected through cellular interactions. We have recently reported that both normal and thioglycollate-activated peritoneal macrophages are capable of elaborating prostaglandin (PG) as determined by a sensitive radioimmunoassay (Kurland and Bockman, manuscript in preparation). This is in agreement with the observations of Bray et al. (4) that purified protein derivative (PPD) primed guinea pig peritoneal exudate cells release PG. The adherent macrophage, and not the nonadherent components of the exudate (i.e., lymphocytes and polymorphonuclear leukocytes), is the source of PG, and though prostaglandin $F_{2\alpha}$ ($PGF_{2\alpha}$) is capable of being produced, PGE is by far the principal prostaglandin found in macrophage supernatants (Kurland and Bockman, manuscript in preparation). The efficacy of indomethacin, a known PG synthesis inhibitor (12), at concentrations as low as $10^{-8} M$, to completely inhibit PGE production by macrophages, makes this agent a useful tool in examining the possible influence of macrophage-derived prostaglandin on hemopoietic proliferation in vitro.

This chapter deals with our investigations into the regulatory role of the macrophage in normal and neoplastic hemopoietic proliferation, and the involvement of the prostaglandins in effecting such responses.

MATERIALS AND METHODS
MICE

Female C57BL/6J mice, obtained from Jackson Laboratories, Bar Harbor, Me., were used when 2 to 3 months of age.

PREPARATION OF CELL SUSPENSIONS

Mouse bone marrow suspensions were prepared in ice-cold serum-free McCoy's 5A Modified Medium as previously described (21, 27).

Human bone marrow was obtained from consenting normal, healthy, adult volunteers by iliac crest aspiration, and treated as previously described (26).

Activated peritoneal exudates were induced with thioglycollate medium (Difco Laboratories, Detroit, Mich.) as described previously (21). This procedure yielded $30-50 \times 10^6$ peritoneal exudate cells (PEC)/mouse, of which approximately 75% were morphologically typical macrophages (Mϕ). The harvested PEC were washed in ice-cold, serum-free medium, and the number of nucleated cells counted and appropriately diluted.

PREPARATION OF ADHERENT PERITONEAL MACROPHAGES

Various concentrations of washed PEC were allowed to adhere to 35 mm Petri dishes (Lux Scientific) in McCoy's 5A Medium containing 15% fetal calf serum (Flow, Rockville, Md.) for 1.5 hr at 37°C, at which time the nonadherent cells of the exudate (comprising mostly lymphocytes and polymorphonuclear leukocytes) were vigorously washed off. The adherent population consisted mainly of macrophages, as determined by their ability to stain with the vital lysosomal stain Neutral Red (37), and comprised approximately 20–30% of the initial PEC population. This was determined by counting the number of adherent cells in arbitrarily chosen 1 mm fields and then multiplying this count by the ratio of the area of the dish to the area counted. Thus, by varying the number of PEC initially plated, the number of adherent macrophages could be varied.

PREPARATION OF ADHERENT MACROPHAGE UNDERLAYERS

To the macrophages remaining adherent to 35 mm dishes was added a 1 ml cell-free 0.5% agar layer in complete McCoy's Medium which constituted a physical separation between the macrophages and those target cells suspended in a 1 ml 0.3% agar overlay whose colony-forming ability was to be investigated. The adherent macrophages, the 0.5% cell-free agar layer, and the 0.3% target cell layer will be referred to as the "two layer soft agar culture system."

COLONY-FORMING ASSAYS

Bone marrow CFU-c Assay This assay was a slight modification of the technique described previously (21). Briefly, one part of 3.0% Bacto-agar (Difco) was added to nine parts prewarmed McCoy's 5A Modified Medium containing 15% fetal calf serum supplemented with additional essential and nonessential amino acids, glutamine, asparagine, and sodium pyruvate. Mouse bone marrow cells were cultured at a concentration of 5×10^4 or 7.5×10^4 cells/ml, and normal human bone marrow at a concentration of 2.0×10^5 cells/ml. One ml suspensions of the agar-bone marrow cell mixture were added to adherent macrophage underlayers or onto cell-free underlayers. Exogenous colony-stimulating factor (CSF) was provided by the addition of concentrated conditioned medium from cultures of WEHI-3 leukemic cells (21) in the case of mouse CFU-c, and 10% concentrated human monocyte-conditioned medium (courtesy of Dr. R. G. Shah) in the case of human CFU-c. Mouse active CSF was standardized in units, as previously described (26). The cultures, incubated for 7 days at 37°C in a humidified atmosphere of 10% CO_2 in air, were scored for the presence of colonies (greater than 40 cells) and clusters (3–40 cells) at $25\times$ magnification.

Hemopoietic Tumor Cell Line Cloning The following cell lines were cultured in the two layer soft agar culture system, as in the CFU-c assay: Wil-2, a human B-cell lymphoma; CCRF-CEM, a human T-cell lymphoma; K562 (Lozzio), derived from a patient with Philadelphia positive chronic myeloid leukemia; EL_4, a T-cell lymphoma of the C57BL/6 mouse; and WEHI-3, a myelomonocytic leukemia of the BALB/c mouse. Briefly, $1-5 \times 10^3$ viable cells in 0.3% agar in McCoy's Medium containing 15% FCS were added either to adherent macrophage underlayers or to control cell-free underlayers. Colony-forming ability was determined after 5–7 days of culture.

LIQUID CULTURE OF HEMOPOIETIC TUMOR CELL LINES

Cells of the five hemopoietic tumor cell lines described above were cultured in the absence or presence of varying numbers of adherent macrophages, as previously described (22). Briefly, 2×10^5 viable cells were added to adherent macrophages from various concentrations of PEC to establish an adherent macrophage-to-hemopoietic tumor cell (AM:HTC) ratio of 1:1 to 10:1. After 72 hr of coculture in McCoy's 5A Modified Medium containing 15% FCS, the HTC were removed for viable cell counts and cytofluorometric analysis. In addition, 5 μCi/ml of ^{3}H-TdR (specific activity, 6.7 Ci/mM; New England Nuclear, Boston, Mass.) was added to duplicate cultures for the final 3 hr of incubation. The incorporation of radioactive precursor into DNA was determined by scintillation counting.

CYTOFLUORIMETRY

Cell cycle analysis of hemopoietic tumor cell lines following periods of liquid culture in contact with adherent macrophages was performed by flow cytofluorimetry (Cytofluorograf, Biophysics Systems, Mahopec, N.Y.) in conjunction with Drs. F. Traganos and Z. Darzynkiewica, Department of Automated Cytology, Sloan-Kettering Institute, New York, according to the method previously described (8, 22, 37). Briefly, detailed analysis of the distribution of individual cells in the G_1, S, and G_2 + M phases of the cell cycle was possible using acridine orange, a metachromatic fluorescent stain, which differentially stains DNA and RNA (9).

INDOMETHACIN

Indomethacin was a generous gift of Dr. Fred Kuehl, Merck and Co., Rahway, New Jersey. Stock solutions of 1×10^{-2} Molar were made up when needed in absolute ethanol and serially diluted, in serum-free McCoy's Medium, to the desired concentrations. Control diluent was ethanol diluted in serum-free medium.

RESULTS
EFFECT OF MACROPHAGES ON THE CLONAL PROLIFERATION OF MOUSE BONE MARROW CFU-c IN THE ABSENCE OF EXOGENOUS CSF

The effect of murine macrophages on the proliferation of mouse bone marrow CFU-c was investigated using a two layer soft agar culture system designed to prevent cell contact but to allow the passage of diffusible factors between the macrophage and bone marrow target cells.

In the absence of exogenous CSF, the macrophage is capable of stimulating the growth of CFU-c with maximal colony number observed with adherent macrophages derived from 0.25×10^6 PEC (Fig. 1). With increasing macrophage number a decline in colony number is observed, with total absence of CFU-c at the macrophage equivalent of 2.5×10^6 PEC. The addition of indomethacin, an inhibitor of prostaglandin synthesis, to the adherent macrophages before the addition of the bone marrow target cells, significantly increased the ability of macrophages to stimulate CFU-c, and allowed proliferation of CFU-c even at a totally suppressive concentration of macrophages. Thus, indomethacin, which by

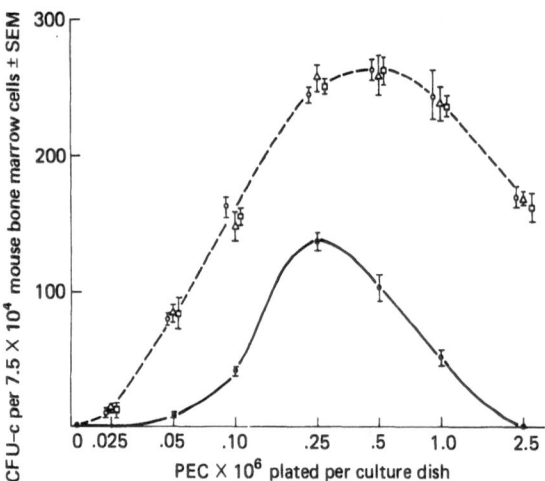

FIGURE 1. The effect of diffusible factors from under-layers of adherent macrophages (Mφ) on the clonal proliferation of syngeneic mouse bone marrow CFU-c. Various concentrations of adherent macrophages were obtained by varying the numbers of peritoneal exudate cells allowed to adhere to 35 mm dishes. A cell-free 0.5% agar layer separated Mφ from the bone marrow cells, preventing cell contact but allowing the passage of diffusible factors. The results are expressed in terms of the capacity of Mφ to stimulate the proliferation of CFU-c in a 0.3% agar overlay, as denoted on the ordinate. Indomethacin, at concentrations of $1.4 \times 10^{-6} M$ (o), $1.4 \times 10^{-7} M$ (Δ), and $1.4 \times 10^{-8} M$ (□), was added in volumes of 0.05 ml to the adherent Mφ immediately prior to the addition of the 0.5% cell-free agar underlayer (-----). Control cultures contained the appropriate indomethacin diluent (●——●). Triplicate cultures were scored for each point.

itself did not stimulate CFU-c in the absence of macrophages, was capable of increasing the net stimulatory ability of a given number of macrophages by suppressing their production of prostaglandin.

EFFECTS OF MACROPHAGES ON THE CLONAL PROLIFERATION OF MOUSE BONE MARROW CFU-c IN THE PRESENCE OF HIGH CONCENTRATIONS OF CSF

In order to examine further the ability of macrophages to decrease colony formation by CFU-c and the possible direct effects of indomethacin on the proliferative ability of the CFU-c, bone marrow cultures were exogenously stimulated by a CSF concentration (1270 units) capable of stimulating maximal colony formation (230 CFU-c/7.5×10^4 mouse bone marrow cells).

As shown in Fig. 2, the presence of relatively high numbers of macrophages was capable of suppressing colony formation by CFU-c. Inhibition of colony formation was evident in those cultures containing adherent macrophages from peritoneal exudate cell (PEC) concentrations greater than 0.1×10^6; such numbers of macrophages were capable of providing only submaximal stimulation in the absence of exogenous CSF (Fig. 1). Indomethacin, which by itself had no effect on the response of the CFU-c to a high CSF concentration, was able to enhance colony formation only in the presence of macrophages, and was capable of partially preventing the suppression of CFU-c by high numbers of macrophages. For example, the adherent macrophages from 1.0×10^6 PEC inhibited CFU-c by 80%; however, in the presence of the prostaglandin synthesis inhibitor, no inhibition was observed.

EFFECT OF MOUSE MACROPHAGES ON THE CLONAL PROLIFERATION OF NORMAL HUMAN BONE MARROW CFU-c

The capacity of the macrophage to release both CSF and PGE complicates the exclusive analysis of the ability of macrophage-derived

FIGURE 2. The effect of Mφ on the proliferation of mouse CFU-c stimulated by a high concentration of CSF. Cultures were prepared in the same manner as in Fig. 1, with the addition of 1270 units of CSF, a concentration capable of stimulating a maximal number of CFU-c in a given bone marrow cell suspension. Indomethacin (----) ($1.4 \times 10^{-6} M$) or control diluent (——) was added to the adherent Mφ, before the addition of the 0.5% agar layer. Results are expressed as a percentage of the maximal number of colonies stimulated by the CSF, in the absence of Mφ. Triplicate cultures were scored for each point.

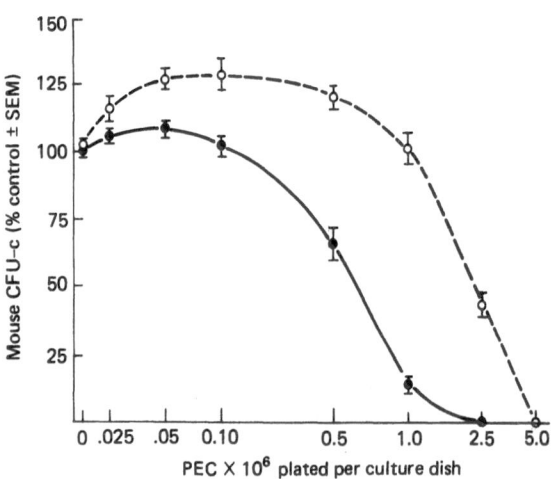

prostaglandin to inhibit CFU-c proliferation, since an increase in the number of macrophages would not only increase the concentration of inhibitory factor(s) per culture but also increase the concentration of CSF. However, because of the species specificity of CSF, due to which human CFU-c are not stimulated by CSF of murine origin, it was possible to examine exclusively the inhibitory action of mouse macrophages uncomplicated by macrophage production of stimulatory factors. Therefore, human bone marrow CFU-c, though not stimulated by mouse macrophage CSF, should theoretically be inhibited by murine macrophage-derived prostaglandin, since PG is known to effect most, if not all, species. Figure 3 shows that normal human bone marrow CFU-c, stimulated by a CSF source of human monocyte-conditioned medium, was in fact inhibited by murine peritoneal macrophages. Indomethacin, which in the absence of macrophages had no effect on colony formation, prevented this inhibition. Thus, in the case of human bone marrow target cells, only the inhibitory activity of murine macrophages was observed, and the ability of indomethacin to prevent inhibition suggests a role for prostaglandin.

FIGURE 3. The effect of varying numbers of adherent mouse macrophages on normal human bone marrow CFU-c stimulated by 10% v/v human monocyte CSF-conditioned medium. Cultures were prepared using 2×10^5 nucleated human bone marrow cells as the target cells in the 0.3% agar overlay. Indomethacin, at concentrations of $1.4 \times 10^{-6} M$ (o), $1.4 \times 10^{-7} M$ (Δ), and $1.4 \times 10^{-8} M$ (□), or control diluent (●) was added, as with the mouse CFU-c cultures. The results are expressed as the number of colonies developing after 7 days of culture. Triplicate cultures were scored for each point.

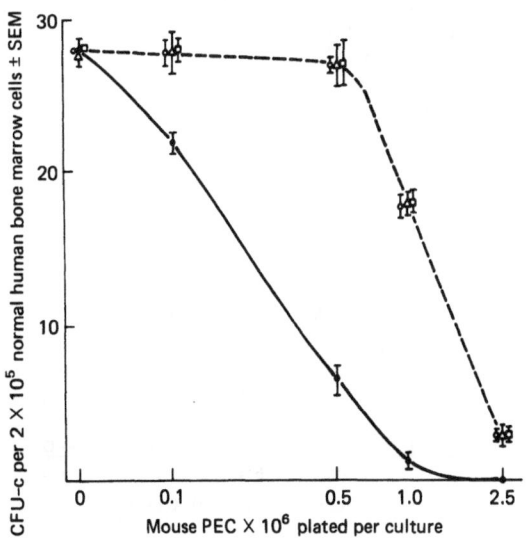

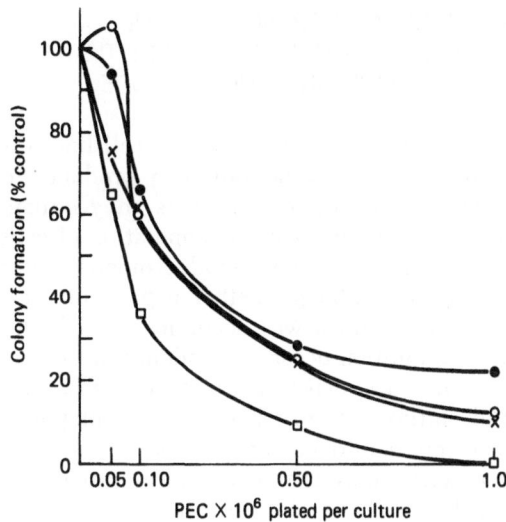

FIGURE 4. The effects of macrophages on the clonal proliferation of four hemopoietic tumor cell lines: WIL-2 (o——o), EL4 (●——●), K562 (Lozzio) (X——X), and WEHI-3 (□——□). In the two layer soft agar assay, $1–5 \times 10^3$ viable cells were cultured, in the absence or presence of adherent Mϕ from various numbers of PEC, and in the absence of exogeous CSF. Mean colony formation was determined at 7 days of culture, and the results expressed as a percentage of colony formation in the absence of Mϕ. Note the slight enhancement of WIL-2 colony formation by low numbers of Mϕ. Triplicate cultures were scored for each point.

EFFECTS OF MACROPHAGES ON THE COLONY-FORMING ABILITY OF HEMOPOIETIC TUMOR CELL LINES

The effects of a soluble factor on the clonal proliferation of leukemic hemopoietic tumor cells were determined in the two layer soft agar culture system used in the quantitation of CFU-c. As shown in Fig. 4, the leukemic cell lines WIL-2, K562 (Lozzio), EL₄, and WEHI-3 were extremely sensitive to the diffusible inhibitory product(s) of adherent macrophages with colony-forming ability, becoming inhibited at lower macrophage numbers than those required for comparble suppression of normal CFU-c. In addition, the ability of macrophages to inhibit clonal proliferation of these four different cell lines suggests that the effects were independent of the species, cell type of origin, and growth characteristics of the tumor cell. Slight enhancement of WIL-2 colony formation was observed with low numbers of macrophages.

EFFECTS OF MACROPHAGES ON THE PROLIFERATION OF HEMOPOIETIC TUMOR CELL LINES IN LIQUID CULTURE

Since the four hemopoietic cell lines displayed sensitivity to the inhibitory product(s) of adherent macrophages, and as they consist primarily of a homogeneous population of cells, they were used in a flow-cytofluorometric determination to establish whether a particular cell cycle-specific block was occurring, or whether growth inhibition occurred randomly in the cell cycle. The basic experimental design required a liquid culture system whereby the hemopoietic target cells could be easily removed for cytofluorometric analysis and for determination of viable cell numbers and ^{3}H-TdR incorporation.

Figures 5A and 5B show that the growth of the four leukemic cell lines was inhibited in liquid culture as a function of the adherent macrophage-to-hemopoietic tumor cell (AM:HTC)

ratio. These effects were not dependent on cell death as determined by trypan blue exclusion, thus suggesting cytostasis rather than cytotoxicity. The inhibition of cell growth in liquid culture was similar to the macrophage inhibition of colony-forming ability in soft agar. There was a parallel inhibition of both tumor cell growth (Fig. 5A) and tritiated thymidine incorporation (Fig. 5B), the magnitude of which was dependent upon the number of macrophages present. However, in contrast to the strong inhibition of EL$_4$ cell growth and colony formation by high numbers of macrophages, there was observed only a moderate decrease in ^{3}H-TdR incorporation at equivalent macrophage numbers. As will be shown later, this discrepancy can be explained by the particular phase of the cell cycle in which EL$_4$ cells were inhibited. The growth of the human B-cell lymphoma, WIL-2, was enhanced by low numbers of macrophages, but was markedly inhibited at AM:HTC ratios of 2.5–10:1.

FIGURE 5. The ability of macrophages to inhibit the growth of hemopoietic tumor cell lines in liquid culture. (A) Suppression of proliferation determined by viable cell counts. Added to 35 mm dishes containing adherent Mφ from various numbers of PEC were 2 × 10^5 viable cells of the four different cell lines. The adherent macrophage-to-hemopoietic tumor cell (AM:HTC) ratio is actually the ratio of PEC initially plated to the number of tumor cells added at the initiation of the culture period. After 72 hr of coculture, viable cell counts of the tumor cells were performed and the results expressed as a percentage of the respective cell numbers of those cultures in the absence of Mφ. Triplicate cultures were used for the determination of each point. (B) Suppression of tritiated thymidine incorporation. To duplicate cultures prepared as in (A) was added 5 μCi/ml ^{3}H-TdR (sp. act., 6.7 Ci/mM) for the final 3 hr of culture; and the incorporation was determined by scintillation counting. Each point represents the mean incorporation expressed as a percentage of the respective cell lines cultured in the absence of macrophages.

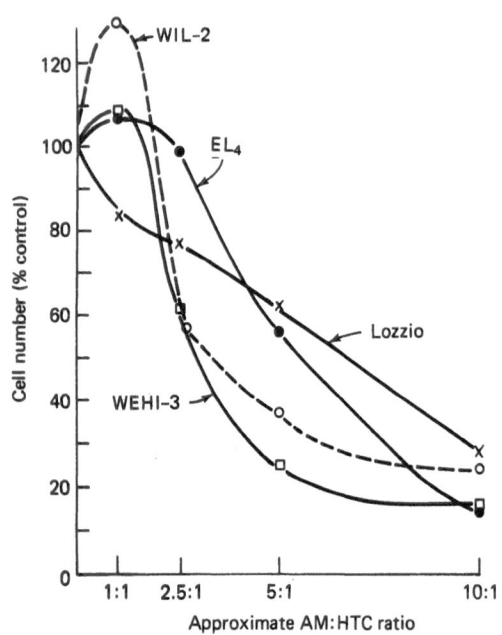

a

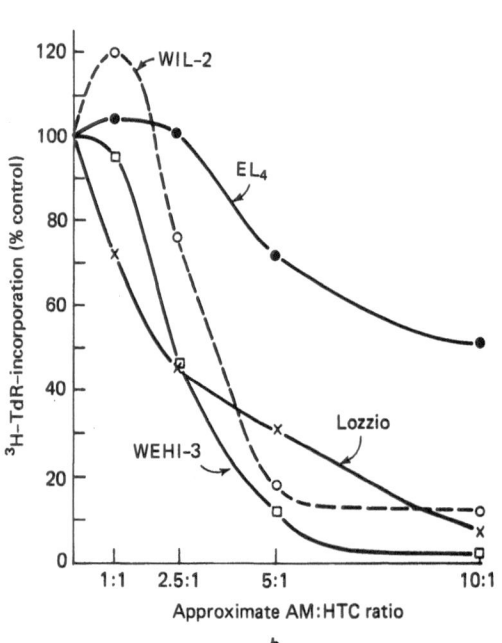

b

CELL CYCLE ANALYSIS OF HEMOPOIETIC CELL LINES CULTURED IN THE PRESENCE OF MACROPHAGES

Cell cycle analysis of the tumor cells, following 72 hr coincubation with macrophages, was performed by flow cytofluorimetry (22). As shown in Table 1, three of the cell lines, WIL-2, CCRF-CEM, and K562 (Lozzio), exhibited a G_1 cell cycle block, which was progressively more intense as the number of macrophages present in the cultures was increased. EL_4 was blocked in the G_2 phase of the cell cycle with an accumulation of cells in S-phase. This explains the apparent discrepancy between the moderate suppression of ^{3}H-TdR incorporation compared with the strong inhibition of cell numbers and colony-forming ability. However, by increasing the macrophage-to-EL_4 cell ratio to greater than 10:1, it was possible to block EL_4 cells in the G_1 phase of the cell cycle. This is in agreement with the fact that EL_4 was relatively resistant to macrophage inhibition, and reverts to its normal cell cycle distribution within 48 hr after being removed from macrophages. WEHI-3, the most sensitive cell line studied, showed a G_1 block as well as a major decrease in red (RNA) and green (DNA) fluorescence. This observed decrease in red fluorescence is a measure of decreased RNA accumulation, which may be attributed to a suppression of RNA synthesis. Work in progress suggests that the dramatic decrease in green (DNA) fluorescence is due to a change in the structure of nuclear chromatin associated with suppressed genome activity, and not attributable to cell death.

EFFECT OF INDOMETHACIN ON THE ABILITY OF MACROPHAGES TO INHIBIT COLONY-FORMATION MURINE MYELOMONOCYTIC LEUKEMIC (WEHI-3) CELLS

As the WEHI-3 cell line was extremely sensitive to the inhibitory activity of the macrophage, we elected to examine a possible role of prostaglandin in the mediation of growth inhibition. WEHI-3 cells were cultured in the two layer soft agar assay in the presence both of untreated macrophages, and of macrophages treated with indomethacin. Figure 6 shows that the macrophage-mediated inhibition of WEHI-3 colony formation can be effectively reduced by this prostaglandin synthesis inhibitor in a manner identical to that observed with human bone marrow CFU-c (Fig. 4). Colony formation by both

FIGURE 6. The effect of indomethacin on the inhibition of myelomonocytic leukemic colony formation by Mϕ in soft agar. WEHI-3 cells were cultured in the two layer soft agar assay, as described in the Materials and Methods section, at a viable cell concentration of 1×10^3 cells/plate in the absence or presence of adherent Mϕ, derived from various concentrations of PEC. Results are expressed as a percentage of WEHI-3 colony formation in the absence of Mϕ (452 colonies/1×10^3 cells). Indomethacin, at a concentration of 1.4×10^{-6} M (o----o), or control diluent ($\bullet$———$\bullet$), was added to adherent Mϕ prior to the addition of the 0.5% cell-free agar layer. Triplicate cultures were scored at 5 days of culture for each point.

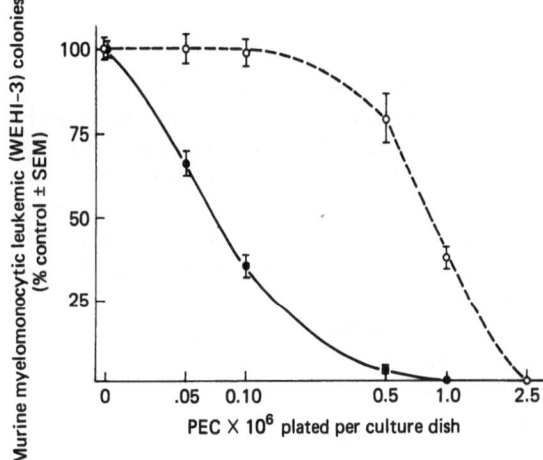

TABLE 1 Nature of Hemopoietic Cell Cycle Inhibition by Macrophages

CELL LINE	CELL TYPE AND ORIGIN	CELL CYCLE BLOCK
WIL-2	Human B-cell lymphoma	G_1
CCRF-CEM	Human T-cell lymphoma	G_1
K562 (Lozzio)	Human; derived from a patient with Ph[1+] Chronic Myeloid Leukemia in blast crisis	G_1
EL_4	Murine T-cell lymphoma	G_2
WEHI-3	Murine myelomonocytic leukemia	G_1 (G_0?)

human CFU-c (Fig. 3) and WEHI-3 cells was uncomplicated by the stimulatory efforts of murine macrophage-derived CSF.

DISCUSSION

Numerous reports attest to the capacity of the macrophage to elaborate factors which can either augment (2, 7, 11, 13–16, 29, 32) or suppress (10, 18, 22, 31, 34) the proliferation of hemopoietic cells. In addition, considerable evidence is available that phagocytic mononuclear cells are capable of exerting a cytostatic influence (1, 5, 17, 19, 20, 24, 25) on neoplastic cell proliferation, although an enhancement of hemopoietic tumor cells has been reported (6, 30). In some cases cell contact is required (1, 17), though activities have been shown to exist in macrophage supernatants (5, 24, 25). The question of whether normal and neoplastic hemopoietic cell proliferation can be regulated by macrophages has been very difficult to answer because of the limited use of normal control cell populations. The use of normal hemopoietic progenitor cells capable of clonal proliferation in soft agar medium provides an ideal control population which can be compared with a number of hemopoietic tumor cell lines. The selection of methods to quantitate the effects of macrophages on tumor cell proliferation is also important, since the results of using ^{3}H-TdR incorporation may not necessarily reflect a true index of proliferation. This has been shown to be due to the release of endogenous macrophage thymidine capable of competing with the incorporation of tritiated precursor and thus leading to an underestimation of cell proliferation (33). Therefore, a multiparameter approach to the possible effect of the macrophage in the regulation of cell proliferation was undertaken.

This study clearly documents the capacity of macrophages to either stimulate, enhance, or suppress the growth of both normal and neoplastic hemopoietic cells. Such effects were independent of cell contact, suggesting the elaboration of at least two opposing biologically active principles, one stimulatory, the other inhibitory, the latter of which is capable of exerting its influence upon cells irrespective of the species, cell type of origin, and growth characteristics. These inhibitory, and, in some cases, stimulatory effects have been shown to occur whether hemopoietic tumor cells were cultured in liquid or in semisolid medium and whether end-point determinations were viable cell numbers, colony formation or ^{3}H-TdR incorporation. The suppression of proliferation was cytostatic in nature, rather than

cytolytic or cytotoxic, as cell viability was maintained throughout the culture period in the presence of macrophages.

If such inhibitory effects are indeed cytostatic, is there a particular phase of the cell cycle in which such hemopoietic cells are blocked, or does inhibition occur randomly throughout the cell cycle? As determined by flow cytofluorimetry (8, 22, 37), the phase of the cell cycle in which the tumor cells were blocked correlated with their susceptibility to macrophage-mediated inhibition. For example, the three cell lines WIL-2, K562 (Lozzio), and CCFR-CEM are blocked in the G_1 phase, and these cells were intermediately sensitive to macrophage inhibition. The EL_4 cell line was blocked in the G_2 phase, which accounted for only the moderate suppression in the incorporation of ^{3}H-TdR. However, when cultured at higher macrophage concentrations it is possible to inhibit EL_4 cells in the G_1 phase of their cell cycle. The ability of the cell line to revert to normal cycle status shortly after being removed from macrophages suggests the relative resistance of EL_4 to macrophage cytostasis.

The murine myelomonocytic leukemic cell line WEHI-3, a possible leukemic counterpart of the normal murine CFU-c, was blocked in a subG$_1$ (G_0) phase of the cell cycle, characterized by both decreased RNA and DNA contents. Work in progress indicates that, in the presence of macrophages, the WEHI-3 cell moves into this restricted cell cycle phase, which is characterized by a decreased ability of the metachromatic fluorescent dye acridine orange to bind to DNA, suggesting a change in nuclear chromatin leading to a decreased RNA synthesis (Kurland and Traganos, manuscript in preparation). The rapidity with which WEHI-3 cells move into this cell cycle phase correlated with its particular sensitivity to macrophage inhibition in both liquid and soft agar culture.

The enhancement of cell proliferation of the human B-cell lymphoma WIL-2 by low numbers of macrophages deserves particular attention in light of recent observations in this laboratory that colony formation by normal mouse B-lymphocytes is also enhanced by, if not dependent upon, low numbers of macrophages (Kurland, unpublished observation). This latter effect is characterized by the ability of diffusible factors of macrophages to reconstitute the poor colony-forming response of adherent cell-depleted mouse spleen or lymph node cultures. Colony formation by WIL-2 became increasingly dependent upon macrophages when the cell line was cultured at low cell numbers. However, both normal B-lymphocyte and WIL-2 colony formation were inhibited in a similar fashion by relatively high numbers of macrophages, suggesting that the neoplastic hemopoietic cell retains some responsiveness to regulatory influences.

The macrophage is capable of producing CSF and thereby of stimulating the proliferation of normal granulocyte-macrophage progenitor cells. The magnitude of stimulation (or the concentration of CSF) increased with the number of adherent macrophages. Increasing the number of macrophages further caused a decline in colony formation to a point at which no CFU-c were detectable. This ability of the macrophage to both stimulate and inhibit the proliferation of CFU-c suggests the production of opposing influences, with a favoring of inhibition at the higher number of macrophages. Macrophage inhibition of normal progenitor cell proliferation was more clearly evident when an exogenous concentration of CSF, capable of stimulating a maximal number of CFU-c, was added to mouse bone marrow cultures. The macrophage concentration at which inhibition of such exogenously stimulated colony formation was first evident corresponded to the concentration of macrophages beyond which colony formation progressively declined in the absence of exogenous CSF. These results indicate that, at *low* macrophage concentrations, promotion of colony formation (or enhancement of exogenously stimulated CFU-c proliferation) occurs, whereas *increasing* macrophage numbers leads to production of an inhibitor which counteracts the action of CSF whether produced endogenously by macrophages or provided exogenously.

Likely candidates for such an inhibitor(s) of macrophage origin are the prostaglandins, in light of (1) our previous observations that the E-series prostaglandins (PGE) are extremely efficacious inhibitors of both CSF-dependent and CSF-independent hemopoietic proliferation (21, 28), and (2) our recent finding that the macrophage is capable of elaborating PGE in culture (Kurland and Bockman, manuscript in preparation). This is further indicated by the ability of the prostaglandin synthesis inhibitor indomethacin(12) to suppress macrophage PG production and to effectively reduce the ability of the macrophages to exert such profound inhibition upon clonal proliferation of CFU-c. Furthermore, the dramatic macrophage inhibition of the myelomonocytic leukemic cell WEHI-3 which is prevented by indomethacin, suggests that prostaglandin production by macrophages may modulate both normal and neoplastic hemopoietic proliferation.

The manner by which synthetic PGE in-

hibits the clonal proliferation of CFU-c is by decreasing CFU-c responsiveness to the stimulatory actions of CSF, so that, in the presence of PGE, the CFU-c require an 8–9-fold greater concentration of CSF in order to proliferate to the same extent as they would in the absence of PGE (21). The ability of low numbers of macrophages to provide greater net stimulation of mouse CFU-c in the presence of indomethacin may be due to the inhibition of prostaglandin synthesis by the macrophages, with a resultant unmasking of colony-stimulating activity. In addition, the decline in colony formation at the higher number of macrophages, presumably due to an increased concentration of prostaglandin with the ability to decrease CFU-c responsiveness to CSF, can be effectively reduced by indomethacin. The species nonspecificity of prostaglandin-mediated effects is clearly documented by the capacity of indomethacin to decrease the inhibition of human CFU-c by murine macrophages.

The regulation of proliferation and differentiation within the hemopoietic system must involve a means of balancing the levels of controlling factors in order to promote and maintain homeostasis. The role of CSF in the humoral regulation of granulopoiesis and macrophage production has led to suggestions that the factor may be a granulopoietin. Although in the mouse many tissues can be found to contain CSF, the principal cell responsible for elaboration of CSF is the monocyte (7, 29) or tissue macrophage (11, 14, 29). Since the elaboration of CSF is increased as a function of macrophage activation (11), it has been proposed that granulocyte and monocyte-macrophage production is controlled by a positive feedback mechanism involving CSF. The CSF-dependent proliferation of the two hemopoietic progenitor cells, CFU-c and CFU-pm (21, 27), each of which generates differentiated macrophage progeny, is therefore capable of recruiting additional cells which synthesize and release CSF. This increased CSF level is then in turn capable of stimulating the proliferation of additional macrophage precursors, unless particular events intervene which limit the proliferative stimulus of CSF.

By a sensitive radioimmunoassay, we have been able to demonstrate that peritoneal macrophages release PGE into their bathing medium, when cultured for brief periods *in vitro*. The adherent macrophage—and not the nonadherent components of the exudate (lymphocytes and polymorphonuclear leukocytes)—is the principal prostaglandin-producing cell, and thus the amount of PGE released into the medium is a function of the number of macrophages (Kurland

and Bockman, manuscript in preparation). These findings thus suggest that the macrophage is able to produce both a stimulator of proliferation and differentiation of the CFU-c and CFU-pm (CSF), and also a molecule which can reduce the ability of these progenitor cells to respond to CSF (PGE).

The elaboration of CSF and PGE by the macrophage suggests a central role in the control of steady-state production of granulocytes, monocytes, and macrophages, as well as the potential to re-establish homeostasis in response to granulopoietic perturbation. Such a proposition would implicate the macrophage as a surveillance cell able to adjust or balance the levels of these opposing regulators in response to a variety of stimuli. The well-documented ability of endotoxin to elevate serum levels of CSF (36) constitutes a macrophage response resulting in increased CSF production (11); and it is possible that similar stimuli exist which may modulate PGE production by the macrophage. We have recently observed that CSF, when added to liquid cultures of macrophages, results in a dramatic increase in the production of prostaglandin (Kurland, unpublished observation). This response by the macrophage was dependent upon the concentration of CSF added; addition of a concentration of CSF which could well stimulate maximal CFU-c pro-

FIGURE 7. A schematic representation of the production of CSF and PGE by the macrophage.

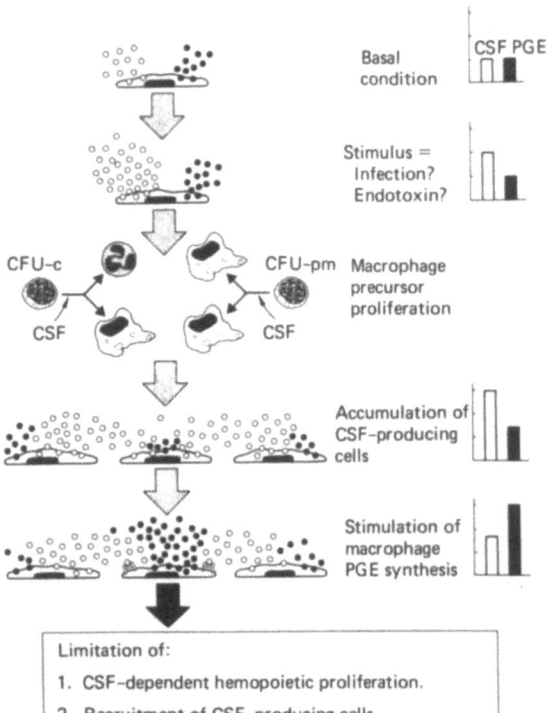

liferation produced a remarkable 16-fold increase in PGE levels. This finding therefore indicates a feedback mechanism whereby the macrophage can control the synthesis of PGE by sensing the CSF concentration in its external milieu. The likelihood that macrophages possess receptors of CSF is based upon their ability to exhibit a proliferative response to exogenous CSF (23).

The central role of the macrophage in the control of granulopoiesis and macrophage production is shown in Fig. 7, which is a model depicting those regulatory influences defined *in vitro*, and which may have *in vivo* significance. In this model, baseline production of CSF by circulating monocytes and by fixed tissue macrophages in hemopoietic tissues, and in lung, liver, and peritoneal cavities provides a constant stimulus for CFU-c and CFU-pm proliferation and differentiation into the mature granulocyte and monocyte-macrophage compartments. A baseline production of PGE by macrophages serves to limit inappropriate myeloid precursor cell proliferation, ensuring steady-state conditions. Not included in this model, but presumably of biologic relevance, are the documented heterogeneity of CFU-c to CSF stimulation (26) and the functional heterogeneity of the different molecular forms of CSF (26). The response of the host to bacterial infection or transient endotoxemia causes an increase in macrophage production of CSF, a high circulating level of which in turn promotes increased granulopoiesis and macrophage production. It should be noted that, in addition to recruitment of granulocytes and monocyte macrophages from the bipotential stem cell compartment (CFU-c), high CSF concentrations promote proliferation of a subpopulation of macrophage precursors in the peritoneal cavity (CFU-pm) and presumably elsewhere. In this context, macrophages can condition their own environment and promote active macrophage proliferation within foci of infection and inflammation. Thus the proliferative stimulus of CSF on these macrophage precursor cells allows for further recruitment of additional CSF-producing cells. Possible perturbations secondary to infection or endotoxin-induced alterations in granulopoiesis and macrophage proliferation are eventually limited by an increased PGE production of macrophages in response to high CSF levels. The capability of PGE to modulate the responsiveness of both the granulocyte-macrophage progenitor cell and the peritoneal macrophage precursor cell to CSF is limited by the lability of PGE, but its presence within hemopoietic milieus constitutes an important control in the prevention of persistant CSF-independent proliferation.

SUMMARY

Nonimmune mouse peritoneal macrophages have the capacity to either stimulate, enhance, or suppress the clonal proliferation of both normal and neoplastic hemopoietic cells. Such effects are independent of cell contact, suggesting the elaboration of at least two opposing biologically active principles, one stimulatory, the other inhibitory, the latter of which is capable of exerting its influence upon hemopoietic cells irrespective of the species, cell type of origin, or growth characteristics.

Macrophages are capable of producing Colony-Stimulating Factors (CSF), and thereby of stimulating the proliferation and/or differentiation of normal granulocyte-macrophage progenitor cells (CFU-c). At low macrophage numbers, species-specific promotion of CFU-c clonal proliferation (or enhancement of exogenously stimulated CFU-c proliferation) occurs, whereas increasing macrophage numbers leads to accumulation of an inhibitor which counteracts the action of CSF, both produced endogenously by macrophages or provided exogenously.

Likely candidates for such an inhibitor(s) of macrophage origin are the prostaglandins, as indicated by the ability of the prostaglandin synthetase inhibitor indomethacin to suppress macrophage prostaglandin production and effectively reduce macrophage-mediated inhibition. This is in agreement with our recent observations that synthetic prostaglandins of the E-series decrease the responsiveness of CSF-dependent hemopoietic progenitor cells to the proliferative stimulus of CSF.

The elaboration of colony-stimulating factor and prostaglandin E by the macrophage indicates a central role of the macrophage in the steady-state production of granulocytes, monocytes, and macrophages. It also has the potential to reestablish homeostasis in response to granulopoietic perturbations.

ACKNOWLEDGMENTS

The authors would like to thank Dr. Peter Ralph for providing the hemopoietic cell lines, Ms. Sabariah Schrader for excellent technical assistance, and Ms. Linda Wilkins for preparation of the manuscript.

This work was supported by NIH Grants CA-17085, CA-17353, CA-19052, and NIH-08748-11A, and by American Cancer Society Grant CH-3.

REFERENCES

(1) Alexander, P., and Evans, R. Endotoxin and double stranded RNA renders macrophages cytotoxic. *Nature (London), New Biol.*, *232*:76, 1971.

(2) Bach, F. H., Alter, B. J., Solliday, S., Zoschke, D. C., and Janis, M. Lymphocyte reactivity in vitro. II. Soluble reconstitution factor permitting response of purified lymphocytes. *Cell. Immunol.*, *1*:219, 1970.

(3) Bradley, T. R., and Metcalf, D. The growth of mouse bone marrow cells in vitro. *Aust. J. Exp. Biol. Med. Sci.*, *44*:287, 1966.

(4) Bray, H. A., Gordon, D., and Morley, J. Role of prostaglandins in reactions of cellular immunity. *Br. J. Pharm.*, *52*:453pp., 1974.

(5) Calderon, J., Williams, R. T., and Unanue, E. R. An inhibitor of cell proliferation released by cultures of macrophages. *Proc. Natl. Acad. Sci. U.S.A.*, *71*:4273, 1974.

(6) Calderon, J., and Unanue, E. R. Two biological activities regulating cell proliferation found in cultures of peritoneal exudate cells. *Nature (London)*, *253*:359, 1975.

(7) Chervenick, P. A., and LoBuglio, A. F. Human blood monocytes: Stimulators of granulocyte and mononuclear colony formation in vitro. *Science*, *178*:164, 1972.

(8) Darzynkiewicz, Z., Traganos, F., Sharpless, T., and Melamed, M. R. Lymphocyte stimulation: A rapid multiparameter analysis. *Proc. Natl. Acad. Sci. U.S.A.*, (in press).

(9) Darzynkiewicz, Z., Traganos, F., Friend, C., Sharpless, T., and Melamed, M. R. Nuclear chromatin changes during erythroid differentiation of Friend virus-induced leukemic cells. *Exp. Cell Res.*, *99*:301, 1976.

(10) Diener, E., Shortman, K., and Russel, P. Induction

of immunity and tolerance in vitro in the absence of phagocytic cells. *Nature (London)*, *225*:731, 1970.

(11) Eaves, A. C., and Bruce, W. R. In vitro production of colony stimulating activity. I. Exposure of mouse peritoneal cells to endotoxin. *Cell Tissue Kinet.*, *7*:19, 1970.

(12) Ferreira, S. H., Moncada, S., and Vane, J. R. Indomethacin and aspirin abolish prostaglandin release from the spleen. *Nature (London), New Biol.*, *231*:237, 1971.

(13) Grey, I., and Waksman, B. H. Potentiation of the T-lymphocyte response to mitogens. II. The cellular source of the potentiating mediator(s). *J. Exp. Med.*, *136*:143, 1972.

(14) Golde, D. W., Finley, T. N., and Cline, H. J. Production of colony stimulating factor by human macrophages. *Lancet*, *ii*:1397, 1972.

(15) Hersch, E. M., and Harris, J. E. Macrophage-lymphocyte interactions in the antigen induced blastogenic response of human peripheral blood leucocytes. *J. Immunol.*, *100*:1184, 1968.

(16) Hoffman, M., and Dutton, R. W. Immune response restoration with macrophage culture supernatants. *Science*, *172*:1047, 1971.

(17) Holtermann, O. A., Klein, E., and Casale, G. P. Selective cytotoxicity of peritoneal leucocytes for neoplastic cells. *Cell. Immunol.*, *9*:339, 1975.

(18) Ichikawa, Y., Pluznik, D. H., and Sachs, L. Feedback inhibition of the development of macrophage and granulocyte colonies. I. Inhibition by macrophages. *Proc. Natl. Acad. Sci. U.S.A.*, *58*:1480, 1967.

(19) Keller, R. Modulation of cell proliferation by macrophages: A possible function apart from cytotoxic tumor rejection. *Br. J. Cancer*, *30*:401, 1974.

(20) Kirchner, H., Holden, H. T., and Herberman, R. B. Inhibition of in vitro growth of lymphoma cells by macrophages from tumor bearing mice. *J. Nat. Cancer Inst.*, *55*:971, 1975.

(21) Kurland, J., and Moore, M. A. S. Modulation of hemopoiesis by prostaglandins. *Exp. Hematol.* (in press).

(22) Kurland, J., Traganos, F., Darzynkiewicz, C., and Moore, M. A. S. Macrophage mediated suppression of neoplastic hemopoietic proliferation. *Nature (London)* (submitted for publication).

(23) Lin, H., and Stewart, C. C. Colony formation by mouse peritoneal exudate cells in vitro. *Nature (London), New Biol.*, *243*:176, 1973.

(24) McIvor, K. L., and Weiser, R. S. Mechanisms of target cell destruction by alloimmune peritoneal cells. *Immunology*, *20*:315, 1971.

(25) Melson, H., Kearney, G., Gruca, S., and Seljelid, R. Evidence for a cytolytic factor released by macrophages. *J. Exp. Med.*, *140*:1085, 1974.

(26) Metcalf, D., and Moore, M. A. S. *Hemopoietic Cells.* Amsterdam: North-Holland Publishing Company, 1971.

(27) Moore, M. A. S., and Kurland, J. Regulation of granulopoiesis. In Viza, D., and Muller-Berat, N., eds., *Progress in Differentiation Research.* Amsterdam: Elsevier/North-Holland Biomedical Press, p. 483, 1976.

(28) Moore, M. A. S., Kurland, J., and Broxmeyer, H. The granulocytic-monocytic stem cell. In Cairnie, A. B., Lala, P. K., and Osmond, D. G., eds., *Stem Cells of Renewing Cell Populations.* New York: Academic Press, p. 181, 1976.

(29) Moore, M. A. S., and Williams, N. Physical separation of colony stimulating cells from in vitro colony forming cells in hemopoietic tissue. *J. Cell. Physiol.*, *80*:195, 1972.

(30) Nathan, C. F., and Terry, W. D. Differential stimulation of murine lymphoma growth in vitro by normal and BCG-activated macrophages. *J. Exp. Med.*, *142*:887, 1975.

(31) Nelson, D. S. Production by stimulated macrophages of factors depressing lymphocyte transformation. *Nature (London)*, *246*:306, 1973.

(32) Oppenheim, J. J., Leventhal, B. G., and Hersch, E. M. The transformation of column purified lymphocytes with nonspecific and specific antigenic stimuli. *J. Immunol.*, *101*:262, 1968.

(33) Opitz, H. G., Niethammer, D., Lemke, H., Flad, H. D., and Huget, R. Biochemical characterization of a factor released by macrophages. *Cell. Immunol.*, *16*:379, 1975.

(34) Parkhouse, R. M. E., and Dutton, R. W. Inhibition of spleen DNA synthesis by autologous macrophages. *J. Immunol.*, *97*:663, 1966.

(35) Pike, B. L., and Robinson, W. A. Human bone marrow colony growth in agar. *J. Cell. Physiol.*, *76*:77, 1970.

(36) Quesenberry, P., Morley, A., Stohlman, F., Rickard, K., Howard, D. and Smith, M. Effect of endotoxin on granulopoiesis and colony stimulation factor. *New Eng. J. Med.*, *286*:227, 1972.

(37) Traganos, F., Darzynkiewicz, Z., Sharpless, T., and Melamed, M. R. Simultaneous staining of ribonucleic and deoxyribonucleic acids in unfixed cells using acridine orange in a flow cytofluorometric system. *J. Histochem. Cytochem.* (in press).

(38) Warner, N. L., Moore, M. A. S., and Metcalf, D. A transplantable myelomonocytic leukemia in Balb/c mice: Cytology, karyotype and muramidase content. *J. Nat. Cancer Inst.*, *43*:953, 1969.

(39) Weir, D. H. *Handbook of Experimental Immunology.* Philadelphia: Davies Pub. Company, p. 1019, 1967.

Studies on the humoral regulation of granulocytopoiesis have, for several reasons, met with frustration since the beginning of experimentation on this problem. Stimulators for the initiation of granulocytosis are ubiquitous, resulting in uncontrolled and undesirable variables. Increased medullary granulocytopoiesis can usually be demonstrated following leukemoid reactions, regardless of the stimulus. Techniques for the reproducible quantitation of meaningful parameters of granulocytopoiesis have not been available in the past.

Advances in recent years in techniques for controlled growth of defined populations of hemopoietic cells both *in vivo* and *in vitro* have facilitated such studies. Definition of murine hemopoietic stem cells by transplantation into radiated animals (35), growth of marrow cells within diffusion chambers (3, 4), and the development of *in vitro* techniques for the growth of granulocyte and monocyte precursor cells in semisolid nutrient media (5, 27) and fluid culture systems (18, 34) have led to extensive developments in our understanding of granulocytopoiesis and its regulation. However, these assay systems have similarly led to considerable variability with regard to the relationship between factors and moieties able to stimulate such proliferation and differentiation in these systems (22, 29, 33).

In an attempt to clarify many of these problems, we have chosen a single granulocyte population, which by its very nature is limited in its reactivity, and which represents a minor percentage of the marrow granulocyte population. The eosinophil granulocyte was first identified as a separate granulocyte cell type nearly 100 years ago by Ehrlich in his original observations on morphology of blood cells (10, 11). In the ensuing 100 years little has been added to the understanding of the function of this cell in the defense mechanisms of the body. Granules are carriers of potent enzymes and it is known that the eosinophil granulocyte is involved in most parasitic and chronic allergic conditions (1, 14).

Several aspects of eosinophils make them a suitable model system in the study of granulocytopoiesis. They represent a minor population of granulocytes within normal bone marrow (2% to 5%). Their proliferation and accumulation is regulated independently of the other granulocyte cell types, resulting in a stable population for study. In order to achieve an artifactual doubling

Humoral Regulation of Eosinophil Granulocytopoiesis

7

M. P. McGarry, A. M. Miller, and D. G. Colley

of the percentage of eosinophil granulocytes, a full halving of total marrow cellularity would be necessary; although fully 8–10-fold increases can occur within the normal range of marrow granulocyte cellularity. Furthermore, in the murine system, total femoral marrow cellularity can be measured.

LYMPHOID CELL DEPENDENCE OF EOSINOPHIL ACCUMULATIONS IN RESPONSE TO ANTIGEN

An interesting and significant aspect of eosinophil granulocytes is that their participation in immune-inflammatory responses and granulocyte exudations is restricted. It has been demonstrated that the accumulation of eosinophil granulocytes in *Trichinella* infections (2, 37), at the site of secondary challenge with tetanus toxoid in mice (21, 28, 32), and in egg granulomas in experimental murine schistosomiasis is dependent on the involvement of thymus derived lymphoid cells (12).

EOSINOPHILOPOIESIS

Similarly, increased peripheral accumulation of eosinophils is followed by increased incidence of eosinophil granulocytes within bone marrow (17, 20, 21). This increase in the number of marrow eosinophil granulocytes is not explained by the release of noneosinophil cells from the bone marrow, nor by eosinophilia of marrow. These increases in absolute numbers, as measured by differential staining and increases in eosinophilic precursors may account for the accumulation of marrow eosinophils.

With tetanus toxoid-induced eosinophil responses in mice it has been possible to demonstrate that elevations in the numbers of bone marrow eosinophils are associated with increases in the activity of eosinophil hemopoietic inductive microenvironments (36). By doing a spleen colony assay in mice experiencing secondary tetanus toxoid-induced eosinophilia it was possible to demonstrate that spleen and bone marrow tissue contain increased capacity for the support of eosinophil granulocytopoiesis. This is shown by the numbers of eosinophil-containing colonies, and represents a real, rather than artifactual, increase in eosinophilopoietic potential of these hemopoietic tissues (15). This is strong evidence demonstrating the increase in eosinophil granulocytopoiesis in such mice. Should the elevations in marrow eosinophils be attributable to

eosinophilia of marrow, no such change in activity of eosinophilopoietic inductive microenvironments would be expected.

EOSINOPHIL GRANULOCYTOPOIETIN

Several investigators have presented evidence indicating the existence of a humoral factor capable of stimulating eosinophil granulocytopoiesis (8, 16, 31). Our own studies have led to similar conclusions (20). Additional experiments, directed to a definition of the cell of origin of the factor(s), were done using a specially constructed diffusion chamber apparatus. This consists of four lucite rings with millipore membrane filters which permits cocultivation of four separate cell populations, pairs of which are separated only by a Millipore filter membrane (the Quadrachamber Diffusion Assembly, QDA). It was found that prior exposure to tetanus toxoid was necessary for the production of the factor by spleen cells (25). This is identical to the requirements found for the *in vivo* eosinophilopoietic response associated with tetanus toxoid-induced eosinophil exudation.

A similar, more impressive, marrow eosinophilopoietic response is manifest in experimental murine schistosomiasis (Table 1). Peripheral eosinophilia is not a consistent observation in such animals although impressive increases in femoral marrow eosinophils occur as the infection progresses. The marrow increase can be correlated, temporally, with the accumulation of eosinophils around schistosome eggs. Egg production in the mouse begins about 4–5 weeks following infection.

Using spleen cells from *Schistosoma mansoni*-infected mice and normal bone marrow in QDA, it was possible to show that such spleen cells also produce an eosinophilopoiesis-stimulating factor (EoSF) in response to stimulation with a soluble schistosomal egg antigen (SEA) preparation. This was measured as eosinophil granulocyte production in the *in vivo* diffusion chambers (24). Normal spleen cells did not produce a detectable EoSF.

We suggested, based on these studies, that the factor demonstrated is the *in vivo* stimulus responsible for the increased eosinophilopoiesis in femoral marrow of such animals.

HYPOTHESIS

These studies have led to the following hypothesis (Fig. 1): Antigen-stimulated lymphoid cells release (or are necessary for the production

TABLE 1 Peripheral Blood and Femoral Marrow Eosinophil Granulocytes in C57BL/6J and Ha/ICR (Swiss) Mice Following Infection with 125 Cercariae of S. mansoni[a]

| Strain | Week | PERIPHERAL BLOOD[b] | | FEMORAL MARROW[c] | |
		WBC/mm³	% Eosinophils	Nucleated cells × 10⁶	% Eosinophils
C57BL/6J					
	1–2	7590	2.75	22.2	3.2
	3–4	7732	3.75	20.0	6.8
	5–6	24377	3.75	20.9	25.7
	7–8	15642	1.50	21.3	32.0
	9–10	20561	3.50	25.9	38.0
Ha/ICR					
	1–2	12345	3.25	21.2	5.5
	3–4	18292	6.00	20.8	5.5
	5–6	7465	2.50	18.6	9.0
	7–8	22374	11.00	28.1	36.2
	9–10	10837	24.50	25.7	35.2
	11–12	38279	9.00	22.6	17.7

[a]Each value represents the average obtained from determinations made on four separate mice.

[b]Values obtained from aliquots electronically counted and May-Gruenwald Giemsa stained smears of samples obtained from lateral tail vein.

[c]Femoral marrow expelled from bone shaft with 10 ml physiologic saline. Suspensions were aspirated, diluted in saline, and counted electronically. Contralateral femurs were used to obtain such smears for differential counts following May-Gruenwald Giemsa stain.

FIGURE 1. Proposed schema for the production of an eosinophilopoiesis-stimulating factor (EoSF) during immune-inflammatory responses to antigens characterized by the involvement of eosinophil granulocytes.

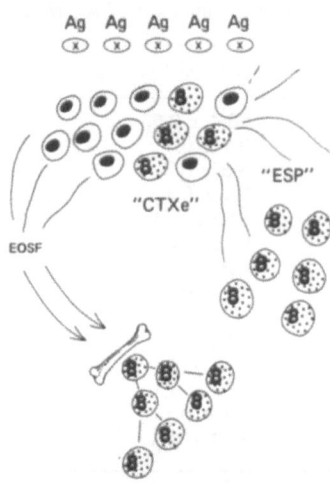

of) a factor which, in some way, stimulates cells either in the eosinophil granulocyte series or in precursors to these cells to increase by proliferation and/or differentiation the number of eosinophil granulocytes within hemopoietic tissues. These cells are released and made available for participation in an ongoing immune-inflammatory response. Such release and production of the eosinophilopoietic factor continues for such time as there is continued antigen stimulation of the sensitized lymphoid cell population. Loss of appropriate antigen challenge or inhibition/cessation of immune-responsiveness results in termination of EoSF production.

PRODUCTION OF HUMORAL FACTOR(S) *IN VITRO*
COLONY-STIMULATING FACTOR

Spleen cells from *S. mansoni*-infected mice had already been demonstrated to release a lymphokine-eosinophil stimulation promoter (ESP) *in vitro* in response to a soluble SEA preparation (9). ESP activity is measured by increases in *in vitro* migration of eosinophil granulocytes, and is believed to facilitate migration of these cells *in vivo*. It seemed possible that the techniques used for the elaboration by spleen cells, *in vitro*, of a factor affecting the function of eosinophils might be useful for the production of the EoSF.

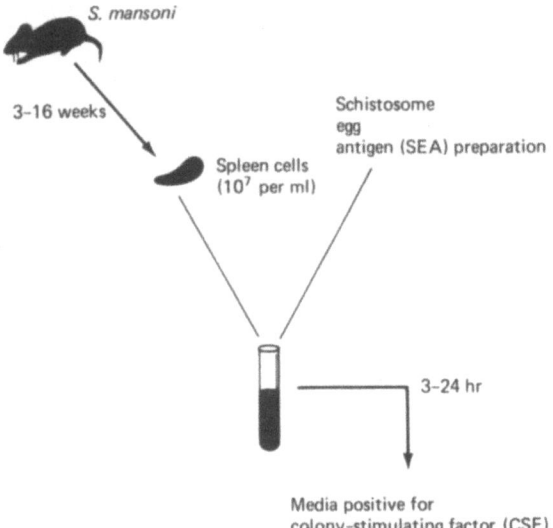

FIGURE 2. Two-milliliter cultures of 10^7 spleen cells from mice infected with cercariae of *S. mansoni* contain colony-stimulating factor (CSF) when cultures contain a soluble schistosome egg antigen (SEA) preparation and mice have been infected at least 3 and not more than 16–20 weeks.

Initial studies were done to establish if such culture fluids, positive for ESP, were positive for colony-stimulating factor, CSF, a myelopoietic factor active *in vitro* (22). Supernatant culture fluids were obtained from 2-ml cultures of 10^7

spleen cells per ml in RPMI 1640 without serum. To these were added 2.2 μg/ml SEA (Fig. 2). CSF was first detectable between 3 and 6 hr after initiation of the cultures and did not increase over those values established between 18 and 24 hr. The production is dependent on the presence of SEA and spleen cells from mice infected minimally 3, and maximally 16–20, weeks with *S. mansoni* (Table 2). Spleen cells from infected C57Bl/6Ja, CBA/J, and Ha/ICR mice have been shown capable of producing CSF in response to SEA.

Further studies have demonstrated that, like ESP (13), the production of the CSF is dependent on a thymus-derived lymphocyte population (Table 2). Eight weeks following infection with 20–40 cercariae of *S. mansoni*, spleen cells from mice release CSF in response to SEA (2.2 μg/ml), phytohemagglutinin (PHA) (0.25 μg/ml), and concanavalin A (Con A) (2.0 μg/ml). [CSF production *in vitro* by PHA and Con A is lymphocyte-dependent (7, 26, 30).] Treatment of 8-week-infected CBA/J mice with daily injections of a rabbit anti-mouse thymocyte serum (ATS) on the 3 days prior to initiation of 24 hr cultures (13) eliminates a population of cells required for the CSF production. Treatment of control-infected mice with normal rabbit serum (NRS) has no deleterious effect on CSF production in response to these antigens. Congenitally athymic, hairless, "nude" (nu/nu) mice, 8 weeks after infection with *S. mansoni*, do not contain cells in their spleen

TABLE 2 Characteristics of *in Vitro* Production of CSF by Spleen Cells from *S. mansoni*-infected CBA/J Mice[a]

ORIGIN OF SPLEEN CELLS	ANTIGEN	NO. OF COLONIES/10^5 BM CELLS[b]
8-week-infected	SEA	60.4 ± 2.58
	—	0.1 ± 0.1
< 3-week-infected	SEA	3.5 ± 0.3
> 20-week-infected	SEA	0.1 ± 0.1
8-week-infected	PHA	31.3 ± 3.4
	Con A	90.5 ± 4.0
8-week-infected + ATS[c]	SEA	0.2 ± 0.2
	PHA	1.3 ± 0.9
	Con A	1.2 ± 0.4
8-week-infected + NRS[c]	SEA	59.5 ± 2.7
	PHA	30.5 ± 2.1
	Con A	81.7 ± 4.1
8-week-infected Balb/c nu/nu	SEA	0.1 ± 0.1
8-week-infected Balb/c nu/+	SEA	19.7 ± 3.5

[a]Spleen cell cultures were done in RPMI 1640 without serum (3% Penicillin-Streptomycin, BBL-Cockeysville, Md.). Two-milliliter cultures were used, with 10^7 cells/ml. SEA was added at 2.2 μg/ml. Media was harvested after centrifugation of cultures. Assays for CSF were done according to previously published techniques[19].

[b]CSF as measured in 0.1 ml culture fluid.

[c]Spleen cells obtained for cultures with SEA, PHA, or Con A from CBA/J mice 8 weeks after infection with 30 cercariae *S. mansoni*. Three daily i.p. injections of a rabbit anti-mouse thymocyte serum (ATS) or normal rabbit serum (NRS) were given prior to sacrifice to eliminate T-cells from the donor[13].

which will produce CSF on challenge *in vitro* with SEA. Normal (nu/+) littermates do contain such cells. Together, these data suggest that the cell of origin of the CSF (or a cell on which production is dependent) is thymus-derived. A similar dependence exists for the production of ESP (13). Also, as was noted previously, *in vivo* eosinophil responses are dependent on a thymus-derived cell population.

Although encouraging and suggestive, these data imply only the possibility that the material being detected by its ability to stimulate proliferation of some cells within a bone marrow population *in vitro* has specificity for eosinophils. We ourselves have not attempted to do granulocyte-type classifications of the colonies produced *in vitro*. Eosinophil-like granulocytic colonies have been reported (23), although differential quantitation of colonies in cultures would be required to evaluate eosinophil-specific effects of CSF.

Media from cultures of lymphocytes from *Trichinella spiralis*-infected mice stimulated with a trichinella antigen have been shown to contain colony-stimulating factor (31). Liquid cultures of bone marrow cells from either *T. spiralis*-infected or normal mice to which such media had been added contained increased numbers of eosinophil granulocytes (31). The implication is apparent that the CSF is reflective of eosinophil-specific myelopoietic activity.

EOSINOPHILOPOIESIS-STIMULATING FACTOR (EoSF)

Rather than attempt to demonstrate directly the eosinophil granulocyte specificity of the CSF, or study eosinophil-specific granulocyte stimulation using other culture techniques, we

have endeavored to investigate whether the media from SEA-stimulated spleen cell populations from *S. mansoni*-infected mice contain an eosinophilopoiesis-stimulating factor (EoSF) active *in vivo*.

One complicating feature of attempts to demonstrate granulocytopoietic activity *in vivo* has been the superimposed response of the host to the antigenic/inflammatory effect of *any* injected material. These aspects are minimized by studying the eosinophil, since its reactions are restricted. They do not normally exudate in response to nonspecific inflammatory stimuli. In order to reduce the possibility that specific antigen stimulation might occur, resulting in eosinophil responses, especially to contaminating SEA, we rendered experimental male B6D2F1 mice immunologically unresponsive by a single subcutaneous injection of 2.0 mg hydrocortisone acetate (HCA). Peripheral blood eosinophils are virtually eliminated in such mice (1). Also, such animals are severely lymphopenic. Bone marrow eosinophils are not decreased during this time (Table 3) (6).

Experimental media (EM) (supernatant harvested from RPMI 1640 without serum + SEA with 10^7 spleen cells from infected donors cultured 24 hr) and Control media (CM) (RPMI 1640 without serum ± SEA) were concentrated 20-fold using Polyethylene glycol-6000 (Carbowax, Union Carbide, Buffalo, N.Y.) (Fig. 3). EM and CM concentrates were dialyzed 24 hr against physiologic saline. Three intraperitoneal injections of 0.5 ml were given at daily intervals, beginning 6 hr after a single, subcutaneous injection of HCA. Six hr after the final injection of media concentrate femoral marrow was evaluated for eosinophil granulocytes by differential and total cellularity. Five such media concentrates have been tested. The data presented in Table 4 illustrate that in

TABLE 3 Femoral Marrow Eosinophil Granulocytes in B6D2F$_1$ Mice following 2.0 mg Subcutaneous Hydrocortisone Acetate (HCA)[a]

TIME FOLLOWING HCA[b]	TOTAL CELLS	% EOSINOPHILS	EOSINOPHILS
3 hr	23.7 ± 0.8	2.6 ± 0.5	0.60 ± 0.12
6 hr	22.4 ± 0.9	2.8 ± 0.5	0.59 ± 0.09
1 day	21.6 ± 0.6	5.1 ± 0.8	1.08 ± 0.17
2 days	21.9 ± 1.0	3.7 ± 0.4	0.81 ± 0.10
3 days	21.7 ± 1.4	1.9 ± 0.7	0.40 ± 0.14
7 days	21.8 ± 0.6	2.4 ± 0.8	0.51 ± 0.16

[a]Total and differential percentages determined as for data in Table 1.

[b]Data for days following HCA obtained from animals sacrificed uniformly between 9:00 AM and 10:00 AM.

TABLE 4 Femoral Marrow Eosinophil Granulocytes in HCA-Treated B6D2F₁ Mice Given 20 × Concentrated Media from SEA-Stimulated Spleen Cells from *S. mansoni*-infected Mice[a]

	TOTAL CELLS ($\times 10^6$)	% EOSINOPHILS	TOTAL EOSINOPHILS ($\times 10^6$)	% $\dfrac{\text{EOSINOPHILS}}{\text{TOTAL GRANULOCYTES}}$
1. Control (2)	20.0	4.0	0.80	8.3
Experimental (2)	19.8	9.6	1.94	20.5
2. Control (3)	17.3	4.1	0.70	8.4
Experimental (3)	16.5	7.1	1.17	12.5
3. Control (2)	19.4	3.0	0.57	5.6
Experimental (2)	7.4	12.1	0.88	22.9
Experimental (2)	12.5	9.6	1.19	17.7
4. Control (2)	19.3	3.2	0.62	6.6
Experimental (2)	14.7	6.3	0.90	12.1
Summary				
Controls (9)	19.0 ± 0.6	3.6 ± 0.3	0.68 ± 0.05	7.2
Experimental (11)	14.4 ± 2.2	8.9 ± 1.0	1.22 ± 0.19	16.6
	$p < 0.005$	$p < 0.05$	$p < 0.005$	

[a]Media collected from 24-hr, 2-ml cultures of 10^7/ml spleen cells from *S. mansoni*-infected Ha/ICR or CBA/J mice with 2.2 μg/ml SEA. Individual sets of cultures were initiated with spleen cells from three to eight *isogenic* mice infected with 40 to 60 *S. mansoni* cercariae. Media were not pooled from separate experiments. Pools of media from separate experiments were concentrated 20-fold using Carbowax (polyethylene glycol-6000), followed by dialysis against physiologic saline for 24-hr.

FIGURE 3. One-half-milliliter aliquots of 20-fold concentrated media from cultures (2 ml, 10^7 cells/ml) of spleen cells from *S. mansoni*-infected (8–10 weeks) mice to which SEA had been added are injected into HCA-treated mice at daily intervals. Six hours following the final injection, mice are sacrificed and femoral marrow percentage and total eosinophils determined. Control media concentrates consist of media with or without SEA.

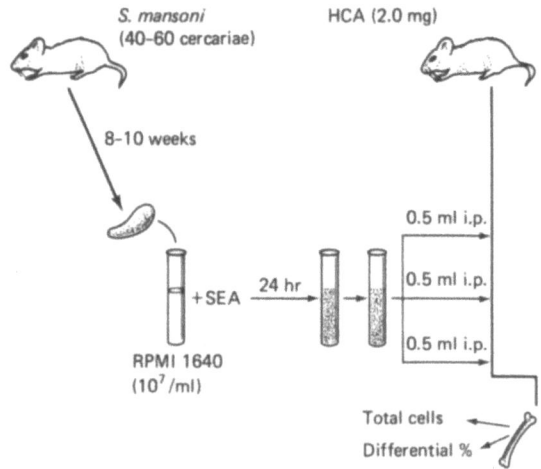

most cases more than twice as many cells in the eosinophil granulocyte series occur in marrow of mice treated with EM concentrate as in those treated with CM concentrate. The marrow of experimental mice indicated active eosinophilopoiesis. When expressed either as the percentage of the differential or by the total number of eosinophils, CM concentrate-treated mice exhibit values similar to those seen in HCA only-treated mice. The parameter:

$$\% \ \frac{\text{Eosinophils}}{\text{Total granulocyte}}$$

reflects more impressively the selective stimulation of eosinophil granulocytes in the EM concentrate-treated mice. This number, calculated from total numbers of femoral marrow eosinophil and neutrophil granulocytes, would not be expected to change substantially if increases in eosinophils were attributable to nonspecific myelopoietic stimulations.

SUMMARY

Information from studies on experimental eosinophilia and eosinophilopoiesis have revealed that: (1) There exists *in vivo* a diffusible factor capable of stimulating eosinophilopoiesis within bone marrow. This factor occurs as a normal constituent of immune-inflammatory reactions characterized by the involvement of eosinophil granulocytes. This factor is specific for eosinophils, and its presence results in increases in numbers of eosinophil granulocytes within myelopoietic tissues. (2) Tissue culture media from spleen cells from animals in which eosinophilopoiesis is actively occurring will produce, on appropriate stimulation, a colony-stimulating factor, a known *in vitro* myelopoietic factor. Conditions associated with the production of this CSF reflect and parallel the prerequisites *in vivo* for the lymphoid cell-dependent stimulation of eosinophilopoiesis. (3) Tissue culture media from SEA-stimulated spleen cells from *S. mansoni*-infected mice can be shown to contain an eosinophilopoiesis-stimulating factor capable of increasing marrow eosinophil granulocytes *in vivo* in appropriately prepared experimental assay mice.

ACKNOWLEDGMENTS

The authors wish to acknowledge the expert technical assistance of Bruce Styles, Rosemary Reszka, Sarah Hieny, George Freeman, Jr., A. Nasrallah, and John Ulatowski, and thank the Museum of Zoology, University of Michigan, for assistance. The materials used in this study were provided by the U.S.-Japan Cooperative Medical Science Program, NIAID. We also wish to acknowledge the excellent assistance of Elena Greco in the preparation of the charts and drawings, and of Mrs. Ann Bean in the preparation of the manuscript.

REFERENCES

(1) Archer, R. K. Regulatory mechanisms in eosinophil leukocyte production, release and distribution. In Gordon, A. S., ed., *Regulation of Hematopoiesis*. New York: Appleton-Century-Crofts, 1970, vol. 2, p. 917.

(2) Basten, A., and Beeson, P. B. Mechanism of eosinophilia. II. Role of the lymphocyte. *J. Exp. Med., 131*:1288, 1970.

(3) Berman, I., and Kaplan, H. S. The cultivation of mouse bone marrow in diffusion chambers. *Blood, 14*:1040, 1959.

(4) Bøyum, A., and Breivik, H. Kinetics of murine hemopoietic cell proliferation in diffusion chambers. *Cell Tissue Kinet., 6*:101, 1973.

(5) Bradley, T. R., and Metcalf, D. The growth of mouse bone marrow cells *in vitro*. *Aust. J. Exp. Biol. Med. Sci., 44*:287, 1966.

(6) Bro-Rasmussen, F. Eosinophils in the bone marrow of normal and cortisol-treated rats. *Acta Pathol. Microbiol. Scand., 81A*:593, 1973.

(7) Cline, M. J., and Golde, D. W. Production of colony-stimulating activity by human lymphocytes. *Nature (London), 248*:703, 1974.

(8) Cohen, N. S., LoBue, J., and Gordon, A. S. Humoral regulation of eosinophil production and release. In LoBue, J., and Gordon, A. S., eds., *Humoral Control of Growth and Differentiation*. New York: Academic Press, 1973, vol. 1, p. 69.

(9) Colley, D. G. Eosinophils and immune mechanisms. I. Eosinophil stimulation promoter (ESP): A lymphokine induced by specific antigen or phytohemagglutinin. *J. Immunol., 110*:1419, 1973.

(10) Ehrlich, P. Beiträge zue Kenntniss der granulisten Bindergewebszellen und der eosinophililen Leukocythen. *Arch. Anat. Physiol., 3*:166, 1879.

(11) Ehrlich, P. Über die specifischen Granulationen des Blutes. *Arch. Anat. Physiol., Physiol. Abt. (Ver. Berl. Physiol. Ges. 16/5), 571, 1879.

(12) Fine, D. P., Buchanan, R. D., and Colley, D. G. *Schistosoma mansoni* infection in mice depleted of thymus-dependent lymphocytes. I. Eosinophilia and immunologic responses to a schistosomal egg preparation. *Am. J. Pathol., 71*:193, 1973.

(13) Greene, B. M., and Colley, D. G. Eosinophils and immune mechanisms. III. Production of the lymphokine eosinophil stimulation promoter by mouse T lymphocytes. *J. Immunol., 116*:1078, 1976.

(14) Gross, R. The eosinophils. In Braunsteiner, H., and Zucker-Franklin, D., eds., *The Physiology and Pathology of Leukocytes*. New York: Grune and Stratton, 1962, p. 1.

(15) Jenkins, V. K., Trentin, J. J., Speirs, R. S., and McGarry, M. P. Hemopoietic colony studies. VI. Increased eosinophil-containing colonies obtained by antigen pretreatment of irradiated mice reconstituted with bone marrow cells. *J. Cell. Physiol., 79*:413, 1972.

(16) Mahmoud, A. A. F., Stone, M. K., Kellermeyer, R. W., and Warren, K. S. Eosinophilopoietic activity in mouse serum. *Clin. Res., 23*:524A, 1975.

(17) Mahmoud, A. A., Warren, K. S., and Graham, R. C., Jr. Anti-eosinophil serum and the kinetics of eosinophilia in Schistosomiasis mansoni. *J. Exp. Med., 142*:560, 1975.

(18) Mauel, J., and Defendi, V. Regulation of DNA

synthesis in mouse macrophages. I. Sources, action and purification of the macrophage growth factor (MGF). *Exp. Cell Res.*, *65*:33, 1971.

(19) McGarry, M. P. Metallic copper-induced granulocyte exudation in the study of granulocytopoiesis. *Cell Tissue Kinet.*, *8*:355, 1975.

(20) McGarry, M. P., and Miller, A. M. Evidence for the humoral stimulation of eosinophil granulocytopoiesis in *in vivo* diffusion chambers. *Exp. Hematol.*, *2*:372, 1974.

(21) McGarry, M. P., Speirs, R. S., Jenkins, V. K., and Trentin, J. J. Lymphoid cell dependence of eosinophil response to antigen. *J. Exp. Med.*, *134*:801, 1971.

(22) Metcalf, D. Regulation of granulocyte and monocyte-macrophage proliferation by colony stimulating factor (CSF): A review. *Exp. Hematol.*, *1*:185, 1973.

(23) Metcalf, D., Parker, J., Chester, H. M., and Kincade, P. W. Formation of eosinophilic-like granulocytic colonies by mouse bone marrow cells *in vitro*. *J. Cell. Physiol.*, *84*:275, 1974.

(24) Miller, A. M., Colley, D. G., and McGarry, M. P. Spleen cells from *Schistosoma mansoni*-infected mice produce diffusible stimulator of eosinophilopoiesis *in vivo*. *Nature (London)*, 1976 (in press).

(25) Miller, A. M., and McGarry, M. P. A diffusible stimulator of eosinophilopoiesis produced by lymphoid cells as demonstrated with diffusion chambers. *Blood*, *48*:293, 1976.

(26) Parker, J. W., and Metcalf, D. Production of colony-stimulating factor in mitogen-stimulated lymphocyte cultures. *J. Immunol.*, *112*:502, 1974.

(27) Pluznik, D. H., and Sachs, L. The cloning of normal mast cells in tissue culture. *J. Cell. Physiol.*, *66*:319, 1965.

(28) Ponzio, N. M., and Speirs, R. S. Lymphoid cell dependence of eosinophil response to antigen. VI. The effect of selective removal of T or B lymphocytes on the capacity of primed spleen cells to adoptively transferred immunity to tetanus toxoid. *Immunology*, *28*:243, 1975.

(29) Price, G. B., Senn, J. S., McCulloch, E. A., and Till, J. E. Heterogeneity of molecules with low molecular weight isolated from media conditioned by human leukocytes and capable of stimulating colony formation by human granulopoietic progenitor cells. *J. Cell. Physiol.*, *84*:383, 1974. ·

(30) Ruscetti, F. W., and Chervenick, P. A. Regulation of the release of colony-stimulating activity from mitogen-stimulated lymphocytes. *J. Immunol.*, *114*:1513, 1975.

(31) Ruscetti, F. W., Cypess, R. H., and Chervenick, P. A. Specific release of neutrophilic- and eosinophilic-stimulating factors from sensitized lymphocytes. *Blood*, *47*:757, 1976.

(32) Speirs, R. S., Gallagher, M. T., Rauschwerger, J., Heim, L. R., and Trentin, J. J. Lymphoid cell dependence of eosinophil response to antigen. II. Location of memory cells and their dependence upon thymic influence. *Exp. Hematol.*, *1*:150, 1973.

(33) Stanley, E. R., Cifone, M., Heard, P. M., and Defendi, V. Factors regulating macrophage production and growth: Identity of colony-stimulating factor and macrophage growth factor. *J. Exp. Med.*, *143*:631, 1976.

(34) Sumner, M. A., Bradley, T. R., Hodgson, G. S., Cline, M. J., Fry, P. A., and Sutherland, L. The growth of bone marrow cells in liquid culture. *Br. J. Haematol.*, *23*:221, 1972.

(35) Till, J. E., and McCulloch, E. A. A direct measurement of the radiation sensitivity of normal mouse bone marrow cells. *Radiat. Res.*, *14*:213, 1961.

(36) Trentin, J. J. Influence of hemopoietic organ stroma (hemopoietic inductive microenvironments) on stem cell differentiation. In Gordon, A. S., ed., *Regulation of Hematopoiesis*. New York: Appleton-Century-Crofts, 1970, vol. 1, p. 161.

(37) Walls, R. S., Bass, D. A., and Beeson, P. B. Mechanism of eosinophils. X. Evidence for immunologic specificity of the stimulus. *Proc. Soc. Exp. Biol. Med.*, *145*:1240, 1974.

INTRODUCTION

Hemopoietic inductive microenvironments, associated with the stroma of blood-forming tissues, demonstrate specificity in regulating blood cell formation (7, 8, 25). The inductive influences of microenvironments are radio-resistant, and their distribution can be mapped in radiated mice or rats injected with pluripotent stem cells (14). Such cells, depending on the microenvironment in which they lodge, give rise to at least three colony types: predominantly erythroid, granulocytic, or megakaryocytic (8). The influence of the microenvironment on the pluripotent stem cell is not sufficient for erythroid colony development. Without erythropoietin (epo), pluripotent stem cells in an erythroid-inductive microenvironment develop into microscopic colonies of apparently undifferentiated cells (4, 8). In the mouse spleen, erythroid colonies outnumber granulocytic colonies 3 to 1 whereas, in the bone marrow. granulocytic colonies outnumber erythroid colonies 2 to 1 (8). The rat spleen supports the growth of erythroid colonies only (18).

Determination of the biochemical basis of hemopoietic microenvironments will require a simple, quantitative assay of their effects on differentiation. We have recently demonstrated that the erythroid-inductive microenvironment can be studied *in vitro* (22). Spleen cells from radiated mice, when combined with marrow cells in culture, caused an increase in baseline hemoglobin (Hb) formation and in the response of the marrow cells to exogenous epo. In the present studies, we show that this method can be used to characterize the erythroid-inductive microenvironment and to probe the biochemical mechanisms of its action.

MATERIALS AND METHODS
ANIMALS

Male Long-Evans rats weighing 250–300 g and female BDF_1, (C57BL/6 × DBA)F_1 mice 12–16 weeks old were used. Ex-hypoxic rats, used in one experiment (Fig. 4B), had been subjected to 0.5 atmosphere in a hypobaric chamber for 5 weeks, followed by 9 days at ambient pressure. They were removed from the chamber for 3 hr, 5 days per week, to allow for replenishing of food and water and for cage cleaning. Drinking water was supplemented with 15 mg of ferrous sulfate/liter to provide iron for extra erythropoiesis. The hematocrits of the rats used were 76% and 72%.

Studies of the Erythroid Inductive Microenvironment in Vitro

8

Gary van Zant
and Eugene Goldwasser,
with the technical assistance
of Nancy Pech

71

RADIATION

Mice were radiated at a dose rate of 64 R/min, as measured with a Victoreen meter, with x-rays from a 250 kVp Maxitron, with filtration of 0.25 mm Cu plus 1.0 mm Al and a half-value thickness of 1.05 mm Cu, at a distance of 77 cm. Rats were radiated at a dose rate of 69 R/min with filtration of 0.5 mm Cu plus 1.0 mm Al and a half-value thickness of 1.25 mm Cu, at a distance of 66 cm. The animals were rotated under the x-ray source during radiation. Cell suspensions, in an ice bath, were radiated at a dose rate of 640 R/min with filtration of 0.25 mm Cu plus 1.0 mm Al and a half-value thickness of 1.05 mm Cu, at a distance of 14.5 cm.

PREPARATION OF CELL SUSPENSIONS

Spleen, lung, thymus, and liver cell suspensions were prepared by aseptic removal of the organs from animals killed by cervical dislocation or with carbon dioxide. The organs were minced with scissors and forced through a sterile stainless steel screen (60 mesh) into medium composed of 70% NCTC-109 and 30% fetal calf serum. Cell clumps were dispersed by repeated passage of the suspension first through a 19 gauge, then through a 22 gauge, needle. The cell suspension was then filtered through a 100 mesh screen, and the cells were washed once in phosphate-buffered saline (PBS) (0.14 NaCl, 0.01 M phosphate buffer, pH 7.2) before appropriate dilution in culture medium (described below). Cell counts were performed with a hemocytometer.

CULTURE METHODS

Marrow cells aspirated from tibiae and femora were cultured as previously described (11) in a medium at pH 6.9 consisting of 65% NCTC-109 (Microbiological Associates, Bethesda, Md.) containing 30 mM morpholino-propane sulfonic acid (Sigma Chemical Co., St. Louis, Mo.) to maintain the pH, 30% fetal calf serum (International Scientific Industries, Cary, Ill.), 5% rat serum containing 4.1 μg/ml of iron as ferric nitrate, and 0.05 mg/ml of Gentamicin (Schering Corp., Port Reading, N.J.). Cysteine HCl (Sigma Chemical Co., St. Louis, Mo.) (390 μg/ml) was added to the medium because we have found that Hb synthesis (both in the presence and absence of epo) was increased by cysteine at an optimal concentration of 559 μg/ml (NCTC-109 contains 169

μg/ml) (23). All sera were heat-treated (30 min at 56°C) prior to use. Two-tenths milliliter aliquots of the cell suspension were pipetted into the wells of Disposo trays (Linbro Chemical Co., New Haven, Conn.); six replicates were used per group. The wells were sealed with the sterile plastic sheets provided and incubated at 37°C for 24 hr, when 20 μl of rat serum labeled with ^{59}Fe (0.2 μCi) was added and incubation allowed to proceed for 5 more hr. The cells were transferred to tubes, kept at ice-bath temperature, and washed with PBS, then twice with 5% trichloroacetic acid. Heme was extracted into cyclohexanone, and radioactivity was determined in an aliquot of this extract as previously described (11). In this system, essentially all of the heme is derived from Hb, allowing the use of this simple method to quantify synthesis of Hb. Marrow cells were always cultured at a cell density of 15 $\times$ 10^6 nucleated cells/ml.

Spleen cells, when cultured, were prepared as described and diluted to a cell density of 15 $\times$ 10^6 nucleated cells/ml in complete medium. The cell suspension was cultured in 25 cm^2, 75 cm^2, or 150 cm^2 screw-top tissue culture flasks (Falcon Plastics, Oxnard, Cal., or Corning, Corning, N.Y.) in 5% CO$_2$ in fully humidified air at 37°C. Conditioned medium was centrifuged in the cold first at 1000 RPM, then at 10,000 RPM, before it was stored frozen. Medium incubated under the same conditions without cells served as a control.

EPO PREPARATIONS

Epo with a potency of 200 u/mg of protein, prepared in this laboratory from anemic sheep plasma and diluted to 16 u/ml in 0.1% bovine serum albumin (Pentex Fraction V, Miles Laboratories, Elkhart, Ind.) in PBS, was added to cultures in microliter volumes. In two experiments (Figs. 4A and 4B), epo with a potency of 23,000 u/mg of protein (< 50% pure), prepared from human urine in this laboratory (18), was employed.

ANTI-EPO

In one experiment (Table 4), antibody to epo was used (it was a gift of Drs. J. Garcia and J. Schooley, University of California, Berkeley, Cal.). Antibody was added to epo or spleen cell medium and incubated at 37°C for 1 hr and then overnight at 4°C before assay with marrow cells.

RESULTS

CELL SPECIFICITY

The data summarized in Table 1 show the following: Normal mouse marrow cells in culture alone responded to added epo with a 5-fold increase in Hb synthesis. When normal mouse spleen cells (NS) were added to the marrow, baseline Hb synthesis increased by about 80% and the response to epo was essentially unchanged. In contrast, the same number of spleen cells taken from radiated mice (XS) caused both a 5-fold increase in the baseline and a 50% increase in response to epo. Neither XS nor NS by themselves synthesize measurable amounts of Hb nor respond to added epo. Normal rat spleen cells, when added to mouse marrow, also increase baseline Hb formation and response to epo. At the same cell density, rat spleen cells alone do not detectably synthesize Hb nor respond to epo.

Lung cells from either normal or radiated mice caused a modest increase in baseline Hb synthesis by marrow cells, but had essentially no effect on their response to epo. Marrow cells from radiated mice were tested at a low cell density because the number of nucleated cells fell by 87% in the 3 days after radiation. At this low concentration, the radiated cells by themselves did not measurably synthesize Hb, but they did cause an increase in both baseline and epo-stimulated Hb formation by normal marrow.

In a series of experiments (Table 2) similar to those reported in Table 1 but with rat marrow cells, we found that rat NS and XS increased baseline Hb synthesis about 3-fold and increased the response to epo by 69% and 49% respectively. When marrow cells from radiated rats (3 days after 500 R) (XM) were added to normal marrow, baseline Hb synthesis increased about 3-fold, and the response to epo was essentially unchanged. XM synthesized a small but detectable amount of Hb when cultured alone, and gave a small response to epo. The results from cocultures were corrected for Hb synthesis by XM alone. Similar controls were carried out for other cell types appearing in Table 2, but only XM synthesized measurable amounts of Hb.

In contrast to the capacity of both rat NS and XS to increase Hb synthesis by marrow cells, a clear difference existed between normal lung cells (NL) and lung cells from radiated rats (XL). NL increased the baseline by 260% and increased the response to epo by 12%, whereas XL increased the baseline about 6-fold and the response to epo by about 70%. Similarly, normal thymus cells (NT) increased baseline Hb synthesis 2-fold and significantly potentiated the response to epo (by 13%). Thymus cells from radiated rats (XT) increased Hb synthesis in the absence of epo by 3–4-fold and increased the response to epo by about 60%.

Normal rat liver cells, when added to marrow, had essentially no effect on the baseline, but almost completely abolished (−94%) the response to epo. Mouse NS increased Hb synthesis by rat marrow cells by about 50%, but had essentially no effect on their response to epo. Mouse XS increased the baseline about 2-fold, but substantially decreased (−36%) the response by rat marrow cells to epo.

TABLE 1 Mouse Marrow Cell Hemoglobin Synthesis: Effect of Other Cell Types in Coculture

CELL TYPES	NO. OF CELLS/ML ($\times 10^{-6}$)	NO. OF EXPERIMENTS	MEAN % OF CONTROL	
			− epo	+ epo
Marrow	15	25	100	500
Marrow + normal spleen	15 9	7	179	525
Marrow + X-radiated spleen	15 9	6	496	755
Marrow + Normal lung	15 5	3	125	520
Marrow + X-radiated lung	15 5	2	145	465
Marrow + X-radiated marrow	15 1.5	1	138	630
Marrow + Normal rat spleen	15 9	2	361	930

TABLE 2 Rat Marrow Cell Hemoglobin Synthesis: Effect of Other Cell Types in Coculture

CELL TYPES	NO. OF CELLS/ML (× 10⁻⁶)	NO. OF EXPERIMENTS	MEAN % OF CONTROL	
			− epo	+ epo
Marrow	15	33	100	396
Marrow +	15	11	419	668
Normal spleen	15			
Marrow +	15	7	392	592
X-radiated spleen	15			
Marrow +	15	1	398	412
X-radiated marrow	8			
Marrow +	15	2	360	444
Normal lung	15			
Marrow +	15	2	718	666
X-radiated lung	15			
Marrow +	15	4	319	448
Normal thymus	15			
Marrow +	15	3	457	636
X-radiated thymus	15			
Marrow +	15	1	93	24
Normal liver	8			
Marrow +	15	1	149	391
Normal mouse spleen	9			
Marrow +	15	2	298	255
X-radiated mouse spleen	9			

FIGURE 1. *In vivo* radiation dose: effect of spleen cells on hemoglobin synthesis by marrow cells. Spleen cells (9 × 10⁶/ml) were harvested from mice 3 days after they had received 350, 450, 600, or 900 R and were combined with mouse marrow cells (15 × 10⁶/ml). Hemoglobin synthesis was measured during the last 5 hr of a 29-hr culture period. ○, cocultures without epo; ●, cocultures to which 16 mU of epo were added.

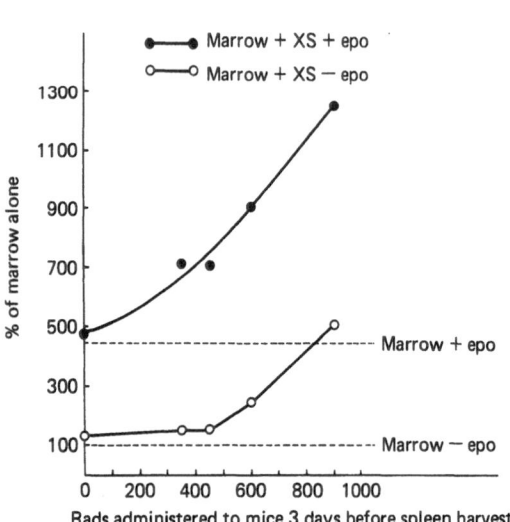

RADIATION DOSE

Mouse In an experiment in which NS increased Hb synthesis in mouse marrow cells by 30% and had no effect on their response to epo, XS from mice given four different doses of radiation were combined with marrow cells, and the effects on Hb synthesis were measured (Fig. 1). XS from mice given 350 or 450 R 3 days previously had little effect on baseline Hb synthesis when compared with NS. However, an appreciable increase was observed in cells from mice given 600 R, and an even greater increase at 900 R. With the 900 R dose, the increase in baseline Hb synthesis was 4-fold, i.e., more than that caused by the addition of 16 mU of epo.

Potentiation of the response to epo by marrow cells was found with XS cells from animals given 350 R. These cells increased the response to epo by 220% over marrow alone and by 220% over cocultures of marrow and NS. XS cells from mice given 450 R showed essentially the same results as found with 350 R, but XS cells from mice given 600 R and 900 R gave successively larger increases in the response to epo. The latter dose more than doubled the response by marrow cells in cocultures with NS.

Rat Normal spleen cells and XS from rats given up to 700 R, when added to marrow cells, increased Hb synthesis to the same extent.

Effect of in vitro radiation We previously showed that radiation *in vitro* of mouse XS cells, just prior to their addition to marrow cells, enhanced their ability to potentiate the response by mouse marrow cells to epo (23). When rat NS cells were radiated *in vitro* just prior to their addition to rat marrow cells, their capacity to augment Hb synthesis was impaired (Fig. 2). Loss of activity increased with increasing dose up to 10,000 R; at this dose, which also enhanced the effect of mouse XS, epo-induced Hb synthesis was reduced by 54% and the effect on the baseline was reduced by 59%, compared to unradiated NS.

CELL NUMBER

Mouse Increasing numbers of mouse XS increased baseline Hb synthesis by mouse marrow cells in a dose-dependent manner, with a plateau beginning at about 15×10^6 cells/ml. However, no optimum cell density was found for potentiating the marrow response to epo. In three experiments, it appeared that the broad range from 4 to 10×10^6 XS/ml provided a somewhat greater response to epo than cell densities above 10×10^6/ml, where potentiation declined (data not shown).

Rat Both rat NS and XS increased baseline HB synthesis and potentiated the cellular response to epo in a dose-dependent manner. Results of an experiment in which up to 20×10^6 NS/ml were added to marrow cells are shown in Fig. 3. Both baseline and potentiation of the response of epo increased steadily with increasing numbers of added NS. In an experiment employing larger numbers of cells (XS), the increase in baseline continued up to 43×10^6 cells/ml; potentiation of the response to epo, however, reached a peak at 22×10^6 cells/ml and declined at 32 and 43×10^6/ml (data not shown).

TIME AFTER RADIATION

Mouse Mouse XS, when compared with NS, generally caused an increase in Hb synthesis by marrow cells 2 days after 900 R, although activity first appeared sometimes as early as 1 day or as late as 3 days. XS harvested 3 days after 900 R consistently increased Hb synthesis, but XS 4–6 days after radiation gave inconsistent results.

Rat Rat spleen cells showed little change in their capacity to increase marrow cell Hb synthesis for up to 17 days after 500 R.

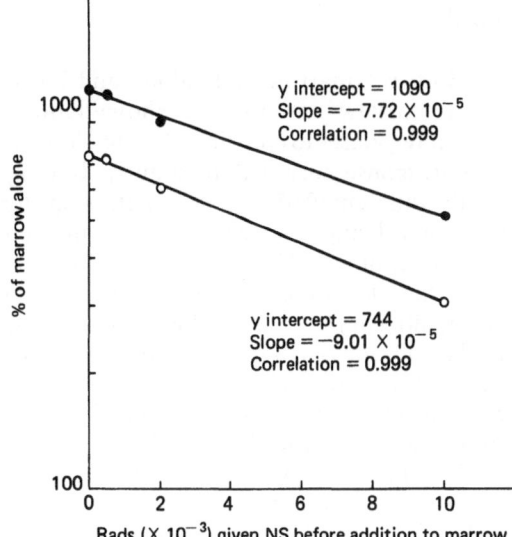

FIGURE 2. Effect of spleen cell number on hemoglobin synthesis by marrow cells in cocultures. Normal rat spleen cells were combined with rat marrow cells. Hemoglobin synthesis was measured during the last 5 hr of a 29-hr culture period. ○, cocultures without epo; ●, cocultures to which 16 mU of epo were added.

FIGURE 3. *In vitro* radiation of spleen cells: effect on marrow cell hemoglobin synthesis in cocultures. Rat spleen cells (15×10^6/ml) were radiated *in vitro* and combined with rat marrow cells. Hemoglobin synthesis was measured during the last 5 hr of a 29-hr culture period. ○, cocultures without epo. Characteristics of the curve are: intercept, 744; slope, -9.01×10^{-5}; correlation, 0.999. Characteristics of the upper curve (with 16 mU of epo) are: intercept, 1,090; slope, -7.72×10^{-5}; correlation, 0.999.

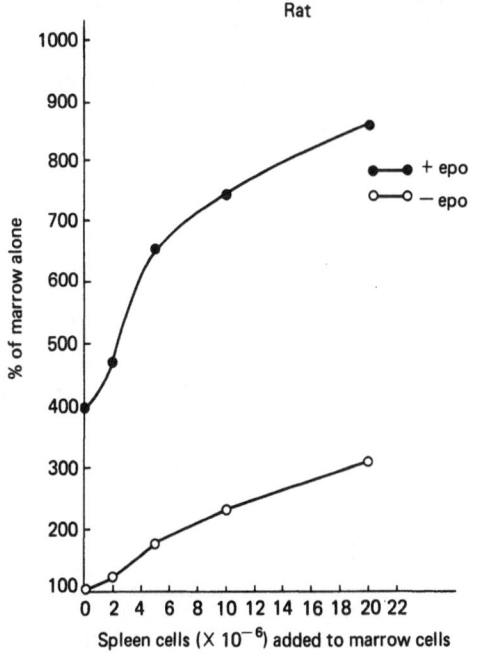

EPO DOSE

Both rat marrow cells alone and cocultures of marrow and rat NS showed dose-dependent responses to epo (Fig. 4A). The slopes of the dose-response curves were slightly, but significantly, different ($0.05 > p > 0.02$), the slope for marrow alone being greater than that for marrow and NS cocultures (1.28 vs. 1.01). The difference between the elevations of the two curves was highly significant ($p < 0.001$). Under these·conditions, the addition of NS increased the sensitivity of marrow to epo by an average of 89%. In a parallel experiment, the same preparation of NS was combined with marrow from ex-hypoxic rates (Fig. 4B). This marrow retains more primitive erythropoietic precursors while lacking erythroblasts late in the developmental sequence, a population upon which epo is known to act (12, 13, 17). The slopes of the two dose-response curves did not differ significantly, but the elevations were different ($0.01 > p > 0.001$). The addition of NS increased the sensitivity of ex-hypoxic marrow to epo by an average of 70%.

SPLEEN CELL-CONDITIONED MEDIUM

To determine whether the presence of spleen cells was required to stimulate marrow-cell Hb synthesis, or whether medium from spleen

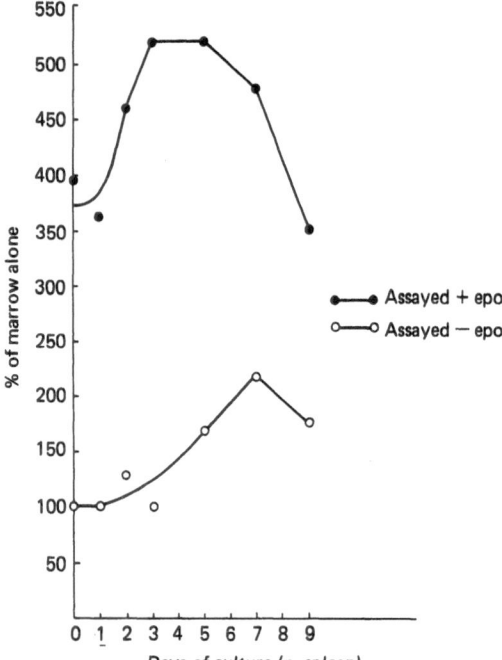

FIGURE 5. Effect of spleen cell medium on marrow cell hemoglobin synthesis. Rat spleen cells were cultured at 15×10^6 cells/ml. Medium was collected from cultures at 1, 2, 3, 5, 7, or 9 days and combined with marrow cells to make up 17% of the marrow cell medium. Hemoglobin synthesis was measured during the last 5 hr of a 29 hr culture period. ○, medium assayed without epo; ●, medium assayed in the presence of epo (5 mU).

FIGURE 4. Epo dose response of cocultures of rat marrow and rat spleen cells. Hemoglobin synthesis was measured in cultures of rat marrow alone (15 × 10⁶/ml) (○) or cocultures of marrow and rat spleen cells (15×10^6/ml) (●) to which 0.6, 1.2, 2.4, or 4.8 mU of epo were added. Normal marrow was used in panel A and ex-hypoxic marrow was used in panel B.

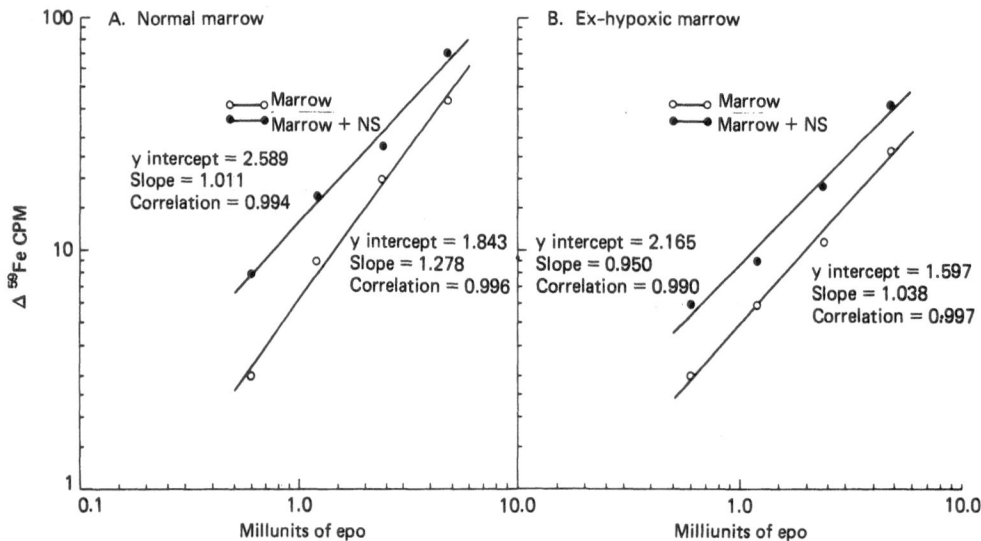

cells cultured alone could also stimulate marrow cells, we carried out the following experiment. Rat NS was cultured for up to 9 days, after which medium freed of cells was assayed by being added to marrow cells. Medium incubated under the same conditions, but containing no cells, served as a control. Medium from 7-day spleen cultures caused a more than 2-fold increase in baseline Hb synthesis by marrow cells (Fig. 5). This effect was first detectable at 5 days and, by 9 days, had fallen off slightly from the peak at 7 days. Spleen cell medium (SCM) also effected an increase in epo-induced Hb synthesis in marrow cells. Potentiation was detectable at 2 days, reached a peak at 3–5 days, and declined to control levels by 9 days.

SCM, when added to marrow cells, caused increased baseline Hb synthesis and response to epo in a dose-dependent manner (Table 3). Significant stimulation was obtained when SCM comprised as little as 10% of the marrow cell medium. The effect of SCM on the baseline could be described by a log-log curve with the following characteristics: intercept, −0.94; slope, 1.20; correlation coefficient, 0.97. Its stimulation of epo-induced Hb synthesis was described by a similar curve, with an intercept of −0.98, a slope of 1.38; and a correlation coefficient of 1.00.

To rule out the possibility that epo in SCM was responsible for stimulating Hb synthesis, we exposed medium to anti-epo before adding it to marrow cells (Table 4). Under conditions where antibody neutralized 70% of the epo (4.3 mU), it had no significant effect on the 3-fold stimulation of Hb synthesis caused by SCM.

DISCUSSION

A fundamental difference exists between the erythropoietic response of mouse and rat marrow to the addition of cells from different tissues. Of the cell types tested, only three had a major effect

TABLE 3 Effect of Spleen Cell Medium on Marrow Cell Hemoglobin Synthesis

% MADE UP BY SPLEEN CELL MEDIUM[a]	EPO[b]	INCREASE OVER CONTROL MEDIUM (^{59}FE CPM)
3	−	2
	+	2
10	−	4
	+	8
20	−	16
	+	27
33	−	29
	+	44

[a]Each concentration of spleen cell medium was compared to the appropriate concentration of medium incubated under the same conditions (6 days), but with no cells (control medium).

[b]Five milliunits.

on the response of mouse marrow to epo: mouse XS, rate NS, and possibly mouse XM (Table 1). Significantly, all three are from tissues that are potentially erythropoietic. All of the cell types tested caused increased baseline Hb synthesis by marrow cells, but mouse XS and rat NS clearly differed from the others.

Unlike mouse marrow, rat marrow Hb synthesis was increased by cells from both erythropoietic and nonerythropoietic tissues (Table 2). In the latter category, lung and thymus cells potentiated the response of marrow cells to epo, but only if they were harvested from previously radiated rats. XL appeared to be the most active cell type tested. Normal lung and thymus cells potentiated the response to epo only slightly. Liver cells almost completely suppressed the response of marrow to epo, possibly by inactivating epo. The fact that they did not appreciably affect baseline Hb synthesis argues against a toxic effect on the marrow cells.

All of the cell types listed in Table 2 which

TABLE 4 Effect of Anti-epo on Spleen Cell Medium

GROUP	ADDITION	^{59}FE CPM (±S.D.)	p VALUE[a]
1	Control Medium (CM)[b,c]	28 ± 4	
2	CM + anti-epo	26 ± 5	0.5 > p > 0.4
3	CM + epo (4.3 mU)	63 ± 4	0.001 > p
4	CM + epo + anti-epo	36 ± 4	0.001 > p
5	Spleen cell medium (SCM)[c]	111 ± 6	0.001 > p
6	SCM + anti-epo	100 ± 17	0.3 > p > 0.2

[a]The following groups were compared to yield the p value: 2 with 1, 3 with 1, 4 with 3, 5 with 1, 6 with 5.

[b]CM is medium incubated under the same conditions (6 days) as spleen cell medium (SCM), but with no cells.

[c]CM or SCM comprised 33% of the marrow cell medium.

were harvested from previously radiated rats, with the exception of bone marrow, potentiated the response by marrow cells to epo. In contrast, only one of these cell types from unradiated rats, namely, spleen cells, possessed the same activity.

The capacity of NS to increase marrow cell Hb synthesis was abrogated in a dose-dependent manner by radiation (Fig. 2). The fact that the capacities to increase the baseline and to increase epo-induced Hb formation were lost in parallel suggests that the same radiosensitive cell population may be responsible for both effects. Alternatively, two cell subpopulations, each with the same sensitivity to radiation, could be required for both effects. Since mouse XS, subjected to further radiation (10,000 R) before addition to marrow cells, was more effective in potentiating epo-induced Hb synthesis (23), different cell subpopulations in mouse XS and rat NS may be responsible for their effects on marrow cells.

Rat NS, in addition to its effects on rat marrow cells, was also a potent stimulator of Hb synthesis in mouse marrow. In the converse experiment, however, mouse XS, which was a potent stimulator of Hb synthesis in mouse marrow, suppressed the response by rat marrow cells to epo. It appears that a unidirectional species-specificity characterizes the potentiation of marrow cell erythropoiesis by rat and mouse spleen cells.

All of the cell types tested, with the exception of liver cells, increased baseline Hb synthesis by rat marrow cells. No clear difference was apparent between the amount by which a certain cell type increased the baseline and its capacity to potentiate the response to epo. For example, XM and NL increased the baseline about as much as XS, but neither had a major effect on the marrow response to epo. No cell type which increased the baseline 3-fold, however, failed to potentiate the response to epo in the same cocultures. It is possible that part of the increase in baseline was due simply to increasing the cell density in cocultures, thus approximating more closely the marrow cell density. Krantz and Goldwasser (15) found that Hb synthesis per cell increased in marrow cell cultures of increasing density. Spleen cells harvested from mice 3 days after exposure to various doses of radiation, when combined with marrow cells at the same cell density, increased Hb synthesis to various degrees (Fig. 1). These results argue against nonspecific cell contact in cocultures playing a major role in increasing Hb synthesis by marrow cells.

Our results here and in previously reported work (23) have established that the concept of the hemopoietic inductive microenviron-ment, originally proposed by Curry and Trentin (7), can be studied in vitro. The cell type or types responsible for the augmentation of Hb synthesis are unknown. Our findings, however, permit several conclusions regarding their distribution: (a) in the mouse, these cells are found in XS and XBM and not in lung; (b) in the rat, they are present in NS as well as in XS, XL, and XT; and (c) their relative frequency in XL is apparently higher than in NS, XS, and XT.

Tissue stromal cells have been implicated in a number of important aspects of hemopoiesis. Several workers have described a large cell in erythropoietic tissue surrounded by developing erythroblasts making up a unit called an eryth-roblastic islet (2, 3, 25). The central cell resembles a macrophage and may be involved in the transfer of ferritin to erythroblasts (3). A radioresistant cell occurring in spleen and resembling a macrophage is involved in processing antigen in the immune response (20). Cellulose acetate membranes implanted into the peritoneal cavities of radiated recipients supported the growth of hemopoietic colonies after development of a layer of macrophages and fibroblasts (21). Fibroblasts obtained from serially cultured marrow cells, when injected under the kidney capsule, differentiated into stroma which supported the growth of bone and hemopoietic tissue (10). Wilson et al. (26) have developed an assay for marrow cells forming plaques of fibroblasts in culture. These cells are relatively radioresistant, and the plaques of resulting fibroblasts are a source of colony-stimulating factor (CSF). These cells may be related to the endosteal-associated cells described by Chan and Metcalf (5, 6), which were radioresistant and were a potent source of CSF.

It is tempting to speculate that a subpopulation of lung and spleen macrophage-like cells is responsible for increased Hb synthesis obtained when lung or spleen cells are added to marrow cell cultures. It is not clear why XL, and not NL, possesses this capacity. We have observed that the cellularity of lungs, thymus, and spleen from radiated rats is greatly reduced by prior radiation. If the active cells are radioresistant, as are fibroblast plaque-forming cells (26) and cells responsible for processing antigen in the immune response (20), radiation may simply serve to increase the relative frequency of active cells through the loss of cells sensitive to radiation. This explanation fits the data for mouse spleen where XS, but not NS, was active. It does not, however, explain the results for rat spleen, where NS was more active on a per-cell basis than was XS.

Thymus cells have been shown to influence hemopoiesis in several respects. Lord and Schofield (16) found that thymus cells injected into radiated recipient mice, together with an inoculum of bone marrow cells which had been radiated to reduce the number of colony-forming cells to about 20% of normal, enhanced spleen colony formation 2-fold. Thymus cells had no effect on colony formation by normal marrow or spleen cells. Endogenous spleen colony formation in sublethally radiated mice was enhanced about 3-fold by the injection of thymocytes. Thymus cells given 1600 R prior to injection did not have these effects. Dexter and Lajtha (9) found that the survival of spleen colony-forming cells was enhanced by the addition of thymocytes to marrow cell cultures. When injected together with bone marrow cells into radiated mice, thymus cells partially overcame the "poor growth" of splenic colonies that is characteristic of parental marrow in F_1 hybrid mice in particular strain combinations (1). In our hands, thymus cells harvested from radiated rats stimulated Hb synthesis by marrow cells in the presence and in the absence of epo. Normal thymocytes increased baseline Hb synthesis, but did not have a major effect on the response to epo.

It has been shown that the radiated mouse spleen supports the growth of at least three colony types: erythroid, granulocytic, and megakaryocytic (7). The radiated rat spleen, on the other hand, supports the growth of erythroid colonies only (19). The rat splenic microenvironment thus appears to be exclusively erythroid, and may provide a clue as to why NS as well as XS augmented marrow cell erythropoiesis in the rat, but not in the mouse.

The presence of radioresistant cells associated with bone endosteum which produce CSF, a candidate granulopoietin (22), may be related to the fact that the mouse marrow microenvironment supports the growth of about twice as many granulocytic as erythroid colonies. In the mouse spleen, erythroid colonies outnumber granulocytic colonies by about 3 to 1 (8). We have found that rat NS produces a substance *in vitro* which has the same effect as the cells themselves, namely, that of augmenting marrow cell Hb synthesis. When added to marrow cells, it increased both baseline Hb synthesis and epo-induced hemoglobin formation in a dose-dependent manner. Since this substance was not inactivated by anti-epo, it does not appear to be epo. The identification of this substance and its source are the subject of current studies.

SUMMARY

Spleen cells from radiated mice (XS), but not from normal mice (NS), when combined with mouse marrow cells *in vitro*, had two effects on hemoglobin synthesis: (1) increased baseline, and (2) augmented erythropoietin-induced, hemoglobin formation. Spleen cells from normal and radiated rats caused increased hemoglobin synthesis by rat marrow cells in a dose-dependent manner. In addition, lung and thymus cells from radiated rats, but not from normal rats, caused increased hemoglobin synthesis by rat marrow cells. Rat NS augmented hemoglobin formation by mouse marrow, but neither mouse NS nor mouse XS had an effect on rat marrow.

Medium from rat NS cultures stimulated hemoglobin synthesis when added to rat marrow cells. The increases in baseline, and in the response to erythropoietin, were dependent on the amount of "conditioned medium." The substance in the medium responsible for these effects was not erythropoietin.

ACKNOWLEDGMENTS

We are deeply indebted to Drs. J. Garcia and J. Schooley of the University of California, Berkeley, California, for the anti-epo, and to Mr. James Bland for radiating the animals and cell suspensions.

Supported in part by Grants CA-14599 and CA-18375 from the National Cancer Institute. G. van Zant held a postdoctoral research fellowship from the National Institutes of Health.

REFERENCES

(1) Basford, N. L., and Goodman, J. W. Effects of lymphocytes from the thymus and lymph nodes on differentiation of hemopoietic spleen colonies in irradiated mice. *J. Cell. Physiol.*, *84*:37, 1974.

(2) Ben-Ishay, Z., and Yoffey, J. M. Reticular cells of erythroid islands of rat bone marrow in hypoxia and rebound. *J. Reticuloendothelial Soc.*, *10*:482, 1971.

(3) Bessis, M., and Breton-Gorius, J. Iron metabolism in the bone marrow as seen by electron microscopy: A critical review. *Blood*, *19*:635, 1962.

(4) Bleiberg, I., Liron, M., and Feldman, M. Reversion by erythropoietin of the suppression of erythroid clones caused by transfusion-induced polycythaemia. *Transplantation*, *3*:706, 1965.

(5) Chan, S. H., and Metcalf, D. Local production of colony stimulating factor within the bone marrow: Role of non-hematopoietic cells. *Blood*, *40*:646, 1972.

(6) Chan, S. H., and Metcalf, D. Local and systemic control of granulocytic and macrophage progenitor cell regeneration after irradiation. *Cell Tissue Kinet.*, *6*:185, 1973.

(7) Curry, J. L., and Trentin, J. J. Hemopoietic spleen colony studies. I. Growth and differentiation. *Dev. Biol.*, *15*:395, 1967.

(8) Curry, J. L., Trentin, J. J., and Wolf, N. Hemopoietic spleen colony studies. II. Erythropoiesis. *J. Exp. Med.*, *125*:703, 1967.

(9) Dexter, T. M., and Lajtha, L. G. Proliferation of hemopoietic stem cells *in vitro*. *Brit. J. Haematol.*, *28*:525, 1974.

(10) Freidenstein, A. J., Chailakhyan, R. K., Latsinik, N. V., Panasyuk, A. F., and Keiliss-Borok, I. V. Stromal cells responsible for transferring the microenvironment of the hemopoietic tissues. *Transplantation*, *17*:331, 1974.

(11) Goldwasser, E., Eliason, J. F., and Sikkema, D. An assay for erythropoietin *in vitro* at the milliunit level. *Endocrinology*, *97*:315, 1975.

(12) Gross, M., and Goldwasser, E. On the mechanism of erythropoietin-induced differentiation. IX. Induced synthesis of 9S ribonucleic acid and of hemoglobin. *J. Biol. Chem.*, *246*:2480, 1971.

(13) Hodgson, G., and Eskuche, I. Study of the effects of chemotherapeutic agents on the "early" and "late" responses to erythropoietin. *Proc. Soc. Exp. Biol. Med.*, *127*:328, 1968.

(14) Jenkins, V. K., Trentin, J. J., and Wolf, N. S. Radioresistance of the splenic hemopoietic inductive microenvironments (HIM). *Radiat. Res.*, *43*:212, 1970.

(15) Krantz, S. B., and Goldwasser, E. On the mechanism of erythropoietin-induced differentiation. IV. Some characteristics of erythropoietin action on hemoglobin synthesis in marrow cell culture. *Biochim. Biophys. Acta*, *108*:455, 1965.

(16) Lord, B. I., and Schofield, R. The influence of thymus cells in hemopoiesis: Stimulation of hemopoietic stem cells in a syngeneic, *in vivo*, situation. *Blood*, *42*:395, 1973.

(17) Matoth, Y., and Ben-Porath, E. Effect of erythropoietin on mitotic rate of erythroblasts in bone marrow cultures. *J. Lab. Clin. Med.*, *54*:722, 1959.

(18) Miyake, T., Kung, C. K. -H., and Goldwasser, E. Unpublished.

(19) Rauchwerger, J. M., Gallagher, M. T., and Trentin, J. J. Role of the hemopoietic inductive microenvironments (HIM) in xenogeneic bone marrow transplantation. *Transplantation*, *15*:610, 1973.

(20) Roseman, J. X-ray resistant cell required for the induction of *in vitro* antibody formation. *Science*, *165*:1125, 1969.

(21) Seki, M. Hematopoietic colony formation in a macrophage layer provided by intraperitoneal insertion of cellulose acetate membrane. *Transplantation*, *16*:544, 1973.

(22) Stanley, E. R., Hansen, G., Woodcock, J., and Metcalf, D. Colony stimulating factor and the regulation of granulopoiesis and macrophage production. *Fed. Proc.*, *34*:2272, 1975.

(23) van Zant, G. E., Goldwasser, E., and Baron, J. M. Study of hemopoietic microenvironment *in vitro*. *Nature (London)*, *260*:609, 1976.

(24) van Zant, G. E., and Goldwasser, E. Unpublished.

(25) Weiss, L. The structure of bone marrow: Functional interrelationships of vascular and hematopoietic compartments in experimental hemolytic anemia: An electron microscope study. *J. Morphol.*, *117*:467, 1965.

(26) Wilson, F. D., O'Grady, L., McNeill, C. J., and Munn, S. L. The formation of bone marrow derived fibroblastic plaques *in vitro*: Preliminary results contrasting these populations to CFU-C. *Exp. Hematol.*, *2*:343, 1974.

(27) Wolf, N. S., and Trentin, J. Hemopoietic colony studies. V. Effect of hemopoietic organ stroma on differentiation of pluripotent stem cells. *J. Exp. Med.*, *127*:205, 1968.

INTRODUCTION

As a result of a genetic defect expressed in the stroma of the tissues supporting hematopoiesis rather than in the hematopoietic cells themselves, mice of genotype Sl/Sl^d (Steel, Steel-Dickie mutant mice) suffer a chronic macrocytic anemia, and are extremely sensitive to ionizing radiation (11). Previously, it was hypothesized that the genetic defect disturbs erythropoiesis very early in the erythron, perhaps at the point of commitment of *in vivo* colony-forming units (CFU) to the erythrocytic cellular line of differentiation (8, 19). In an earlier study (9), this hypothesis was tested by measuring and comparing, in Sl/Sl^d mice and their congenic +/+ littermates, population sizes of high self-renewal potential and low self-renewal potential CFU. It was reasoned that a block in stem cell differentiation occurring early in the erythron would result in a deficiency of the latter but not of the former. However, our study did not bear this hypothesis out. Rather, it led to the unexpected observation that all the stem cell populations in Sl/Sl^d mice, with the exception of the splenic CFU population, were reduced in size.

However, anemia and other forms of hematopoietic stress are known to initiate substantial increases in extramedullary, but not medullary CFU population sizes (7, 10). and, because Sl/Sl^d mice suffer a chronic macrocytic anemia, it was reasoned that a comparison of the effects of Sl and + genes on CFU population sizes might be more meaningful if the comparison were undertaken not only on the same genetic background but also under similar physiologic conditions in which the blood RBC concentrations of +/+ and Sl/Sl^d mice are approximately the same. Therefore, in the present study, anemic Sl/Sl^d mice and normal +/+ mice were rendered polycythemic by hypertransfusion and the sizes of their CFU population determined.

In Vivo Colony Forming Unit Population Sizes in Hypertransfused Sl/Sl^d Mice

9

Kenneth F. McCarthy

METHODS
MICE

WCB6F$_1$-Sl/Sl^d, B6D2F$_1$, C57BL/6J, and WC/Re mice were obtained from the Jackson Laboratory, Bar Harbor, Me. B6WCF$_1$ mice were raised at the Armed Forces Radiobiology Research Institute by mating C57BL/6J females with WC/Re males. The animals were maintained on a 6 AM to 6 PM (light-dark) cycle. Wayne Lab-Blox and acidified (pH 2.5) water were available *ad libitum*. All mice were acclimated to laboratory conditions for 2 weeks. During this time they were certified free of lesions of murine pneumonia complex and of oropharyngeal *Pseudomonas* spr.

RADIATION

Mice were exposed to ^{60}Co wholebody gamma-radiation at a dose rate of 150 rad/min to a total dose of 950 rad.

IN VIVO COLONY FORMING ASSAY (CFU)

The CFU assay of Till and McCulloch (16) was performed as previously described (9). The following donor-recipient combinations were used: B6D2F$_1$ hematopoietic cells were transplanted into B6D2F$_1$ mice and WCB6F$_1$-*Sl/Sl^d* into B6WCF$_1$ mice.

CFU SEEDING EFFICIENCY

The 2-hour seeding efficiencies f of CFU were determined according to the method of Siminovitch et al. (14). Briefly, in the case of femoral CFU, 6×10^6 marrow cells from a pool of one to four donor mice were injected into five intermediate recipients. Two hr later the mice were euthanized, their spleens removed, and 1/16 to 1/12 of a spleen was then injected into 10 secondary recipients. In the case of splenic CFU, $1-5 \times 10^7$ spleen cells from a pool of one to four donor mice were injected into five intermediate recipients. Two hours later 0.25–0.5 of a spleen was injected into four secondary recipients. The CFU

content of the original cell suspensions was determined in primary recipients by the *in vivo* CFU assay.

HYPERTRANSFUSION

Blood for transfusion was collected from normal and heterozygous littermates from the orbital sinus into heparinized phosphate buffered saline and washed 3 times. One-half ml of washed packed red cells was then injected i.p. into each recipient daily on 3 successive days. Six days after the last injection, hematocrit (Hct.) values of the peripheral blood were determined. Mice having a Hct. of at least 55 were considered hypertransfused.

CALCULATIONS

CFU per 10^5 cells was calculated according to the formula: CFU/10^5 cells = observed colonies/10^5 cells $\times 1/f$. CFU per organ was calculated according to the formula: CFU/organ = CFU/cell $\times$ cells/organ.

RESULTS

ORGAN CELLULARITY AND CFU NUMBERS IN NORMAL AND HYPERTRANSFUSED +/+ MICE

Presented in Table 1 and Table 2 are the number and colony-forming potential of nucleated cells from the spleen and femurs respectively of normal and hypertransfused male B6D2F$_1$ of genotype +/+. It was found, as has been

TABLE 1 Colony-Forming Ability of Spleen Cells from Normal and Hypertransfused B6D2F$_1$ − +/+ Mice

Hct. range	48–50	67–76
Colonies/10^5 cells	2.85 ± 0.73[a] [4][b]	3.60 ± 0.51 [3]
Nucleated cells/ spleen ($\times 10^{-7}$)	6.36 ± 0.85 [4]	8.35 ± 2.32 [3]
f (%)	14.8 ± 3.1 [4]	4.2 ± 1.2 [3]
CFU/10^5 cells[c]	19	86
CFU/spleen[d]	12,200	71,500

[a]Mean ± SE.

[b]Figures in brackets refer to number of separate determinations. Each determination consisted of a cell suspension prepared from three to four experimental mice being injected into seven to twelve recipient mice.

[c]Calculated from data for f and colonies/10^5 cells.

[d]Calculated from CFU/10^5 cells and average number of cells per donor spleen.

TABLE 2 Colony-Forming Ability of Marrow Cells from Normal and Hypertransfused B6D2F$_1$ − +/+ Mice

Hct. range	48–50	67–76
Colonies/10^5 cells	20 [1][a]	58 [1]
Nucleated cells/ femur ($\times 10^{-7}$)	1.07 [1]	1.13 [1]
f (%)	17.5 [1]	8.2 [1]
CFU/10^5 cells[b]	115	700
CFU/femur[c]	11,800	75,000

[a]Figures in brackets refer to number of separate determinations. Each determination consisted of a cell suspension prepared from four experimental mice being injected into twelve recipient mice.

[b]Calculated from data for f and colonies/10^5 cells.

[c]Calculated from CFU/10^5 cells and average number of cells per donor marrow.

reported by others (3, 13), that hypertransfusion increases the colony-forming potential of hematopoietic nucleated cells from both the marrow and spleen. Also, it was found that hypertransfusion decreases the CFU seeding efficiency 2–4-fold. Therefore, correcting for changes in both the spleen colony-forming potential and *f*, it was calculated that the femoral and splenic CFU population sizes of hypertransfused mice are approximately 4–6-fold larger on either a per cellular or per organ basis than are comparable CFU population sizes in normal mice.

ORGAN CELLULARITY AND CFU NUMBERS IN NORMAL AND HYPERTRANSFUSED *Sl/Sl*^d MICE

As compared to male mice of genotype +/+, hypertransfusion has exactly the opposite effect on the CFU populations of male *Sl/Sl*^d mice. Hypertransfusion (a) lowers the colony-forming potentials of the hematopoietic nucleated cells rather than increasing them; and (b) increases rather than decreases *f* (Tables 3 and 4). Taking these differences into consideration when calculating the CFU population sizes of hypertransfused *Sl/Sl*^d mice, as compared to anemic *Sl/Sl*^d mice, it was determined that hypertransfusion drastically reduces by about 50-fold the size of the splenic CFU population, and to a lesser extent—about 2-fold—the size of the marrow CFU population.

DISCUSSION

It was the tentative conclusion of a previous study (9) that the factors supporting a normal size splenic CFU population in *Sl/Sl*^d mice were predominantly long-range or systemic in nature, and were produced in response to the macrocytic anemia suffered by these mice (5). This hypothesis was tested in the present study by temporarily eliminating the anemia of *Sl/Sl*^d mice by hypertransfusion and measuring their CFU population sizes. It was found that hypertransfusion reduced the *Sl/Sl*^d splenic CFU population size 50-fold while reducing that of the marrow only 2-fold. In contrast, this same treatment increased both marrow and splenic CFU numbers of +/+ mice 6-fold.

When these findings are viewed in the light of the recent work of Bozzini et al. (1), they offer an insight into the puzzling fact that although erythropoiesis in *Sl/Sl*^d mice is erythropoietin-dependent (12), polycythemic

TABLE 3 Colony-Forming Ability of Spleen Cells from Normal and Hypertransfused WCB6F$_1$ – *Sl/Sl*^d Mice

Hct range	22–33	55–66
Colonies/10^5 cells	2.60 ± 0.82[a] [4][b]	0.27 ± 0.12 [4]
Nucleated cells/ spleen (× 10^{-8})	1.72 ± 0.21 [4]	1.34 ± 0.16 [4]
f (%)	4.4 [1]	14.7 ± 12.0 [2]
CFU/10^5 cells[c]	58.1	1.8
CFU/spleen[d]	102,000	2460

[a]Mean ± SE.

[b]Figures in brackets refer to number of separate determinations. Each determination consisted of a cell suspension prepared from one to two experimental mice being injected into seven to twelve recipient mice.

[c]Calculated from data for *f* and colonies/10^5 cells.

[d]Calculated from CFU/10^5 cells and average number of cells per donor spleen

Sl/Sl^d mice are refractory to exogenous erythropoietin (6). Bozzini and coworkers clearly demonstrated that the early reestablishment of erythropoiesis in polycythemic +/+ mice by exogenous erythropoietin is nearly *in toto* a splenic phenomenon. Therefore, given the stem cell-progenitor cell relationship of CFU to erythropoietin-responsive cells (ERC), the lack of a splenic erythropoietin-responsive compartment in hypertransfused Sl/Sl^d mice is consistent with the present finding of a nearly total absence of splenic CFU in these mice. As such, it might be concluded that the refractory character of hypertransfused *Sl/Sl*^d mice to exogenous eryth-

TABLE 4 Colony-Forming Ability of Marrow Cells from Normal and Hypertransfused WCB6F$_1$ – *Sl/Sl*^d Mice

Hct. range	22–33	55–64
Colonies/10^5 cells	15.00 ± 3.50[a] [2][b]	12.35 ± 1.35 [2]
Nucleated cells/ femur (× 10^{-7})	1.09 ± 0.16 [2]	1.51 ± 0.09 [2]
f (%)	7.3 [1]	15.0 [1]
CFU/10^5 cells	185	82
CFU/femur	20,200[c]	12,400

[a]Mean ± SE.

[b]Figures in brackets refer to number of separate determinations. Each determination consisted of a cell suspension prepared from one to two experimental mice being injected into seven to twelve recipient mice.

[c]Calculated from CFU/10^5 cells and average number of cells per donor femur.

ropoietin is a result, in part, of anomalous cell kinetics at the stem cell level.

In addition to a long-range mechanism regulating CFU proliferations, there is considerable evidence for a local one. For example, it is known that the selective depopulation of a hematopoietic cell maturation compartment, such as the erythron, results in the recruitment of CFU into cell cycle (3, 18). However, in the Sl/Sl^d mouse, suppression of the erythron by hypertransfusion does not appear to stimulate CFU proliferation. This might suggest that the stromal tissues of Sl/Sl^d mice are incapable of producing an effective local CFU proliferative factor in response to a depleted erythron. Indeed, similar observations on the absence of a local CFU proliferative mechanism in Sl/Sl^d mice have been reported by others (4, 15).

Given the concepts that (a) commitment of hematopoietic stem cells to the erythrocytic cellular line of differentiation is regulated by local stromal tissue, i.e., the hematopoietic inductive microenvironment (HIM), and (b) the Sl element is an integral part of this microenvironment (17), it might follow from the present work that the erythrocytic HIM can be described, in part, as a feedback loop between the erythron and the multipotent CFU compartment via the specialized stromal tissues supporting hematopoiesis. The mechanism would operate in such a fashion that CFU proliferation would be stimulated by a stromal tissue-CFU interaction in response to a depleted erythron. The exact nature of this stromal-supported CFU proliferation is, of course, not known. However, it could be speculated that this proliferative mechanism by itself or in unison with other factors generates and/or, amplifies a CFU subpopulation with a high capacity for erythroid differentiation. It is known that CFU of Sl/Sl^d origin have considerably less potential for establishing erythroid colonies in radiated recipient mice than do CFU of +/+ origin (17, 19) and, further, what erythropoiesis that does take place in Sl/Sl^d mice does so at erythropoietin concentrations characteristic of *in vitro* rather than *in vivo* systems (2, 6).

SUMMARY

The effect of hypertransfusion on the colony-forming unit (CFU) population size of normal and mutant Sl/Sl^d mice was determined. The main finding was that hypertransfusion reduced the splenic CFU population of Sl/Sl^d mice nearly 50-fold while increasing that of normal mice 6-fold. Hypertransfusion also reduced the marrow CFU population of Sl/Sl^d mice, but the reduction was only 2-fold. In normal mice, hypertransfusion resulted in a 6-fold increase in the marrow CFU population. Two tentative conclusions were drawn from the present study: (a) the refractoriness of the polycythemic Sl/Sl^d mice to exogenous erythropoietin is a result of anomalous stem cell kinetics characterizing the hypertransfused Sl/Sl^d mouse; and (b) the hematopoietic inductive microenvironment can be described, in part, as a feedback loop between the erythron and multipotent CFU compartment via the specialized stromal tissues which support hematopoiesis.

REFERENCES

(1) Bozzini, C. E., Martinez, M. A., Alvarez-Ugarte, C. A., Montangero, V., and Soriano, G. The importance of the spleen on proliferation of erythropoietin-responsive cells induced by erythropoietin. *Exp. Hematol.*, 2:93, 1974.

(2) Gregory, C. J. Erythropoietin sensitivity as a differentiation marker in the hemopoietic system: Studies of three erythropoietic colony responses in culture. *J. Cell. Physiol.*, 89:289, 1976.

(3) Guzman, E., and Lajtha, L. G. Some comparisons of the kinetic properties of femoral and splenic haemopoietic stem cells. *Cell Tissue Kinet.*, 3:91, 1970.

(4) Fried, W., Chamberlin, W., Knospe, W. H., Husseini, S., and Trobaugh, F. E., Jr. Studies of the defective haematopoietic microenvironment of Sl/Sl^d mice. *Br. J. Haematol.*, 24:643, 1973.

(5) Harrison, D. E., Malath, V. G., and Silber, R. Elevated erythrocyte nucleoside deaminase levels in genetically anemic W/W^v and Sl/Sl^d mice. *Blood Cells*, 1:605, 1975.

(6) Harrison, D. E., and Russell, E. S. The response of W/W^v and Sl/Sl^d anaemic mice to haemopoietic stimuli. *Br. J. Haematol.*, 22:155, 1972.

(7) Hodgson, G. S., Bradley, T. R., and Telfer, P. A. Haemopoietic stem cells in experimental

haemolytic anemia. *Cell Tissue Kinet.*, 5:283, 1972.

(8) McCarthy, K. F., and MacVittie, T. J. Erythrocytic committed hematopoietic stem cells in the peripheral blood of mice. *Acta Haematol.*, 53:226, 1975.

(9) McCarthy, K. F., Ledney, G. D., and Mitchell, R. A deficiency of hematopoietic stem cells in steel mice. *Cell Tissue Kinet.*, in press, 1976.

(10) Rencricca, N. J., Rizzoli, V., Howard, D., Duffy, P., and Stohlman, F., Jr. Stem cell migration and proliferation during severe anemia. *Blood*, 36:764, 1970.

(11) Russell, E. S., and Bernstein, S. E. Blood and blood formation. In Green, E. L., ed., *Biology of the Laboratory Mouse*. New York: McGraw-Hill, 1966, 2nd ed., p. 315.

(12) Schooley, J. C. as quoted in Harrison and Russell 1972.

(13) Schooley, J. C. Studies of stem cell kinetics utilizing the spleen colony technique and the erythropoietin sensitivity test. In Bond, U. P., and Sugahara, T., eds., *Comparative Cellular and Species Radiosensitivity*. Tokyo: Igaku Shoin Ltd., 1969.

(14) Siminovitch, L., McCulloch, E. A., and Till, J. E. The distribution of colony-forming cells among spleen colonies. *J. Cell. and Comp. Physiol.*, 62:327, 1963.

(15) Sutherland, D. J. A., Till, J. E., and McCulloch, E. A. A kinetic study of the genetic control of hemopoietic progenitor cells assayed in culture and *in vitro*. *J. Cell. Physiol.*, 75:267, 1970.

(16) Till, J. E., and McCulloch, E. A. A direct measurement of the radiation sensitivity of normal mouse bone marrow cells. *Radiat. Res.*, 14:213, 1961.

(17) Trentin, J. J. Determination of bone marrow stem cell differentiation by stromal hemopoietic inductive microenvironments (HIM). *Am. J. Pathol.* 65:621, 1971.

(18) Vassort, F., Winterholer, M., Frindel, E., and Tubiana, M. Kinetic parameters of bone marrow stem cells using *in vivo* suicide by tritiated thymidine or by hydroxyurea. *Blood*, 41:789, 1973.

(19) Wolf, N. S. Dissecting the hematopoietic microenvironment. I. Stem cell lodgment and commitment, and the proliferation and differentiation of erythropoietic descendants in the *Sl/Sl^d* mouse. *Cell Tissue Kinet.*, 7:89, 1974.

PART III Physiology of Committed Stem Cells
(CFU-e and CFU-m)

J. E. Till, Ph.D.

The revolution in experimental hematology brought about by the introduction of colony assays for hemopoietic progenitor cells is still running its course. Colony assays have now been described for pluripotent stem cells (20), granulocyte and macrophage progenitors (3, 10, 13, 16), erythropoietic progenitors (1, 18), megakaryocyte precursors (12, 15), B- and T-lymphocyte precursors (5, 14, 17), and cells in leukemic marrow (4). For all the above cell types except pluripotent stem cells, the assays involve colony formation in cell culture. There is scope for more advances along these lines as culture methods are refined further, particularly in regard to the development of a culture assay for pluripotent stem cells.

The colony assays have provided new and powerful approaches to the analysis of the initial steps of hemopoietic differentiation. Their major advantages are well known, and can be summarized as follows:

1. Use of assays for colony formation permits the detection of classes of progenitor cells not recognizable at present on the basis of other criteria, such as a characteristic morphology in stained preparations.

2. The number of colonies formed under different experimental conditions can be used to compare the relative numbers of viable progenitor cells present under these different conditions. Such comparisons involve the tacit assumption that the efficiency of colony formation by the viable progenitors remains constant as the experimental conditions are changed.

3. The composition of the colonies provides information about the potentialities for differentiation of the progenitor cells, provided that the conditions used to support colony formation are appropriate for expression of those potentialities.

In addition to uncertainties about detection efficiencies and about the most appropriate culture conditions for expression of potentialities for differentiation, colony assays often have some other limitations which need to be borne in mind. Examples are:

1. Colony assays do not allow the progenitor cell itself to be visualized, but only the colony composed of its progeny. Properties of the progenitors must be deduced from examination of the colonies. Such deductions may yield ambiguous conclusions. For example, it has been known for more than a decade that colonies formed *in vivo* by pluripotent hemopoietic stem cells are very variable in their composition (11, 20). These variations have been interpreted in different

Recognition of Hemopoietic Progenitors

10

J. E. Till

ways; they have been assumed to result from random fluctuations occurring as a function of time during the development of colonies (9, 21), or from spatial variations in the microenvironment (22). The information needed to determine the relative importance of temporal versus spatial sources of variation is still not available. It should be noted, however, that the large variations observed among colonies have provided a useful means for investigating cell lineage relationships, based on the examination of correlations between numbers of different cell types in individual colonies (6, 7, 23).

2. The progenitor cells detected by a given colony assay may prove to be heterogeneous on the basis of other criteria. In order to distinguish more homogeneous subpopulations of progenitors, further refinements of the assay may be needed, such as an initial physical separation of the cells, or modifications in the culture technique. An example is cells on the erythropoietic pathway, where at least two and probably three classes of progenitors can be recognized on the basis of differences in kinetics of colony formation, erythropoietin dose-response curves, and velocity sedimentation rate (6, 8). Recognition of subpopulations is already creating problems about terminology, e.g., CFU-e, and two subclasses of BFU-e (1, 6, 8).

3. Clinical applications of the colony methods are restricted, both because of the time span of 1–2 weeks before results become available, and because a well-equipped cell culture laboratory is essential. Considerable care is often required if the culture methods are to yield reproducible results. In particular, the selection of growth media and sera and the preparation of appropriate growth stimulatory factors require a high level of quality control.

Because of limitations such as these, colony assays should not necessarily be regarded as the optimal means for the study of progenitor cells. The expanding repertoire of colony assays needs to be supplemented by other, more direct, methods for the identification and enumeration of different classes of hemopoietic progenitors. To make this possible, "markers" characteristic of classes of hemopoietic progenitor cells need to be recognized.

The types of markers which might be expected to be present on hemopoietic progenitor cells have been outlined elsewhere (19). Of course, the most useful markers would be those with the greatest specificity for a single class of progenitor cells. However, it may be unwise to assume that such markers are prevalent. Instead, the available markers may be less specific, and perhaps may only vary in relative amounts from one progenitor class to another. To achieve adequate resolution of the different classes of progenitors, it may be necessary to recognize sets of markers which, taken together, are characteristic of a specific progenitor cell class. The feasibility of using multiple markers, together with a fluorescence-activated cell sorter (2), to identify and separate out subpopulations of hemopoietic progenitors is presently being explored.

SUMMARY

The development of criteria for the recognition of the committed progeny of pluripotent hemopoietic stem cells continues to be an important aspect of experimental hematology. Emphasis has been placed on colony assays which permit detection of the capacity of committed progenitors to proliferate and differentiate in culture in response to appropriate stimuli. Such colony techniques make it possible to delineate identifiable stages in early hemopoietic differentiation and to investigate the regulation of transitions between these stages. Current studies on CFU-e and CFU-m provide excellent examples of the value of this approach.

In order to obtain more detailed information about committed progenitor cells and their regulation, another more direct approach to the study of hemopoietic progenitors is needed. In particular, methods need to be developed for the detection of specific markers on pluripotent stem cells and committed progenitor cells. Some possible approaches to this problem are discussed.

REFERENCES

(1) Axelrad, A. A., McLeod, D. L., Shreeve, M. M., and Heath, D. Properties of cells that produce erythrocytic colonies *in vitro*. In Robinson, W. A., ed., *Hemopoiesis in Culture: Second International Workshop*. Washington, D.C.: U.S. Government Printing Office, 1974, p. 226.

(2) Bonner, W. A., Hulett, H. R., Sweet, R. G., and Herzenberg, L. A. Fluorescence activated cell sorting. *Rev. Sci. Instrum.*, 43: 404, 1972.

(3) Bradley, T. R., and Metcalf, D. The growth of mouse bone marrow cells *in vitro*. *Aust. J. Exp. Biol. Med. Sci.*, 44: 287, 1966.

(4) Dicke, K. A., Spitzer, G., and Ahearn, M. J. Colony formation *in vitro* by leukaemic cells in acute myelogenous leukaemia with phytohaemagglutinin as stimulating factor. *Nature (London)*, 259: 129, 1976.

(5) Fibach, E., Gerassi, E., and Sachs, L. Induction of colony formation *in vitro* by human lymphocytes. *Nature (London)*, 259: 127, 1976.

(6) Gregory, C. J. Erythropoietin sensitivity as a differentiation marker in the hemopoietic system: Studies of three erythropoietic colony responses in culture. *J. Cell. Physiol.* (in press).

(7) Gregory, C. J., McCulloch, E. A., and Till, J. E. Erythropoietic progenitors capable of colony formation in culture: State of differentiation. *J. Cell. Physiol.*, 81: 411, 1973.

(8) Heath, D. S., Axelrad, A. A., McLeod, D. L., and Shreeve, M. M. Separation of the erythropoietin responsive progenitors BFU-E and CFU-E in mouse bone marrow by unit gravity sedimentation. *Blood*, 47: 777, 1976.

(9) Korn, A. P., Henkelman, R. M., Ottensmeyer, F. P., and Till, J. E. Investigations of a stochastic model of haemopoiesis. *Exp. Hematol. 1*: 362, 1973.

(10) Lin, H. S., and Stewart, C. C. Colony formation by mouse peritoneal exudate cells *in vitro*. *Nature (London), New Biol.* 243: 176, 1973.

(11) McCulloch, E. A. Les clones de cellules hématopoiétiques *in vivo*. *Rev. Fr. Etud. Clin. Biol.*, 8: 15, 1963.

(12) Metcalf, D., MacDonald, H. R., Odartchenko, N., and Sordat, B. Growth of mouse megakaryocyte colonies *in vitro*. *Proc. Nat. Acad. Sci. U.S.A.*, 72: 1744, 1975.

(13) Metcalf, D., Parker, J., Chester, H. M., and Kincade, P. W. Formation of eosinophilic-like granulocytic colonies by mouse bone marrow cells *in vitro*. *J. Cell. Physiol.*, 84: 275, 1974.

(14) Metcalf, D., Warner, N. L., Nossal, G. J. V., Miller, J. F. A. P., Shortman, K., and Rabellino, E. Growth of B lymphocyte colonies *in vitro* from lymphoid organs. *Nature (London)*, 255: 630, 1975.

(15) Nakeff, A., and Daniels-McQueen, S. *In vitro* colony assay for a new class of megakaryocyte precursor: Colony-forming unit megakaryocyte (CFU-M). *Proc. Soc. Exp. Biol. Med.*, 151: 587, 1976.

(16) Pluznik, D., and Sachs, L. The cloning of normal "mast" cells in tissue culture. *J. Cell. Comp. Physiol.*, 66: 319, 1965.

(17) Sredni, B., Kalechman, Y., Michlin, H., and Rozenszajn, L. A. Development of colonies *in vitro* of mitogen-stimulated mouse T lymphocytes. *Nature (London)*, 259: 130, 1976.

(18) Stephenson, J. R., Axelrad, A. A., McLeod, D. L., and Shreeve, M. M. Induction of colonies of hemoglobin-synthesizing cells by erythropoietin *in vitro*. *Proc. Nat. Acad. Sci. U.S.A.*, 68: 1542, 1971.

(19) Till, J. E. Regulation of hemopoietic stem cells. In Cairnie, A. B., Lala, P. K., and Osmond, D. G., eds., *Stem Cells of Renewing Cell Populations*. New York: Academic Press, 1976, p. 143.

(20) Till, J. E., and McCulloch, E. A. A direct measurement of the radiation sensitivity of normal mouse bone marrow cells. *Radiat. Res.*, 14: 213, 1961.

(21) Till, J. E., McCulloch, E. A., and Siminovitch, L. A stochastic model of stem cell proliferation, based on the growth of spleen colony-forming cells. *Proc. Nat. Acad. Sci. U.S.A.*, 51: 29, 1964.

(22) Trentin, J. J. Hemopoietic inductive microenvironments. In Cairnie, A. B., Lala, P. K., and Osmond, D. G., eds., *Stem Cells of Renewing Cell Populations*. New York: Academic Press, 1976, p. 255.

(23) Wu, A. M., Siminovitch, L., Till, J. E., and McCulloch, E. A. Evidence for a relationship between mouse hemopoietic stem cells and cells forming colonies in culture. *Proc. Nat. Acad. Sci. U.S.A.*, 59: 1209, 1968.

INTRODUCTION

Differentiation in the hemopoietic system is a complex process involving a progression of changes that extend through many cell divisions. In the most mature hemopoietic cells, the results of differentiation can be visualized directly and morphologic criteria can therefore be used to determine the relative positions of cells in the terminal stages of blood cell production. Detection of earlier cell types has relied on the development of assay procedures that exploit the capacity of primitive hemopoietic cells to give rise to colonies of mature progeny under appropriate conditions *in vivo* or in culture. A variety of such colony assays now exist (1, 3, 10, 11, 13, 14, 16, 17), and their use has led to a conceptual model of the hemopoietic system which makes a distinction between *pluripotent* stem cells or spleen colony-forming units (CFU-s), and secondary progenitor cell types still capable of extensive proliferation but *committed* to differentiation along a single pathway (9). Examples of this latter class of progenitor have been identified for bopth the granulopoietic and erythropoietic pathways. These are the progenitors of the granulopoietic colonies and the progenitors

11

of the large erythroid bursts detected by established *in vitro* assay procedures (1, 3, 13), historically referred to as CFU-c (originally the only "colony-forming unit in culture") and BFU-e ("erythropoietin-dependent burst-forming unit"), respectively.

Evidence in support of such a two-stage model has been obtained from analysis of the representation of various types of hemopoietic cells amongst the clonal progeny of individual CFU-s, stimulated to divide and differentiate under conditions that lead to spleen colony formation. Early studies showed that the number of CFU-s found per colony varies widely from one colony to another (18). In addition, the number of CFU-s in any given colony does not correlate strongly with any of the mature hemopoietic cell types recognizable by morphologic criteria (5). This lack of correlation between cells at opposite ends of the system has led to the concept that numerical correlation might provide a measure of relatedness between populations. Thus, for two cell types which are closely related (e.g., parent and daughter cells), it might be expected that a spleen colony containing a large number of cells of the parent type would also contain a large number of cells of the daughter type. Other spleen colonies containing few cells of the parent type would have few cells of the daughter type. On the basis of this

Relationships between Early Hemopoietic Progenitor Cells Determined by Correlation Analysis of Their Numbers in Individual Spleen Colonies

C. J. Gregory
and R. M. Henkelman

type of argument, the strong correlations found between CFU-s and CFU-c (19), and between CFU-s and BFU-e (7), have been taken as evidence of the primitive state of differentiation of CFU-c and BFU-e, respectively.

In the experiments used to provide the CFU-s and BFU-e spleen colony data just referred to (7), simultaneous measurements were also made of N (total number of nucleated cells), CFU-c, and CFU-e (progenitors of the small erythroid colonies recognized after 2 days in culture, (16)). This paper presents an analysis of the complete set of results. For this analysis, Pearson product-moment correlation coefficients and partial coefficients (12) were used to define relationships between the various cell types assayed. These were calculated using the logarithm of the number of cells detected per spleen colony for each determination. The analysis itself provides a rigorous new method for revealing the hierarchy that exists between different types of intermediate cell populations in a system where loss of numerical correlation between distantly related cell types occurs during the processes of differentiation.

EXPERIMENTAL

Spleen colonies were obtained following the injection of $1-2 \times 10^4$ marrow cells from (C57BL/6J × C3H/HeB) F_1 hybrid mice into syngeneic recipients previously radiated with 900 rad (cobalt-60 γ-rays). Eleven or 12 days later, individual colonies were dissected out, cell suspensions prepared in 2 ml of 2% fetal calf serum in α-medium,

and separate aliquots then used to determine the content per colony of each of the six cell types listed in Table 1. As indicated in the table, the details of each of the assay procedures used have been described previously. On the basis of earlier evidence that bursts seen after 3 to 4 days in culture are derived from a different cell population than the bursts that are identified after 8 days (7), the numbers of day 3 BFU-e and day 8 BFU-e have been kept separate in this analysis also.

RESULTS
DATA

One hundred and seventeen spleen colonies were analyzed individually for their content of CFU-s, CFU-c, day 8 BFU-e, day 3 BFU-e, CFU-e, and N, as described above. Figure 1 shows the frequency of occurrence of each of these cell types in the 114 spleen colonies for which complete sets of data were obtained. The open bar in each histogram indicates the proportion of spleen colonies in which no cells of the type indicated could be detected in the aliquot used for that particular assay, i.e., the number of cells of the type indicated was below the detection limit. It can be seen from the shape of the histograms shown in Fig. 1 that all of the distributions extend over several orders of magnitude and are approximately normal when the number of cells detected is plotted on a logarithmic scale.

Figures 2 and 3 show two examples of the 15 scatter plots obtained for all possible pairs of measurements in this series of experiments. Figure 2, which shows the number of CFU-e plotted

TABLE 1 Experimental Design Used to Analyze the Progenitor Cell Content of Individual Spleen Colonies

CELL TYPE ASSAYED	SCORING CRITERIA	FRACTION OF SPLEEN COLONY ASSAYED PER MOUSE OR CULTURE	NO. OF REPLICATES	DETAILS OF ASSAY PROCEDURE
CFU-s	Macroscopic spleen colony	1/8 or 1/80	1–4	Gregory (7)
CFU-c	All colonies with > 20 cells seen after 1 week in CFU-c assay cultures	1/40 or 1/400	2	Gregory et al. (6)
CFU-e	Single or paired clusters of erythroblasts seen after 2 days in CFU-e assay cultures	1/160	2	Gregory (7)
Day 3 BFU-e Day 8 BFU-e	Bursts containing > 3 clusters of erythroblasts seen after 3 or 8 days in BFU-e assay cultures	1/40	2	Gregory (7)
N	Nucleated cells	1/25,000	1	0.05 of each colony diluted in 3% acetic acid for hemocytometer counts

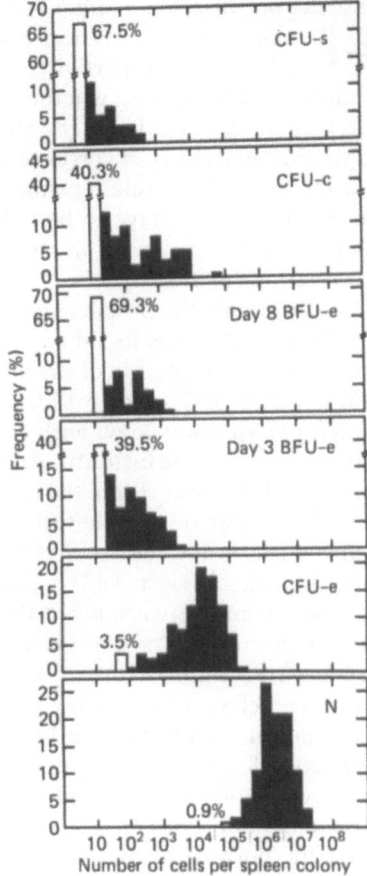

FIGURE 1. Frequency distributions of different hemopoietic cell types in 114 individual spleen colonies assayed after 11–12 days. Open bar indicates the proportion of colonies in which cells of the type shown were below the detectable limit.

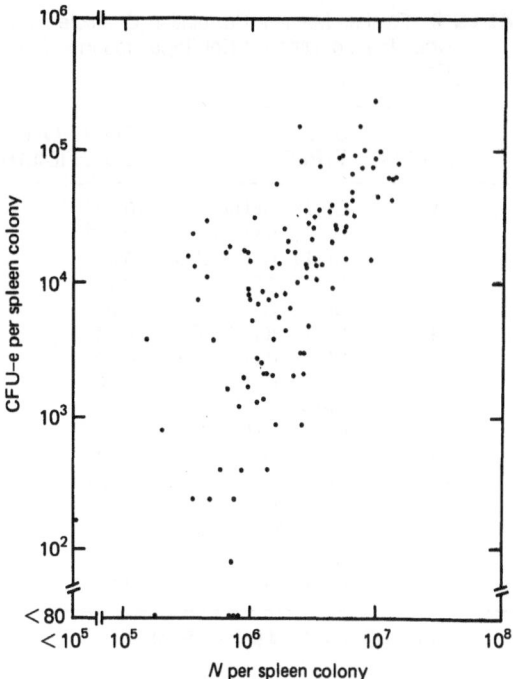

FIGURE 2. Scatter plot of the number of CFU-e versus the total number of nucleated cells (N) in 114 spleen colonies assayed individually.

logarithmic scale rather than a linear scale (Fig. 1). Correlation coefficients were therefore calculated using the logarithms of the numbers of cells of each type detected per spleen colony. In order to get around the problem of zero values, i.e.,

FIGURE 3. Scatter plot of the number of day 8 BFU-e versus the total number of nucleated cells (N) in 114 spleen colonies assayed individually.

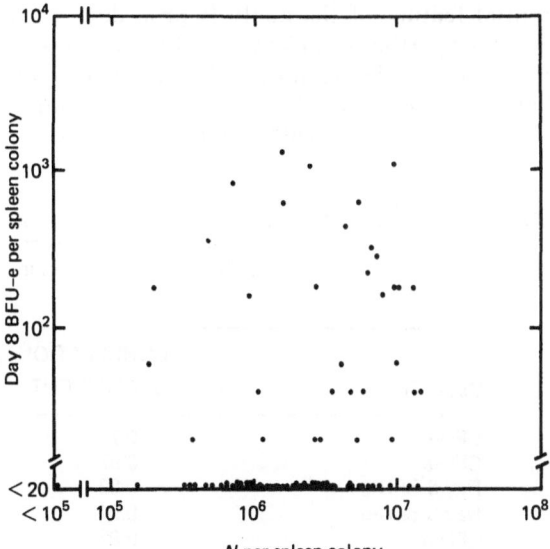

versus the number of N, serves to illustrate one of the strongest correlations found between any two cell types. Figure 3 is a similar plot for day 8 BFU-e versus N. In this case, there was no correlation, and thus, even though 71% of the spleen colonies assayed had no detectable day 8 BFU-e, this could not be explained as a feature particular to small spleen colonies.

CORRELATION ANALYSIS

The degree of relatedness between pairs of cell types was determined using Pearson product-moment correlation coefficients (12). This correlation coefficient requires that the values considered conform to a multivariate normal distribution. For the present study, this type of distribution was better approximated using a

TABLE 2 Product-Moment Correlation Coefficients between Pairs of Different Cell Types found in Spleen Colonies

PAIR NO.	CELL TYPES	CORRELATION COEFFICIENT
1	CFU-s, CFU-c	0.76[a]
2	CFU-s, day 8 BFU-e	0.81[a]
3	CFU-s, Day 3 BFU-e	0.57[a]
4	CFU-s, CFU-e	0.13
5	CFU-s, N	0.20[a]
6	CFU-c, day 8 BFU-e	0.74[a]
7	CFU-c, day 3 BFU-e	0.60[a]
8	CFU-c, CFU-e	0.16
9	CFU-c, N	0.16
10	Day 8 BFU-e, day 3 BFU-e	0.74[a]
11	BFU-e, CFU-e	0.19[a]
12	BFU-e, N	0.18
13	Day 3 BFU-e, CFU-e	0.42[a]
14	BFU-e, N	0.23[a]
15	CFU-e, N	0.68[a]

[a]Statistically significant ($p > 0.95$, (4)).

spleen colonies containing fewer than the minimum detectable number of cells for any of the six assays, these spleen colonies were arbitrarily assigned a number equal to the minimum detectable level. However, the results obtained were not highly dependent on this approximation, since when zero measurements were alternatively assigned a value of 1 the results of the analysis were not appreciably different.

The requirement for normal distributions and hence the selection of an appropriate scale, could have been avoided if rank correlations had been used. However, the usual Spearman rank correlation coefficient (15) does not allow the calculation of partial correlation coefficients, an essential feature of the analysis procedure used here. The partial coefficient provides a measure of the correlation between two cell types that would have been obtained if the numbers of any or all of the others had been held fixed. It thus suppresses

TABLE 3 Product-Moment Correlation Coefficients for Pairs of Measurements of the Same Cell Type using Values from Assay Replicates as Separate Experiments

CELL TYPE	CORRELATION COEFFICIENT
CFU-s	0.74
CFU-c	0.97
Day 8 BFU-e	0.88
Day 3 BFU-e	0.91
CFU-e	0.96

any dependence on variation in other cell populations. The Kendall rank correlation coefficient (15) allows the calculation of partial coefficients, but is deficient for determining whether two correlation coefficients are significantly different (8). Since the present analysis uses correlation coefficients as a measure of the relationship between cell types, it was an advantage to be able to determine significant differences in appropriate pairs of correlation coefficients.

Table 2 presents the product-moment correlation coefficients obtained for all possible pairs of cell types using the logarithm of the number of CFU-s, CFU-c, day 8 BFU-e, day 3 BFU-e, CFU-e, or N detected per spleen colony, with zero measurements set to the minimum detectable number, as described above. From the results presented in Table 2, it can be seen that the strongest correlations were found between CFU-s and CFU-c (pair no. 1), between CFU-s and day 8 BFU-e (pair no. 2), and between day 8 BFU-e and day 3 BFU-e (pair no. 10). Correlation coefficients which are greater than 0.18 are significant ($p > 0.95$ (4)). Therefore, CFU-e do not show a significant correlation with CFU-s, CFU-c, or day 8 BFU-e (pairs no. 4, 8, 11), nor do any of these latter cell types show a significant dependence on colony size (N) (pairs no. 5, 9, 12).

Since the purpose of this analysis was to provide a measure of relatedness between cell types in terms of their correlation coefficients, it was important to assess the extent to which variation in the experimental procedures used might affect the results obtained. This was evaluated by repeating the calculation of correlation coefficients using the data obtained from each assay replicate as a separate determination. Tables 3 and 4 show the results of these calculations for five of the six cell types measured, where at least two replicates per assay were performed (Table 1). Analysis of the pairs of data obtained for a single cell type provides a measure of the internal consistency of the assay procedure used. As can be seen from Table 3, all of the assays showed very strong correlations, the weakest being that found for CFU-s, where most of the counts obtained were low (0–2) and hence subject to large statistical fluctuations. The availability of replicate measurements for five of the cell types allowed the calculation of four sets of correlation coefficients for each pair of different cell types. These values should provide a pessimistic estimate of the kind of variation that might be expected in a completely repeated experiment. The four sets of correlation coefficients obtained are presented in Table 4. The good agreement between these and the correlation coefficients for the averaged data

TABLE 4 Product-Moment Correlation Coefficients between Different Cell Types in Spleen Colonies using Values from Assay Replicates as Separate Determinations

PAIR NO.	CELL TYPES	REPLICATE VALUES CONSIDERED SEPARATELY				REPLICATE VALUES AVERAGED (FROM TABLE 2)
1	CFU-s, CFU-c	0.73[a]	0.73[a]	0.66[a]	0.66[a]	0.76[a]
2	CFU-s, day 8 BFU-e	0.73[a]	0.81[a]	0.62[a]	0.68[a]	0.81[a]
3	CFU-s, day 3 BFU-e	0.57[a]	0.61[a]	0.49[a]	0.52[a]	0.57[a]
4	CFU-s, CFU-e	0.14	0.13	0.06	0.06	0.13
6	CFU-c, day 8 BFU-e	0.71[a]	0.70[a]	0.71[a]	0.71[a]	0.74[a]
7	CFU-c, day 3 BFU-e	0.59[a]	0.61[a]	0.59[a]	0.60[a]	0.60[a]
8	CFU-c, CFU-e	0.16	0.16	0.14	0.14	0.16
10	Day 8 BFU-e, day 3 BFU-e	0.73[a]	0.75[a]	0.68[a]	0.70[a]	0.74[a]
11	BFU-e, CFU-e	0.20[a]	0.16	0.19[a]	0.15	0.19[a]
13	Day 3 BFU-e, CFU-e	0.39[a]	0.34[a]	0.44[a]	0.39[a]	0.42[a]

[a]Statistically significant ($p > 0.95$, (4)).

(results from Table 2) shows the reliability of the averaged data for this type of analysis.

Table 5 presents the partial correlation coefficients for all 15 pairs of cell types conditional on fixed numbers of a third cell type. Several of these are of particular interest. For example, the strong correlation between CFU-e and N (pair no. 15) persists irrespective of the variation in each of the other cell populations. This implies that any relationships with other cell types does not affect the close relationship that exists between CFU-e and N. On the other hand, the moderately strong correlation between CFU-s and day 3 BFU-e (pair no. 3) is lost completely when day 8 BFU-e numbers are held constant. Thus, the apparent relationship between CFU-s and day 3

BFU-e is an artifact explained by the fact that each of these two cell types has an independent close relationship to day 8 BFU-e. The strong correlation between CFU-s and day 8 BFU-e (pair no. 2) is not, however, appreciably changed by fixing the number of day 3 BFU-e. Similarly, the strong correlation between day 8 BFU-e and day 3 BFU-e (pair no. 10) is not highly dependent on variations in any other cell type. This series of results can only be explained by a differentiation sequence in which CFU-s give rise to day 8 BFU-e and thence to day 3 BFU-e. (The other possible explanation, that day 8 BFU-e give rise to both day 3 BFU-e and CFU-s contradicts the property of pluripotency of CFU-s.) It can also be concluded that for all of the progenitor cell types assayed, only day 8 BFU-e

TABLE 5 Partial Product-Moment Correlation Coefficients between Pairs of Different Cell Types in Spleen Colonies with Dependence on a Third Cell Type Suppressed

PAIR NO.	CELL TYPES	CORRELATION COEFFICIENT (FROM TABLE 2)	PARTIAL CORRELATION COEFFICIENT SUPPRESSING THE DEPENDENCE ON:					
			CFU-s	CFU-c	Day 8 BFU-e	Day 3 BFU-e	CFU-e	N
1	CFU-s, CFU-c	0.76	—	—	0.40	0.63	0.75	0.75
2	CFU-s, day 8 BFU-e	0.81	—	0.56	—	0.69	0.80	0.80
3	CFU-s, day 3 BFU-e	0.57	—	0.23	−0.07	—	0.58	0.55
4	CFU-s, CFU-e	0.13	—	0.02	−0.03	−0.15	—	0.00
5	CFU-s, N	0.20	—	0.11	0.09	0.08	0.15	—
6	CFU-c, day 8 BFU-e	0.74	0.34	—	—	0.55	0.73	0.73
7	CFU-c, day 3 BFU-e	0.60	0.31	—	0.11	—	0.60	0.59
8	CFU-c, CFU-e	0.16	0.09	—	0.03	−0.13	—	0.06
9	CFU-c, N	0.16	0.03	—	0.05	0.03	0.08	—
10	Day 8 BFU-e, day 3 BFU-e	0.74	0.58	0.56	—	—	0.75	0.73
11	Day 8 BFU-e, CFU-e	0.19	0.14	0.11	—	−0.21	—	0.09
12	Day 8 BFU-e, N	0.18	0.04	0.09	—	0.01	0.07	—
13	Day 3 BFU-e, CFU-e	0.42	0.43	0.42	0.43	—	—	0.37
14	Day 3 BFU-e, N	0.23	0.15	0.17	0.15	—	−0.08	—
15	CFU-e, N	0.68	0.67	0.67	0.67	0.66	—	—

appear between CFU-s and day 3 BFU-e, as shown by partial correlation coefficient analysis. By similar arguments, the partial correlation coefficients suggest that day 3 BFU-e represent the only progenitor cell type assayed to be located between day 8 BFU-e and CFU-e.

The use of partial correlation coefficients also throws light on the interrelationships that exist between CFU-s, CFU-c, and day 8 BFU-e, although here the situation is more complicated. The strong correlations ($p > 0.9999$) between CFU-s and CFU-c (pair no. 1) and between CFU-s and day 8 BFU-e (pair no. 2) are only slightly affected when the numbers of day 3 BFU-e, CFU-e, or N are held fixed. In contrast, these same two correlations are appreciably reduced when the dependence of day 8 BFU-e for CFU-s versus CFU-c, or CFU-c for CFU-s versus day 8 BFU-e is suppressed. It is also noteworthy that the partial correlation coefficients reveal a significant relationship between CFU-c and day 8 BFU-e (pair no. 6) which exists independently of the correlation of each of these cell types with CFU-s.

An extension of the type of analysis used to calculate the results given in Table 5 is given in Table 6. In this case, correlation coefficients have been calculated for each pair of cell types holding constant the variation in *all* of the remaining cell types. These values provide the most rigorous measure of relatedness between any two cell types. Highly significant ($p > 0.9999$) correlations persist only for the following pairs of numbers: CFU-s and CFU-c (1); CFU-s and day 8 BFU-e (2); day 8 BFU-e and day 3 BFU-e (10); day 3 BFU-e

and CFU-e (13); and CFU-e and N (15). *All* other positive correlations are *not* significant ($p = 0.9$), with the exception of the value obtained for CFU-c versus day 8 BFU-e (pair no. 6) ($p = 0.96$). Evidence of a weak relationship between this pair was already indicated by the partial correlation coefficients shown in Table 5. The negative correlation seen between day 3 BFU-e and N (pair no. 14) ($0.99 > p > 0.95$) is due to the suppression of CFU-e, an intermediate cell type, which correlates very strongly with N.

VALIDATION OF THE ORIGIN OF day 3 BFU-e FROM day 8 BFU-e

One of the conclusions of the correlation analysis described was to locate day 3 BFU-e as a population intermediate between day 8 BFU-e and CFU-e. This suggested that it should be possible to demonstrate the presence of day 3 BFU-e in colonies derived from day 8 BFU-e. Accordingly, normal marrow cells were plated in standard methyl cellulose cultures containing 2.5 units of erythropoietin per ml (7), and after 8 days incubation, 10 isolated colonies identified as bursts were removed and individually resuspended in methyl cellulose culture medium containing 2.5 units per ml erythropoietin. Each resuspended burst was then plated in a new dish and scored at various intervals thereafter, using the number of erythroblast clusters per colony to distinguish between erythroid "colonies" derived from CFU-e and "bursts" derived from BFU-e. The results are

TABLE 6 Partial Correlation Coefficients between Pairs of Different Cell Types in Spleen Colonies with Dependence on all Remaining Cell Types Suppressed

| | | PARTIAL CORRELATION COEFFICIENTS | |
| | | Significant values[a] | Insignificant values |
PAIR NO.	CELL TYPES		
1	CFU-s, CFU-c	0.41[a]	
2	CFU-s, day 8 BFU-e	0.52[a]	
3	CFU-s, day 3 BFU-e		−0.09
4	CFU-s, CFU-e		−0.08
5	CFU-s, N		0.12
6	CFU-c, day 8 BFU-e	0.19[a]	
7	CFU-c, day 3 BFU-e		0.15
8	CFU-c, CFU-e		−0.02
9	CFU-c, N		0.01
10	Day 8 BFU-e, day 3 BFU-e	0.54[a]	
11	Day 8 BFU-e, CFU-e		−0.14
12	Day 8 BFU-e, N		0.05
13	Day 3 BFU-e, CFU-e	0.44[a]	
14	Day 3 BFU-e, N	−0.20[a]	
15	CFU-e, N	0.68[a]	

[a]Statistically significant ($p > 0.95$,(4).

TABLE 7 Number of Erythroid Colonies and Bursts Apparent at Various Times after Replating of 10 Individually Suspended 8-day-old Bursts

REPLATED BURST NO.	COLONY TYPE[a]	SCORING TIME AFTER REPLATING			
		Day 2	Day 4	Day 6	Day 8
1	E	213	40	2	0
	B	6	15	6	1
2	E	42	3	0	0
	B	4	2	0	0
3	E	18	1	0	0
	B	0	1	0	0
4–10	E	0–30	0–3	0	0
	B	0	0	0	0

[a]E = single or paired clusters of erythroblasts (8–50 small cells). B = bursts containing more than 3 erythroid clusters.

shown in Table 7. Substantial numbers of CFU-e could be demonstrated in most of the bursts selected for replating. Three of these replated bursts also contained detectable numbers of BFU-e and these were almost all of the day 3 BFU-e type.

DISCUSSION

Figure 4 shows a schematic representation of the make-up of an "average" spleen colony in terms of its component cell types, as suggested by analysis of spleen colony data, using partial product-moment correlation coefficients. Measurement of the number of hemopoietic progenitors of various types in individual spleen colonies provides information relating to the population size of each cell type assayed (ignoring possible relative differences in plating efficiency), but does not provide information about their interrelationships. In the present analysis, the use of partial coefficients has enabled the definition of each hemopoietic differentiation pathway in terms of a series of related cell types. This was possible because only pairs of adjacent cell types remained significantly correlated when dependence on all other cell types was suppressed. The partial correlation coefficients obtained under these conditions also provide a measure of the relative positions of each cell type along a given pathway. This is also illustrated in Fig. 4, where the length of the arrow between cell types has been drawn inversely proportional to the partial correlation coefficient obtained. It can be seen that, using such a scale to measure differentiation state, presently available colony assays detect populations of erythroid and granulocyte progenitors that are separated from each other and from pluripotent stem cells by roughly equal distances.

Correlations with N provided little information about interrelationships between specific progenitor cell types. However, they were an important inclusion in the analysis as they served to demonstrate that none of the results obtained were influenced by variations in spleen colony size.

The sequence:

day 8 BFU-e → day 3 BFU-e → CFU-e

established by correlation analysis agrees well with previous studies showing a similar progression of changes in proliferative capacity,

FIGURE 4. Diagrammatic representation of the hierarchy of hemopoietic cell types as found in spleen colonies. (For explanation, see discussion.)

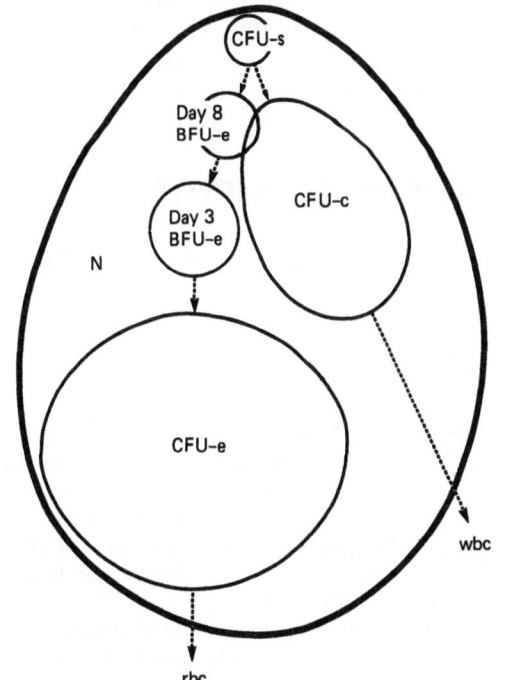

erythropoietin sensitivity, and time to onset of erythroblast differentiation by these three cell types (6). In contrast, the correlation found between CFU-c and day 8 BFU-e was not anticipated. It is interesting that a recent report suggests that BFU-e in man may be responsive to preparations of conditioned media that also stimulate human CFU-c (2). The present data suggest either (a) that primitive (day 8) BFU-e and CFU-c are derived from a common progenitor different from CFU-s, or (b) that in spite of their apparent commitment to two different pathways, some overlap still exists in the cells detected by these two assay procedures.

SUMMARY

Although a number of colony assays have now been devised to detect hemopoietic progenitors apparently committed to either erythroid or granulopoietic differentiation, little is known about their relative states of differentiation or the steps involved in their production from pluripotent stem cells. To approach these questions, 114 spleen colonies were assayed on days 11–12 for their content of: total nucleated cells (N), cells forming spleen colonies in vivo (CFU-s), cells forming granulopoietic colonies in culture (CFU-c), and three types of erythropoietic progenitors: cells forming small erythroid colonies (CFU-e), cells forming small, early-appearing bursts (day 3 BFU-e), and cells forming larger, later-appearing bursts (day 8 BFU-e).

Product-moment correlation coefficients between each possible pair of cell types were calculated using the logarithm of the number of cells of each type per colony. Large correlation coefficients were taken to imply a close relation between the two members of a pair. Partial correlation coefficients were also calculated and used to determine whether strong correlations between two cell types were due to their independent correlation with a third cell type. Using these procedures, the previous findings that CFU-s and CFU-c are closely related, whereas CFU-s and CFU-e are not, have been confirmed. Further, it has now been possible to show the following: no progenitor cell type, except CFU-e, correlates significantly with spleen colony size. CFU-s, day 8 BFU-e, day 3 BFU-e, and CFU-e represent four distinguishable populations, and form a linear sequence in that order, with day 8 and day 3 BFU-e exhibiting a closer relationship than that which exists between day 3 BFU-e and CFU-e. There is an unexpected weak correlation between CFU-c and day 8 BFU-e even when variations with all other cell types including CFU-s are suppressed. This suggests a model of hemopoietic differentiation in which there exists a cell with insufficient self-renewal capacity to be detectable by spleen colony assay but which retains the potential for both erythroid and granulopoietic differentiation.

ACKNOWLEDGMENTS

We are grateful to M. Heppner and C. Smith for expert technical assistance.

REFERENCES

(1) Axelrad, A. A., McLeod, D. L., Shreeve, M. M., and Heath, D. S. Properties of cells that produce erythrocytic colonies in vitro. In Robinson, W. A., ed., Proceedings of the Second International Workshop on Hemopoiesis in Culture. Washington, D.C.: U.S. Government Printing Office, p. 226, 1974.

(2) Aye, M. T. The enhancement of erythroid colony formation by leukocyte conditioned medium (Abstract). Am. Soc. Hemat., 18th Annual Meeting, 1975.

(3) Beyer, W. H., ed. CRC Handbook of Tables for Probability and Statistics. 2nd Edition Cleveland: The Chemical Rubber Company, p. 298, 1974.

(4) Bradley, T. R., and Metcalf, D. The growth of mouse bone marrow cells in vitro. Austr. J. Exp. Biol. Med. Sci., 44:287, 1966.

(5) Fowler, J. H., Wu, A. M., Till, J. E., McCulloch, E. A.,

and Siminovitch, L. (1967) The cellular composition of hemopoietic spleen colonies. *J. Cell. Physiol.*, *69*:65, 1967.

(6) Gregory, C. J., McCulloch, E. A., and Till, J. E. Erythropoietic progenitors capable of colony formation in culture: State of differentiation. *J. Cell. Physiol.*, *81*:411, 1973.

(7) Gregory, C. J. Erythropoietin sensitivity as a differentiation marker in the hemopoietic system: Studies of three erythropoietic colony responses in culture. *J. Cell. Physiol.* In press.

(8) Kendall, M. G. *Rank Correlation Methods.* London: Griffin, Chaps. 4 & 5, 1970.

(9) McCulloch, E. A., & Till, J. E. Regulatory mechanisms acting on hemopoietic stem cells: Some clinical implications. *Am. J. Pathol.*, *65*:601, 1971.

(10) Metcalf, D., MacDonald, H. R. Odartchenko, N., and Sordat, L. B. Growth of mouse megakaryocyte colonies *in vitro*. *Proc. Nat. Acad. Sci. (U.S.A.)*, *72*:1744, 1975.

(11) Metcalf, D., Warner, N. L., Nossal, G. J. V., Miller, J. F. A. P., Shortman, K., and Rabellino, E. Growth of B lymphocyte colonies *in vitro* from mouse lymphoid organs. *Nature (London)*, *255*:630, 1975.

(12) Morrison, D. F. *Multivariate Statistical Methods.* New York: McGraw-Hill Book Company, 1967.

(13) Pluznik, P. H., and Sachs, L. The induction of colonies of normal "mast" cells by a substance in conditioned medium. *Exp. Cell Res.*, *43*:553, 1966.

(14) Rozenszajn, L. A., Shoham, D., and Kalechman, I. Clonal proliferation of PHA-stimulated human lymphocytes in soft agar culture. *Immunology*, *29*:1041, 1975.

(15) Siegel, S. *Nonparametric Statistics for the Behavioral Sciences.* New York: McGraw-Hill, chap. 9, 1956.

(16) Stephenson, J. R., Axelrad, A. A., McLeod, D. L., and Shreeve, M. M. Induction of colonies of hemoglobin-synthesizing cells by erythropoietin *in vitro*. *Proc. Nat. Acad. Sci. (U.S.A.)* *68*:1542, 1971.

(17) Till, J. E., and McCulloch, E. A. A direct measurement of the radiation sensitivity of normal mouse bone marrow cells. *Radiat. Res.*, *14*:213, 1961.

(18) Till, J. E., McCulloch, E. A., and Siminovitch, L. A stochastic model of stem cell proliferation based on the growth of spleen colony-forming cells. *Proc. Nat. Acad. Sci. (U.S.A.)*, *51*:29, 1964.

(19) Wu, A. M., Siminovitch, L., Till, J. E., and McCulloch, E. A. Evidence for a relationship between mouse hemopoietic stem cells and cells forming colonies in culture. *Proc. Nat. Acad. Sci. (U.S.A.)*, *59*:1209.

INTRODUCTION

Regulatory mechanisms of erythropoiesis have been studied mainly *in vivo*. Erythropoietin (EP), thought to be the primary regulator, is defined by its erythropoiesis-stimulating properties in experimental animals. The target cells of EP are operationally termed erythropoietin-responsive cells (ERC) (6). They are considered to be primitive hemopoietic cells, not morphologically recognizable as erythroid, and induced by EP to form hemoglobin (16). ERC, in contrast to the maturing erythroid series, are present in erythropoiesis-suppressed states such as transfusion- or hypoxia-induced polycythemia (11). The ERC seem to be more differentiated than the pluripotent hemopoietic cell population (CFU-s) detected by the *in vivo* spleen colony assay (3, 12). This population of target cells of EP is apparently a class of progenitor cells with a role in hemopoietic differentiation quite similar to that of the *in vitro* myeloid colony-forming cells (CFU-c).

It has recently become possible to clone erythroid progenitor cells *in vitro*. Culture of mouse bone marrow cells in semisolid cultures containing EP detects two distinct populations of erythroid progenitor cells. One, termed CFU-e (erythroid colony-forming unit), forms a cluster of erythroid cells after 2–3 days of incubation (1, 5, 8, 19). The other forms a very large colony at 9–10 days of incubation (1, 10), which is termed a "burst," because of its dispersed appearance and explosive growth; the cells of origin are designated as BFU (burst-forming units). It is the purpose of this chapter to describe some characteristics of BFU and CFU-e, and to discuss the applicability of cloning assays for erythroid progenitor cells as tools in studying the regulation of erythroid differentiation.

METHODS AND MATERIALS
MICE

Female BCBA F_1 mice, 8–10 weeks of age, were used for the experiments, unless otherwise stated.

SPLEEN COLONY ASSAY

CFU-s were estimated as described by Till and McCulloch (20). Briefly, mice were radiated with a supralethal dose of 1025 rad (γ), and in-

Some Characteristics of in Vitro Erythroid Colony and Burst-Forming Units

12

G. Wagemaker, V. E. Ober-Kieftenburg, A. Brouwer, and M. F. Peters-Slough

jected with 2.5 and 5×10^4 bone marrow cells. Macroscopic spleen colonies were counted 10 days later after fixation with Telleyesnicki's solution.

IN VITRO CULTURE OF HEMOPOIETIC PROGENITOR CELLS

Bone marrow cells were flushed from femurs with α-medium (8), and counted by hemocytometer using Türk's solution. The cells were cultured in α-medium, supplemented with 10% horse serum (Flow), 10% fetal calf serum (Flow), and 10% trypticase soy broth (BBL), using methylcellulose (4000 cps, Dow Chemical Int., U.S.A.) as a semisolidifying agent. For erythroid cultures, 10^{-4} M 2-mercaptoethanol was added. Triplicate cultures, containing 1 ml in 3.2 cm diameter tissue culture Petri dishes (Falcon Plastics), were incubated at 37°C, in an atmosphere of 100% humidity, and 10% CO_2 in air. Incubation time was 3 days for CFU-e, 7 days for CFU-c, and 10 days for BFU. Cell aggregates were scored with an inverted microscope (Zeiss) at a magnification of 80× for erythroid colonies and 30× for myeloid colonies and erythroid bursts. Erythroid colonies and bursts may be stained by benzidine (0.2% benzidine, 0.12% H_2O_2, in 0.5 M acetic acid) (J. T. Baker Chemicals B. V., Deventer, The Netherlands) applied directly to the cultures. However, this method also stains some peroxidase-positive myeloid cells.

Colony-stimulating factor (CSF) from pregnant mouse uteri was prepared as described by Bradley et al. (2) and Stanley et al. (18). Whole pregnant mouse uteri, stored at −20°C until processed, were homogenized (Ultra Turrax, 20,000 rpm) in 3 volumes of cold glass-distilled water. The homogenate was centrifuged (15,000 g, 30 min. 4°C), and ammonium sulfate was added to the supernatant until 50% saturation was effected. The suspension was centrifuged (20,000 g, 30 min, 4°C), the sediment was discarded, and the supernatant saturated with ammonium sulfate. Following another centrifugation (20,000 g, 30 min, 4°C), the supernatant was discarded and the precipitate was suspended in 0.25 of the original volume of cold glass-distilled water. The suspension was dialyzed against glass-distilled water for 3 days. The precipitate was removed by centrifugation (20,000 g, 30 min, 4°C), and the supernatant was subjected to heat treatment (60°C, 1 hr). It was then further purified and concentrated by adsorption to calcium phosphate gel. CSF was eluted in 0.5 of the original volume of 0.09 M phosphate buffer, pH 7.4, and filtered using a Mil-

lipore membrane. Characteristics of its biologic activity have been studied in detail by Van den Engh (4).

EP was isolated from the urine of patients with severe bone marrow aplasia or pure red cell aplasia. The urine was collected in the presence of 0.2% phenol, dialyzed against distilled water, and lyophilized. The resulting brown powder was dissolved in a volume of 0.01–0.02 of the original volume of 0.1 M Tris-HCl (pH 8.2), 0.5 M NaCl, 1 mM NaN$_3$, and 10 mM 2-mercaptoethanol. This was subjected to upward flow gel filtration on Sephadex G150 (column dimensions, 5 × 90 cm) in the same buffer solution. The eluate, at a distribution coefficient (13) $K_{av} = 0.43 \pm 0.10$ was collected, and, following dialysis against distilled water and lyophilization, passed through a column containing Con A Sepharose equilibrated with 0.05 M acetate buffer (pH 6.0), 0.5 M NaCl, 5 × 10^{-4} M CaCl$_2$, MgCl$_2$, and MnCl$_2$. The nonadsorbed protein fraction, containing EP, was collected and exhaustively dialyzed against distilled water. It was lyophilized, dissolved in α-medium, and sterilized by filtration through a nucleopore membrane filter.

Erythropoietin bioassay was performed as previously described (21) using female ND2 mice 9–12 weeks of age. Briefly, polycythemia was induced by exposing 90 mice, in a closed cage of 65 × 45 × 25 cm, to an inflow consisting of 8% O_2 and 92% N_2 at a rate of 3 liters/min for 9 days, 8 hr/day. On the 4th and 5th days after termination of this intermittent hypoxia, erythropoietin was injected intraperitoneally in three divided doses at time intervals of 12 hr; 24 hr after the last injection, 1 μCi ^{59}FeCl$_3$ in saline was intravenously injected; the mice were killed 24 hr later and peripheral blood ^{59}Fe uptakes determined. EP was calibrated against the Second International Reference Preparation of erythropoietin, kindly provided by the WHO International Laboratory for Biological Standards, Mill Hill, London, England.

Cell separation by sedimentation rate at unit gravity was performed as described by Miller and Philips (17) against a 1–2% linear BSA gradient in phosphate buffered saline.

RESULTS

Erythroid colonies appearing at days 2 and 3 of incubation contain up to approximately 50 cells, those clusters containing eight or more cells being scored as erythroid colonies. In contrast, the explosive growth of BFU results in the appearance of colonies, approximately half of which can be readily detected macroscopically (Fig. 1). The

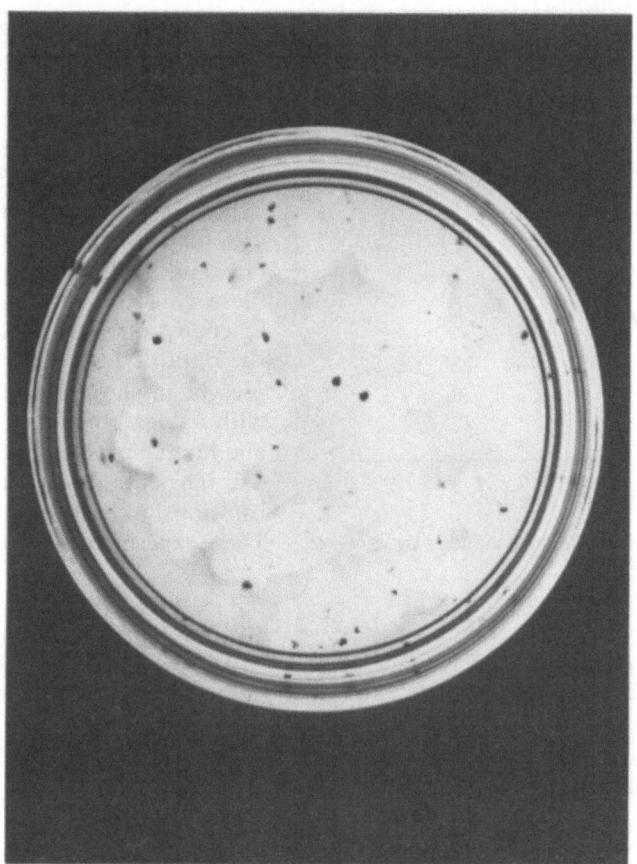

FIGURE 1. *In vitro* macroscopic erythroid burst formation induced by 1.0 I.U. EP. The culture contained 2 × 10⁵ nucleated bone marrow cells and was photographed after·10 days of culture and staining with benzidine.

formation of erythroid colonies and bursts as a function of incubation time is shown in Fig. 2. Separation of colonies and bursts by time is almost complete. There is a small increase in the number of erythroid colonies scored at day 7, which,,presumably, represents burst formation in a very early stage and is being further investigated at present. Both erythroid colonies and bursts appear to lyse approximately 2–3 days after the commencement of hemoglobin synthesis, for unexplained reasons; cloning of erythroid progenitor cells, in contrast to myeloid colony formation, does not result in the formation of colonies consisting of true end cells.

The formation of recognizable bursts is marked by a delay of 6 days. The rapid development of bursts containing up to 2 × 10⁴ cells, indicating at least 14 divisions if bursts originate from single cells, in the period from day 6 to day 9 of incubation, suggests that burst formation is preceded by the formation of a cluster of cells not recognizable as erythroid. So far, however, no

FIGURE 2. Eerythroid colony formation (●) induced by 0.4 I.U. EP, and erythroid burst formation (o) induced by 15 I.U. EP, as a function of incubation time.

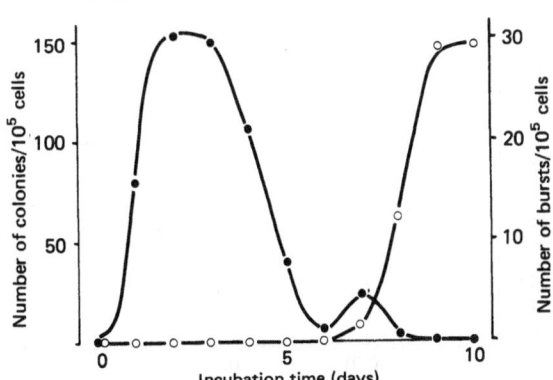

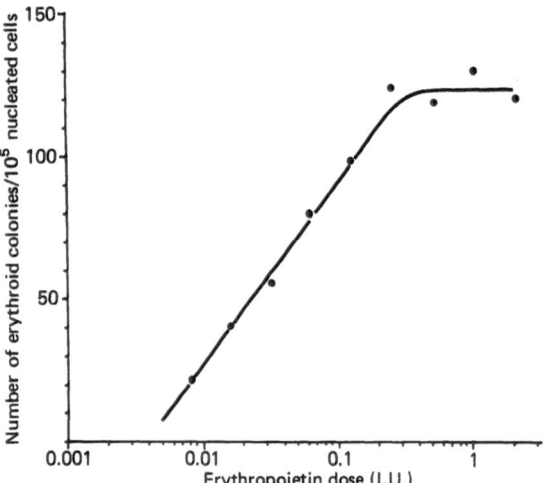

FIGURE 3. Erythroid colony formation induced by varying EP concentrations.

explicit data are available on the early stage of burst formation and its dependency on EP.

Representative dose-response relationships for colony and burst formation in response to human urinary preparations containing EP are shown in Figs. 3 and 4. The number of CFU-e reproducibly detectable was maximal upon stimulation with 0.25–0.50 I.U. EP at an incidence of approximately 150 colonies/10^5, the number of BFU was maximal upon stimulation with 1.0–2.0 I.U. EP at an incidence of 30 bursts/10^5. CFU-e incidence is less than reported by both Axelrad et al. (1) and Iscove et al. (8), but BFU incidence is 10-fold compared to the data of Axelrad et al. (1), and somewhat higher than reported by Iscove et al. (10). It may be inferred that optimal culture conditions apparently differ for BFU and CFU-e.

EP is measured by its ability to stimulate erythropoiesis *in vivo*. The use of relatively impure preparations of EP for *in vitro* studies necessitates some evidence indicating specificity of the observed *in vitro* phenomena to that measured *in vivo*. Such evidence was obtained by comparison of the elution profile of EP obtained by gel filtration of concentrated human urine (Fig. 5), as measured by its effect on ^{59}Fe incorporation in ex-hypoxic polycythemic mice, with the elution profiles for the activities stimulating erythroid colony- and burst-formation. The three elution profiles were identical, possessing a distribution coefficient K_{av} of 0.43. This value is in agreement with the data of Lukowsky and Painter (14) for EP present in sheep plasma and human urine, and with values previously obtained for rat, rabbit, and human plasma, and for rat and human urine (22). The data support the assumption that EP is limiting for both burst- and colony-formation. These experiments also demonstrate separation of

FIGURE 5 Elution profiles obtained by gel filtration of concentrated human urinary proteins on Sephadex G150. The eluate was collected in equal fractions of 250 drops; all fractions were dialyzed against distilled water, lyophilized, and dissolved in 5 ml α-medium. Panel A: absorbancy at 280 nm (---) (broad peak of low molecular weight substances not fully shown) and 24-hr percent ^{59}Fe incorporation (——) in ex-hypoxic polycythemic mice by injection of 50 μl of each fraction. Panel B: erythroid colony formation in response to 10 μl of each fraction. Panel C: myeloid colony formation (---) and erythroid burst formation (——) in response to 50 μl of each fraction.

FIGURE 4 Erythroid burst formation induced by varying EP concentrations.

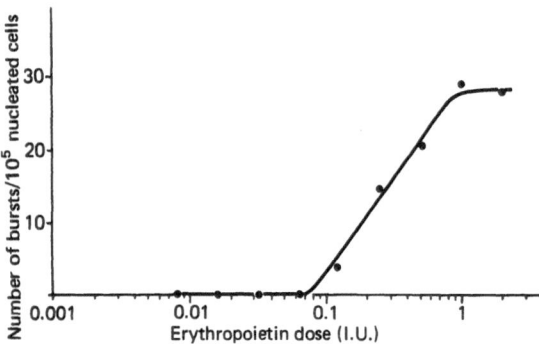

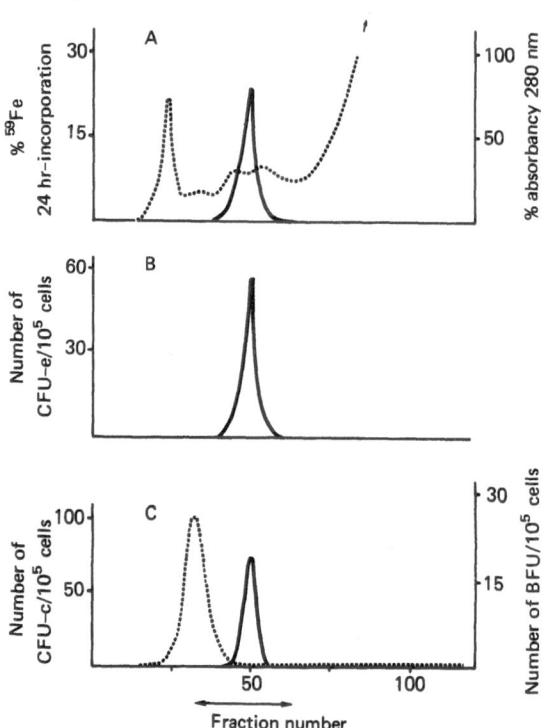

EP and human urinary CSF by gel filtration, and indicate the specificity of the scoring method with regard to distinguishing myeloid and erythroid clones.

Erythropoiesis, in contrast to myelopoiesis, offers the advantage of reproducible *in vivo* manipulation by physiologic means. We have used this attribute of erythropoiesis by studying time courses of hemopoietic progenitor cells following hypoxia-induced polycythemia. After termination of hypoxia, polycythemia results in a virtual absence of erythropoiesis due to profound suppression of EP production. In Fig. 6, time courses after termination of hypoxia are shown for CFU-e, BFU, CFU-s, and CFU-c, using peripheral blood 24-hr percent ^{59}Fe incorporation as an event marker for erythropoiesis. For each cell population measured by an *in vitro* cloning assay, plateau levels of their respective stimulators were employed. Immediately after termination of hypoxia, the incidence of CFU-e was twice normal; it subsequently declined rapidly, preceding and paralleling the fall in peripheral blood 24-hr percent ^{59}Fe incorporation, which reflects the number of mature erythroid cells entering the blood stream. This finding indicates that the CFU-e represent a population of maturing erythroid cells, and suggests that erythropoietin is required for its existence in the bone marrow.

In contrast, BFU incidence was not suppressed by polycythemia, and its time course after termination of hypoxia was characterized by a transient increase. This pattern was also observed for the incidence of CFU-s, and, to a somewhat lesser extent, for CFU-c incidence. These results are not due to a changing cellularity of the bone marrow; expressing incidences of the hemopoietic progenitor cells per femur rather than per 10^5 nucleated cells gave an essentially similar pattern. The observed time course after termination of hypoxia indicates the BFU to be an early erythroid progenitor cell, and demonstrates it to be of a cell population functionally distinct from CFU-e.

Further evidence that BFU and CFU-e are distinct cellular entities was obtained by physical separation by sedimentation velocity. In agreement with data of Iscove and Sieber (9) and of Heath et al. (7), BFU sedimented at a modal rate of 4.0 mm/hr, while the modal sedimentation rate for CFU-e was observed to be 5.6 mm/hr (Fig. 7).

FIGURE 6 Time courses of hemopoietic progenitor cells in ex-hypoxic polycythemia. 24-hr percent ^{59}Fe incorporation (■) and the incidences of CFU-e (●), BFU (○), CFU-s, (△), and CFU-c (▲) as a function of time after termination of hypoxia. Shaded areas indicate normal values.

FIGURE 7 Representative sedimentation rate distributions of total nucleated cells (---), CFU-e (○) (upper panel), and BFU (lower panel) from mouse bone marrow.

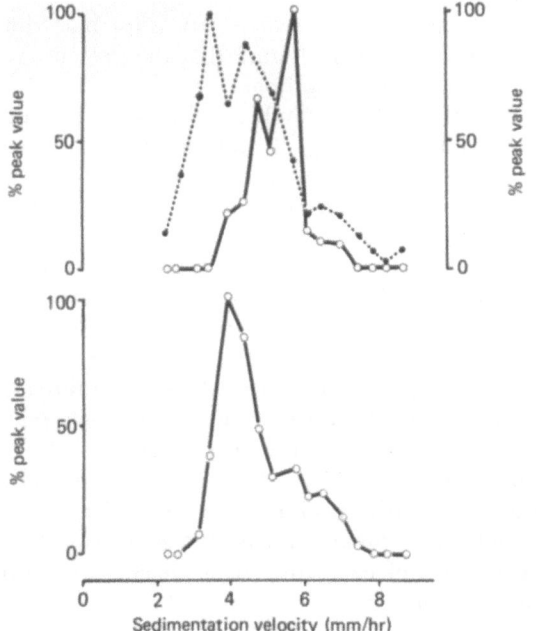

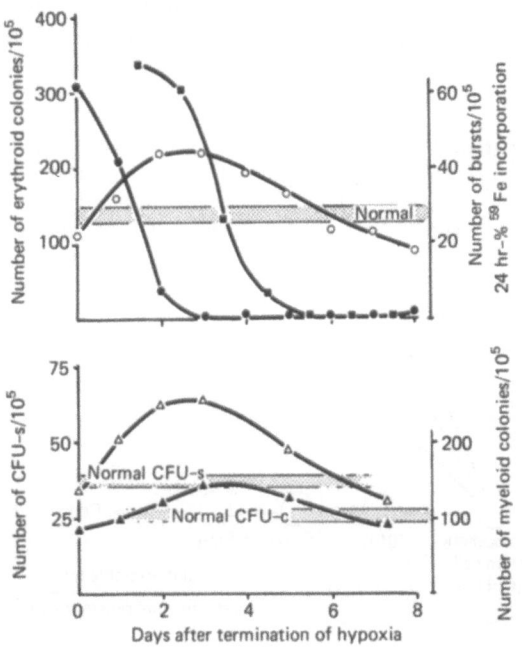

DISCUSSION

The data presented in this report indicate, in line with data from Gregory et al. (5), Axelrad et al. (1). and Iscove et al. (8, 9, 10) that, in the presence of EP, two distinct erythroid progenitor cells may be cloned *in vitro*. They differ in sensitivity to EP, proliferative capacity, incidence, functional behavior *in vivo*, and physical properties.

The characterization of CFU-e as belonging to a maturing erythroid cell population dependent on erythropoietin for its existence in the bone marrow precludes, by definition, its belonging to the ERC population. Rather, CFU-e may belong to the descendants of ERC, thus being of a cell population in which the early initiation of hemoglobin synthesis has already taken place. It is well-known that the maturing erythroid cells *in vivo* do not necessarily require detectable levels of circulating EP to proceed to differentiation towards reticulocytes; in this view, the apparent dependency of CFU-e on EP *in vitro* cannot be readily explained. It might be that EP in intact bone marrow remains available to maturing erythroid elements, this situation no longer holding when the cells are transferred as a single cell suspension to *in vitro* conditions. This hypothesis is the subject of active investigation.

The population of cells developing into erythroid bursts, as judged from its proliferative potential and behavor *in vivo*, apparently forms a class of primitive hemopoietic progenitor cells, the exact place of which in the erythroid differentiation sequence remains to be determined. The pattern of BFU incidence after termination of hypoxia cannot be readily compared to measurements on the ERC compartment, due to the initial high background of peripheral blood ^{59}Fe incorporation. There is evidence, however, that ERC may follow a quite similar time course (15, 21, 22). The occurrence of similar patterns for CFU-s and, notably, CFU-c, resulting from an essentially erythropoietic stumulus, is unexplained. One might speculate about a mechanism apparently operating at the level of the hemopoietic stem cell, but explicit data supporting such a notion are not available.

At present, no information is available on the relationship between BFU and *in vivo*-defined ERC. Data presented by Lajtha et al. (12) indicate the ERC to be in a fast state of turnover, as judged from *in vivo* tritiated thymidine suicide. Quite in contrast, only a relatively low kill by tritiated thymidine *in vitro* was obtained by Iscove and Sieber (9) for BFU. Although one should be very careful comparing *in vivo* and *in vitro* tritiated thymidine kill data, it may be wise not to assume BFU and *in vivo* ERC to be identical cellular entities.

A tentative representation of the place of erythroid progenitor cells assayed by different methods in erythroid differentiation is given in Fig. 8. The position of BFU relative to CFU-s and ERC is purely hypothetical. The sedimentation rate at unit gravity obtained for BFU may indicate a modal size for this progenitor cell population close to that of CFU-s. One striking feature of *in vitro* burst formation is the relatively high level of EP required, as compared to EP levels required, to stimulate erythropoiesis *in vivo*. The incidence of BFU seems to be much lower than that of any other clonable hemopoietic progenitor cell. This may suggest, in relation to the fact that culture conditions seem to be very critical for burst formation (unpublished data), that the present culture method is suboptimal.

It should be emphasized that the recently achieved possibility of cloning erythroid progenitor cells necessarily has a different significance for studying regulatory mechanisms of erythropoiesis than cloning assays have for myelopoiesis. Study of factors regulating myelopoiesis depends heavily on *in vitro* myeloid colony formation, the *in vivo* approach being mainly unsuccessful. In contrast, the humoral regulation of erythropoiesis has been firmly established by *in vivo* studies, leaving the physiologic regulation of erythropoiesis by erythropoietin beyond doubt. Cloning of erythroid progenitor cells, however, with special reference to BFU, may supply additional information concerning the regulation of early initiation of erythroid differentiation.

FIGURE 8 Schematic representation of the position of different erythroid progenitor cells in erythroid differentiation. The position of BFU relative to CFU-s and ERC is purely hypothetical. The abbreviations Ly, My, and Mega stand, respectively, for the lymphatic, myeloid, and megakaryocytic pathways of differentiation.

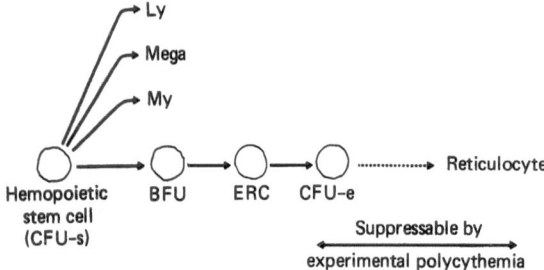

108

SUMMARY

Semisolid culture of mouse bone marrow cells in the presence of a factor assumed to be identical to erythropoietin (EP), as judged from some of its molecular properties, demonstrates two distinct classes of erythroid progenitor cells. The first forms a small cluster of erythroid cells at day 2 of culture and is designated erythroid colony-forming unit (CFU-e). The second gives rise to a large colony of erythroid cells at day 10 of culture; it is termed "burst" because of its dispersed appearance and its explosive growth. The cells of origin are called BFU (burst-forming unit). Erythroid colonies contain 8–50 erythroid bursts up to 10^4 cells. Maximal colony formation occurs at EP concentrations of 0.25–0.50 I.U./ml, maximal burst formation at concentrations of 1–3 I.U./ml. Incidence in methylcellulose culture is approximately $150/10^5$ for CFU-e and $30/10^5$ nucleated cells for BFU. Contrary to erythroid colony formation, burst formation is characterized by a delay of approximately 6 days, which may suggest that BFU does not respond directly to EP. Modal sedimentation rates obtained by velocity sedimentation at unit gravity were 4.0 mm/hr for BFU and 5.6 mm/hr for CFU-e.

To assess some functional characteristics, incidences of BFU and CFU-e were studied in experimental polycythemia. After induction of polycythemia by intermittent hypoxia, CFU-e numbers, initially twice normal, rapidly declined, preceding and paralleling peripheral blood ^{59}Fe incorporation. This finding indicates CFU-e to be a maturing erythroid cell, and suggests that EP is required for its existence in the bone marrow. BFU incidence was not suppressed and its time course during ex-hypoxic polycythemia was characterized by an overshoot, resulting in an incidence of 40–50 $BFU/10^5$ nucleated cells. This is not caused by a changing cellularity of the bone marrow, but reflects a change in the absolute number of BFU/femur. This finding indicates BFU to be a primitive erythroid progenitor cell.

Similar time courses were observed for CFU-s, CFU-c, and ^{59}Fe-incorporation in response to a small standard dose of EP thought to reflect the number of erythropoietin sensitive cells (ERC). The mechanism of this common pattern shown by the hemopoietic progenitor cells following hypoxia is subject to further analysis.

ACKNOWLEDGMENTS

The initial help of Mr. Dries Mulder in performing sedimentation velocity studies is gratefully acknowledged. This investigation is part of a study on the regulation of hemopoiesis which is supported by a program grant of The Netherlands Foundation for Medical Research (FUNGO), which is subsidized by The Netherlands Organization for the Advancement of Pure Research (ZWO).

REFERENCES

(1) Axelrad, A. A., McLeod, D. L., Shreeve, M. M., and Heath, D. A. Properties of cells that produce erythrocytic colonies in vitro. In Robinson, W., ed., *Hemopoiesis in Culture.* Washington, D.C.: U.S. Government Printing Office, p. 226, 1974.

(2) Bradley, T. R., Stanley, E. R., and Sumner, M. S. Factors from mouse tissues stimulating colony growth of mouse bone marrow cells in vitro. *Aust. J. Exp. Biol. Med. Sci., 49*: 595, 1971.

(3) Bruce, W. R., and McCulloch, E. A. The effect of erythropoietic stimulation on the hemopoietic colony forming cells of mice. *Blood, 23*: 216, 1964.

(4) Gregory, C. J., Tepperman, A. D., McCulloch, E. A., and Till, J. E. Erythropoietic progenitor cells capable of colony formation in culture: Response of normal and genetically anemic WW^v mice to manipulation of the erythron. *J. Cell. Physiol., 81*: 411, 1973.

(5) Gurney, C. W., Goldwasser, E., and Pan, C. The regulation of numbers of primitive hemopoietic cells. *Proc. Nat. Acad. Sci. U.S.A., 54*: 1148, 1965.

(6) Health, D. S., Axelrad, A. A., McLeod, D. L., and Shreeve, M. M. Separation of erythropoietin-responsive progenitors BFU-E and CFU-E in mouse bone marrow by unit gravity sedimentation. *Blood, 47*: 777, 1976.

(7) Iscove, N. N., Sieber, F., and Winterhalter, K. H. Erythroid colony formation in cultures of mouse and human bone marrow: Analysis of the requirement for erythropoietin by gel filtration and affinity chromatography on agarose-Concanavalin A. *J. Cell. Physiol., 83*: 309, 1974.

(8) Iscove, N. N., and Sieber, F. Macroscopic erythroid colony formation in cultures of mouse bone marrow cells. *Exp. Hematol., 2*: 278, 1974.

(9) Iscove, N. N., and Sieber, F. Macroscopic erythroid colony formation in cultures of mouse bone marrow cells. *Exp. Hematol., 3*: 32, 1975.

(10) Jacobson, L. D., Gurney, C. W., Plzak, L., and Fried, W. Studies on erythropoiesis. IV. Reticulocyte response of hypophysectomized and polycythemic/rodents to erythropoietin. *Proc. Soc. Exp. Biol. Med., 94*: 243, 1957.

(11) Lajtha, L. G., Pozzi, L. V., Schofield, R., and Fox, M. Kinetic properties of haemopoietic stem cells. *Cell Tissue Kin., 2*: 39, 1969.

(12) Laurent, F. C., and Killander, J. A theory of gel filtration and its experimental verification. *J. Chromatogr., 14*: 317, 1964.

(13) Lukowsky, W., and Painter, R. H. Molecular weight of erythropoietin from anemic sheep plasma and erythropoietin. *Can. J. Biochem., 46*: 731, 1968.

(14) McDonald, T. P., Lange, R. D., and Kretchmar, A. L. Mode of action of erythropoietin in polycythemic mice. *J. Lab. Clin. Med., 77*: 134, 1971.

(15) Metcalf, D., and Moore, M. A. S. *Haemopoietic Cells*. Amsterdam and London: North-Holland Pub. Co., p. 109, 1971.

(16) Miller, H. S., and Phillips, R. A. Separation of cells by velocity sedimentation. *J. Cell. Physiol., 73*: 191, 1969.

(17) Stanley, E. R., Metcalf, D., Maritz, J. S., and Yeo, G. F. Standardized bio-assay for bone marrow colony-stimulating factor in human urine: Levels in normal man. *J. Lab. Clin. Med., 79*: 657, 1972.

(18) Stephenson, J. R., Axelrad, A. A., McLeod, D. L., and Shreeve, M. M. Induction of colonies of hemoglobin synthesizing cells by erythropoietin in vitro. *Proc. Nat. Acad. Sci. U.S.A., 68*: 1542, 1971.

(19) Till, J. E., and McCulloch, E. A. A direct measurement of the radiation sensitivity of normal mouse bone marrow cells. *Radiat. Res., 14*: 313, 1961.

(20) van den Engh, G. J. Quantitative in vitro studies on stimulation of murine haemopoietic cells by colony stimulating factor. *Cell Tissue Kinet., 7*: 537, 1974.

(21) Wagemaker, G., van Eijk, H. G., and Leijnse, B. A sensitive bioassay for the determination of erythropoietin. *Clinica Chim. Acta, 36*: 357, 1972.

(22) Wagemaker, G. Erythropoietine. Acad. Thesis Rotterdam., 1976.

Colony-Forming Unit, Megakaryote (CFU-m): Its Use in Elucidating the Kinetics and Humoral Control of the Megakaryocytic Committed Progenitor Cell Compartment

Alexander Nakeff

13

INTRODUCTION

The stem cell-megakaryocyte-platelet cell system is responsible for the production of blood platelets which are essential to the initiation of hemostasis and maintenance of the integrity of the blood vasculature. Platelet production is dependent upon the steady supply of megakaryocytes, which are themselves incapable of cellular proliferation, being almost exclusively in a state of cytoplasmic maturation (14). The production of megakaryocytes is thus dependent upon progenitor cells that are: (a) morphologically unidentifiable by conventional cytology; (b) committed to megakaryocyte differentiation; and (c) capable of cellular proliferation—including endoreduplication of DNA to produce polyploid cells at various ploidy levels—in addition to cell mitosis. Ultimately, of course, these progenitors must be derived through differentiation or "commitment" of pluripotent hematopoietic stem cells by a process we do not presently understand.

It is clear that the measurement of the extent of proliferative activity in the progenitor cell compartment is important to our understanding of the cell kinetics of megakaryocyte production both in the normal steady state as well as following its perturbation. The development of cell culture methods to support the growth of these progenitors *in vitro* is a useful approach to the assaying for those progenitors capable of cell proliferation, in order to define their physiologic role in megakaryocyte production and, eventually, to study their regulation and control.

We recently reported the establishment of a clonogenic cell assay in plasma culture for a class of progenitors in mouse bone marrow which we entitled the colony-forming unit, megakaryocyte, or CFU-m (8). This report presents some of our most recent findings regarding the physical characterization of the CFU-m, its quantitative response to platelet demand, and its possible humoral regulation.

MATERIALS AND METHODS
ANIMALS

The various strains of mice used were $B_6D_2F_1$ (C57Bl × DBA), C3H/HE, Balb/C, C57Bl, and AKR which have been bred in our own facility (Tyson). Both the male and the female mice used were 8–14 weeks old (approximately 20–25 g).

111

QUANTITATION OF CFU-m

The basic techniques used to culture, strain, and count CFU-m were previously described (8). Briefly, a monodispersed suspension of femoral marrow cells was resuspended in 0.1 ml of L-15 tissue culture medium (Gibco Long Island, N.Y.) at 10^7 cells/ml. This was added to a mixture containing 0.2 ml horse serum (Gibco), 0.1 ml conditioned medium (CM) from pokeweed mitogen (PWM)-stimulated spleen cell cultures (1/320 final dilution of PWM added to cultures of 2×10^6 $B_6D_2F_1$ spleen cells/ml and incubated for 2 days), 0.1 ml bovine embryo extract (Gibco), and 0.4 ml L-15. Following the addition of 0.1 ml of bovine-citrated plasma, 0.1 ml aliquots were cultured in microtiter plates for 4 days in 5% CO_2 in air at 100% humidity. Cultures were transferred to glass slides, fixed with 5% glutaraldehyde, and then stained *in situ* for acetylcholinesterase (AChE) activity as described previously (8). Colonies of four or more AChE-positive cells were counted microscopically as CFU-m and, using the total nucleated cells count, were expressed per femur.

DISCONTINUOUS ALBUMIN DENSITY GRADIENT CENTRIFUGATION

Details for gradient and sample preparation have been presented previously (11). All procedures were performed in a sterile manner. Briefly, bovine albumin powder (Fraction V, Sigma St. Louis, Miss.) was dissolved in distilled water at 4°C to a concentration of 40% (w/w) as determined by refractometry. The pH of this stock solution was 5.2. Its final osmolality was adjusted to 320 mOsm by the addition of sodium chloride (300 mg/ml), and the resulting solution was membrane filtered (0.22 μm). Albumin solutions were then made from 16 to 38%, in 2% steps (density 1.050 to 1.119 g/ml by pycnometry) by dilution with sodium chloride (pH 5.2, 320 mOsm), and 0.5 μl aliquots of each fraction were layered in a glass gradient tube, commencing with the 38% solution. Finally, 0.5 ml of a monodispersed suspension of marrow pooled from the femurs of three mice containing approximately 5×10^7 nucleated cells which had been concentrated by low-speed centrifugation (350 $\times$ g for 10 min) and resuspended in 16% bovine albumin, was layered on the formed gradient. Following centrifugation at 1000 $\times$ g for 30 min at 4°C, each fraction was transferred quantitatively to 10 ml of phosphate-buffered saline (PBS) and washed by low-speed centrifugation. A 0.1 ml sample of the starting cell suspension not placed on the gradient was washed and concentrated in the same manner as the fractions and used to determine the total number of nucleated cells, megakaryocytes, and CFU-m initially placed on the gradient. The fractions were resuspended in a known volume of PBS from which 25 μl samples were withdrawn for determining (1) the total number of megakaryocytes, by quantitative sedimentation onto glass slides and AChE staining (11), and (2) the number of nucleated cells, by electronic counting (Celloscope) (11). The remainder of each cell suspension was adjusted to contain approximately 1×10^7 cells/ml with L-15 and added to the plasma culture system for quantitation of CFU-m as described above.

FLOW MICROFLUOROMETRY (FMF) ANALYSIS

Cell suspensions to be analyzed for DNA content were washed several times by centrifugation in Ca^{2+}- and Mg^{2+}-free Hanks's BSS containing 0.5 mM EDTA. Three volumes of ice-cold 95% ethanol containing 100 μg/ml mithramycin were added to 1 volume of washed cells with constant stirring. Cells in the mithramycin-containing solution were then analyzed in the Los Alamos FMF at a wavelength setting of 457 nm (16), with the generous help of Dr. H. Crissman, University of California at Los Alamos.

IMMUNE-INDUCED THROMBOCYTOPENIA

Mice were rendered thrombocytopenic following the intravenous injection of specific rabbit anti-mouse platelet gamma globulin (RAMP-gG), prepared as described prevously (13). Control mice were either normal (uninjected) or received normal rabbit gamma-globulin (NR-gG) in amounts equivalent to the mg protein of RAMP-gG administered.

THROMBOPOIETIN (Tp)

Serum obtained from immune-induced thrombocytopenic mice served as the *in vivo* source of thrombopoietin, as described previously (13). Control serum was obtained from mice injected with NR-gG.

CELL COUNTING

Total nucleated blood or bone marrow cells were counted on an electronic particle counter (Celloscope, Particle Data) in a solution

(30 mg/ml) of the cytoplasmalytic agent cetrimide (hexadecyltrimethyl ammonium bromide, Baker, Phillipsburg, N.J.). Blood platelet counts were obtained electronically by the method of Nakeff and Ingram (12).

RESULTS

CELLULAR CHARACTERISTICS OF CFU-m

Density Profile As shown in Fig. 1, CFU-m comprises a single population of progenitors with a mean cell density of about 1.075 g/ml. This contrasts with the mean cell density of other nucleated cells of about 1.080 g/ml, and of megakaryocytes of 1.10 g/ml. It thus appears that CFU-m comprise a class of progenitors that are separable on the basis of cell density not only from morphologically recognizable megakaryocytes but also from a class of small, AChE-positive

FIGURE 1. Average density profile of CFU-m (●), nucleated marrow cells (○), and morphologically identifiable megakaryocytes stained for AChE activity (◐) obtained from seven discontinuous albumin gradients at pH 5.2 and 320 mOsm, and expressed as a percentage of their respective numbers intitially placed on the gradient. Errors shown represent ± 1 S.E.

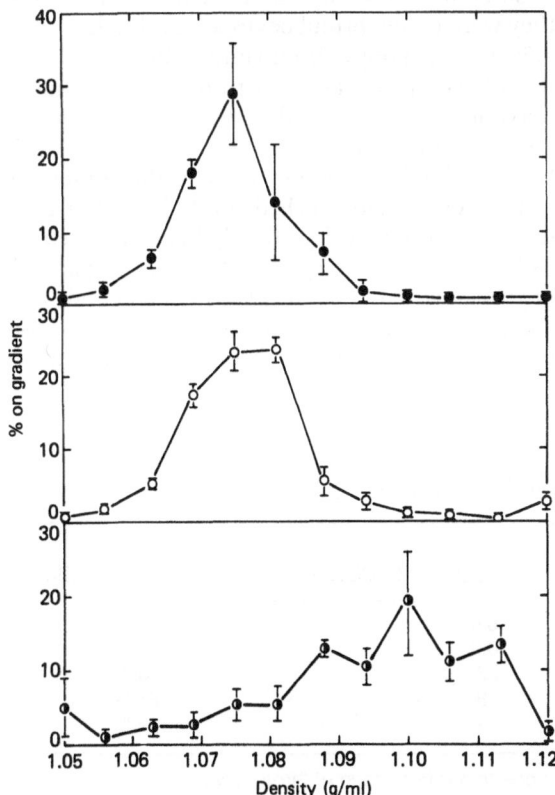

progenitors that have also been shown to exhibit a single density peak at 1.10 g/ml (11). Since the latter two classes are incapable of forming colonies and are thus devoid of proliferative activity, the small (<129 μm^2), AChE-positive progenitors may possibly represent a class of polyploid cells which have been proposed theoretically as being derived from progenitors able to undergo cell mitosis (14); the latter may, by definition, represent the CFU-m population.

Distribution of Polyploidy in Bone Marrow Fractions by FMF Analysis Following Albumin Density Gradient Centrifugation In order to determine the possible ploidy state of CFU-m, we physically separated bone marrow cells on the basis of density as described above and pooled them into two cell suspensions with cell densities either less than or greater than 1.088 g/ml; that is, into fractions that contained the majority of CFU-m and nucleated cells, and fractions that contained megakaryocytes and small, AChE-positive progenitors. These two cell pools were fixed with ethanol, then exposed to mithramycin, which binds quantitatively to their DNA. These populations were analyzed for their DNA content using the Los Alamos FMF as described above. As shown in Fig. 2, cells in the upper (i.e., the lower density) portion of the gradient were characterized by being almost exclusively diploid, whereas cells in the lower (i.e., the higher density) portion of the gradient exhibited ploidy values of 4 and 8N. The latter most probably represent polyploid, morphologically identifiable megakaryocytes, although small, AChE-positive progenitors present in the same fractions may also be polyploid and contribute to the distribution obtained.

Quantitation of CFU-m from Marrow of Different Mouse Strains Since the number of CFU-m from the marrow of $B_6D_2F_1$ mice was relatively low [about $1:10^4$, (8)] when compared to CFU-c or CFU-e, a study was initiated to examine whether this might be a property of this particular strain. Female mice from various strains aged 10–14 weeks were killed by cervical dislocation, and their respective femoral marrow cells were pooled and cultured for 4 days in the plasma culture system for CFU-m using 10% CM collected from $B_6D_2F_1$ spleen cell cultures 2 days after the addition of 12.5 $\mu g/ml$ of PHA. As shown in Table 1, initial experiments have uncovered a clear difference in the number of CFU-m present in the femoral marrow of normal mice of various strains. Possible underlying reasons for this difference are not apparent, since all the mice had normal megakaryocyte and platelet counts (data not shown).

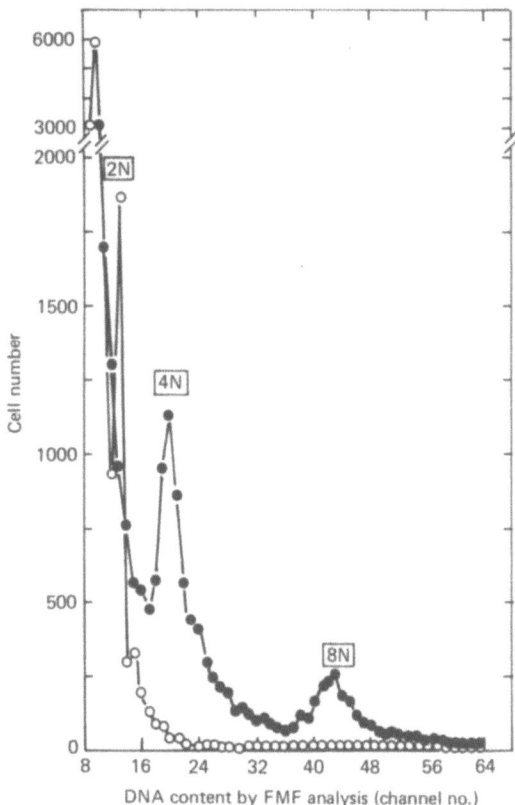

FIGURE 2. DNA profile of $B_6D_2F_1$ femoral bone marrow cells, obtained following density gradient separation, which are present in the upper [<1.088 g/ml (○)] and lower [>1.088 g/ml (●)] portions of the gradient, as analyzed by FMF.

PHYSIOLOGIC ROLE OF CFU-m IN MEGAKARYOCYTOPOIESIS

Response of CFU-m to Platelet Depletion Since the depletion of peripheral blood platelets is known to stimulate an absolute increase in megakaryocytes, which is thought to reflect an increase in progenitor cell proliferation, the response of CFU-m as a class of proliferating

progenitors was studied as a function of the degree of immune-induced thrombocytopenia. Groups of female $B_6D_2F_1$ mice were injected via the tail vein with 0.25 ml of PBS containing either 1, 2, or 4 mg of RAMP-gG. Control mice were untreated. Groups of three mice at each dose level were then assayed 2 hr and 24 hr later, at which time mice were bled from the tail vein into a blood pipette and their blood diluted 1:10 with CPD-PBS and pooled for platelet counting (12). Mice were then killed and their marrow pooled. Nucleated cell counts were determined and an aliquot of bone marrow placed in plasma culture for quantitation of CFU-m as described previously. As shown in Table 2, there was almost a 3–4-fold increase in the number of CFU-m per femur within 2 hr after immune-induced thrombocytopenia, regardless of the amount of RAMP-gG injected and the subsequent level of platelet depletion. By 24 hr, however, the number of CFU-m per femur was similiar to that in normal controls. These data suggests that CFU-m increase in number in response to an acute demand for blood platelets, and that the time course of the response may be rather short.

Time Course of CFU-m Response to Platelet Depletion In order to study the response of CFU-m to immune-induced thrombocytopenia in more detail, CFU-m were cultured from the femoral marrow of mice as a function of time after they were made thrombocytopenic (platelet count <5%) by a single intravenous dose of 2 mg RAMP-gG. It can be seen from Table 3 that a maximum increase in the number of CFU-m of approximately 3-fold was observed as early as 2 hr after platelet depletion. By 6 hr, the number of CFU-m/femur had reached levels less than control, with control values finally being attained by 24 hr. These data suggest that CFU-m respond to physiologic demands for new platelets and may contitute an important class of progenitors whose number can be induced to increase under these conditions.

TABLE 1 Colony-Forming Unit—Megakaryocyte (CFU-m) in Various Mouse Strains

MOUSE STRAIN	CELLS/FEMUR (× 10⁷)	CFU-m[a]/10⁵ CELLS	CFU-m/FEMUR
$B_6D_2F_1$	1.7	6.4	1078
C3H/HE	1.4	0.9	118
BALB/C	2.2	23	5060
C57/BL	2.1	9	2304
AKR	1.8	29	5220

[a]Average of six cultures from two experiments, each pooling one femur from each of three mice.

TABLE 2 Effect of Platelet Depletion on CFU-m

RAMP-gG[a] -INJECTED (MG/MOUSE)	TIME AFTER INJECTION (HR)	PLATELET[b] COUNT ($\times 10^9$/ML)	NUCLEAR[c] CELLS/FEMUR ($\times 10^7$)	CFU-m[d]/10^5 CELLS	CFU-m/FEMUR ($\times 10^7$)
0	2	1.15	2.0	4	0.8
1	2	0.07	1.1	20	2.2
2	2	0.06	1.1	28	3.1
4	2	0.03	1.5	20	3.0
0	24	1.3	1.8	4	0.72
1	24	0.2	1.9	3	0.57
2	24	0.1	1.4	10	1.4
4	24	0.05	1.1	7	0.33

[a]Rabbit anti-mouse platelet gamma globulin.

[b]Blood pooled from three $B_6D_2F_1$ donor mice per dose.

[c]Marrow pooled from one femur obtained from each of three donor mice per dose.

[d]Average of three cultures per pooled marrow sample per dose.

TABLE 3 Time Response of CFU-m to Platelet Depletion

TIME AFTER 2MG RAMP-gG (HR)[a]	NUCLEATED CELLS/FEMUR ($\times 10^7$)[b]	CFU-m/ 10^5 CELLS[c]	CFU-m/FEMUR ($\times 10^3$)
0	1.90	5.4	1.03
2	1.20	21	2.52
4	1.24	11.5	1.43
6	1.20	3.7	0.44
18	1.56	4.3	0.67
24	1.15	11.2	1.29

[a]Rabbit anti-mouse platelet globulin.

[b]$B_6D_2F_1$ marrow pooled from one femur obtained from each of three donor mice per time point.

[c]Average of three cultures at each time point.

Effect of Thrombopoietin (Tp) on Plating Efficiency of CFU-m There is evidence that platelet production is under humoral regulation and that serum from platelet-depleted mice contains a factor(s) (thrombopoietin, Tp) which may be responsible for stimulating new megakaryocyte production (14). We have previously shown that the serum of acutely immune-induced thrombocytopenic mice has the highest titer of Tp some 12 hr after platelet depletion as determined by an *in vivo* radiobioassay (13). To test for the ability of Tp to stimulate CFU-M production *in vitro*, serum was collected from $B_6D_2F_1$ female mice 12 hr after their injection with 2 mg RAMP-gG, and added in increasing concentration to plasma cultures of normal bone marrow. Control cultures either received no mouse serum or serum collected at the same time from control mice injected with 2 mg NR-gG. As shown in Table 4, the addition of up to 20% of Tp resulted in about a doubling of CFU-m per culture. The addition of control serum was generally inhibitory, as was the addition of equivalent volumes of normal serum (data not shown). These preliminary data suggest that there may be a substance specific to thrombocytopoietic serum that is capable of increasing the plating efficiency of CFU-m. The

TABLE 4 Effect of Thrombopoietin (Tp) ON CFU-m

% Tp[a]/CULTURE	CFU-m/10^5 CELLS[b]
0	10
2	5
5	14.3
10	14
20	23.7

[a]Tp as serum 12 hr after 2 mg RAMP-gG.

[b]Average of six cultures from two experiments.

mechanism of induction remains to be determined.

CFU-m AND PLATELET FORMATION IN CULTURE

McLeod et al. (6) have recently published electronmicroscopic evidence of platelet formation in megakaryocyte colonies grown in plasma cultures containing erythropoietin. In optimizing culture conditions for growth of CFU-m, we observed that the substitution of horse serum for fetal calf serum in our original published procedure (8) resulted in an increase in the reproducibility of our assay system. In addition, megakaryocytes both in and among colonies exhibited an improved morphology and seemed

FIGURE 3. Photomicrographs of CFU-m and megakaryocytes *in situ* in plasma cultures. (a) CFU-m of five AChE-positive cells tightly clustered (200×); (b) CFU-m with seven AChE-positive cells spread apart (120×); (c) CFU-m with mitotic figure at arrow (120×); (d) megakaryocyte in mitosis (900×); (e) CFU-m with cells at end stage of platelet formation (200×).

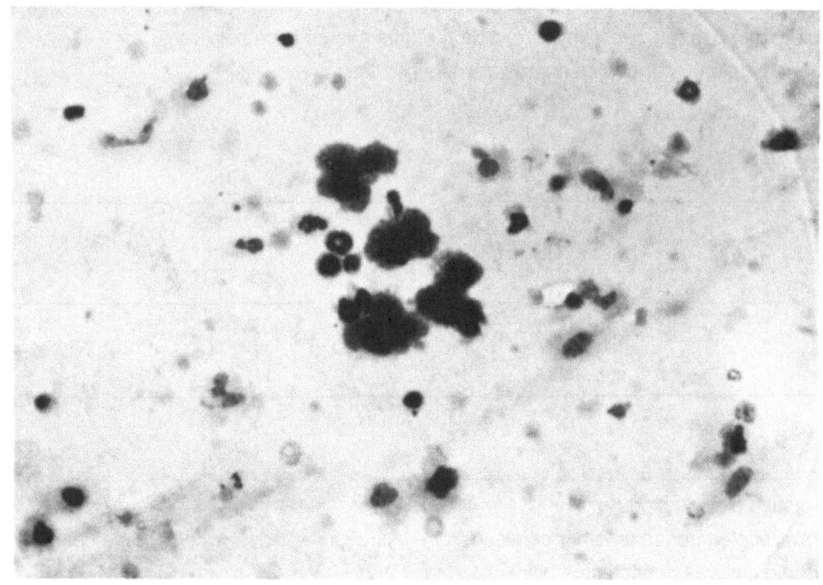

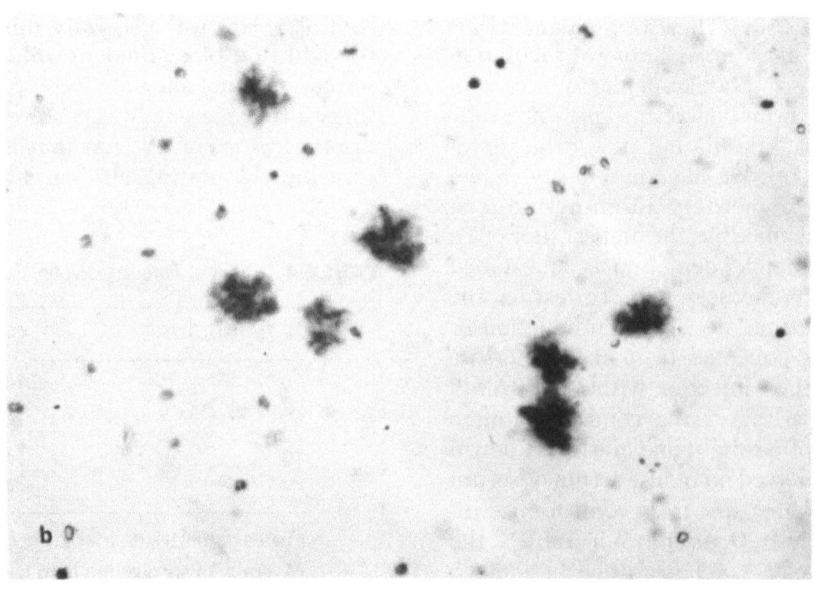

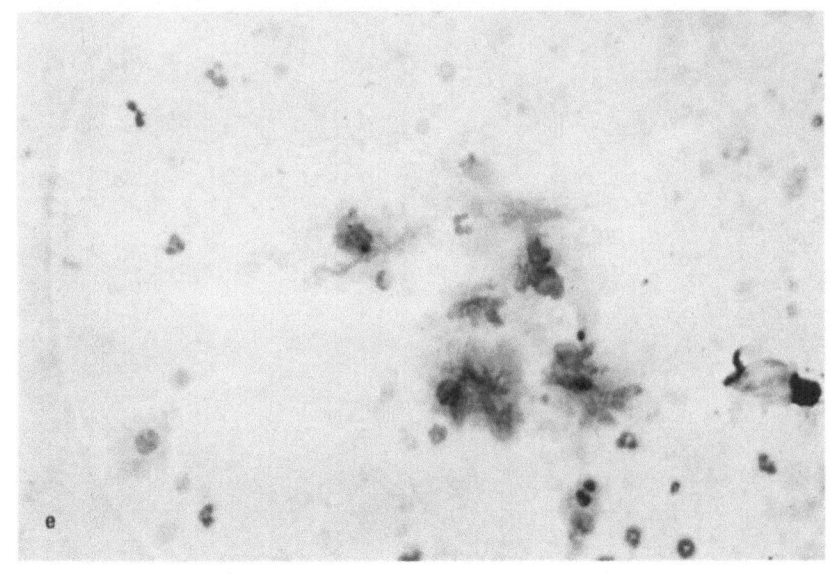

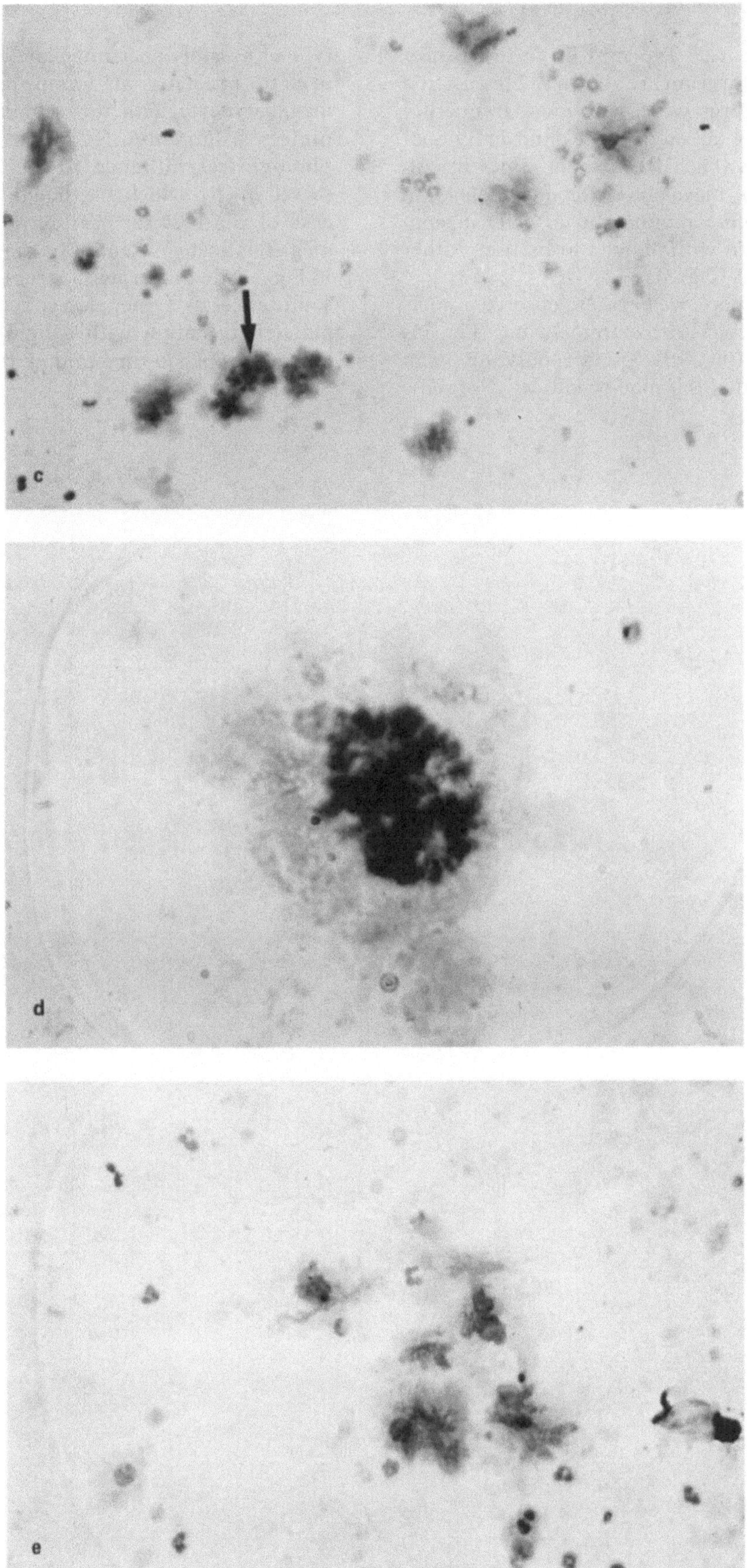

to be more "active." Typical CFU-m in cultures containing horse serum are shown in Figs. 3A and 3B. Mitotic figures were seen more frequently both in colonies (arrow, Fig. 3C) and individual megakaryocytes (Fig. 3D). At later times in culture (>5 days), megakaryocyte morphology in colonies was more reminiscent of cells in end-stage maturation with platelet formation, rather than of cytolysis (Fig. 3E).

Furthermore, we began to observe numerous megakaryocytes with extremely long (Fig. 4A) and numerous (Fig. 4B) "pseudopodia." We can only speculate that this may be indicative of some type of possibly aberrant platelet formation. Figure 4C presents an example of a granular megakaryocyte with its cytoplasm almost completely fragmented. It bears a striking morphologic resemblance to megakaryocytes observed *in vivo* which are thought to be in the process of platelet formation. In particular, the megakaryocyte present *in situ* in a plasma culture in Fig. 4D demonstrates a strong morphologic resemblance to a megakaryocyte at the end of platelet formation with a lightly AChE-positive halo of cytoplasm surrounding the clearly lobular nucleus.

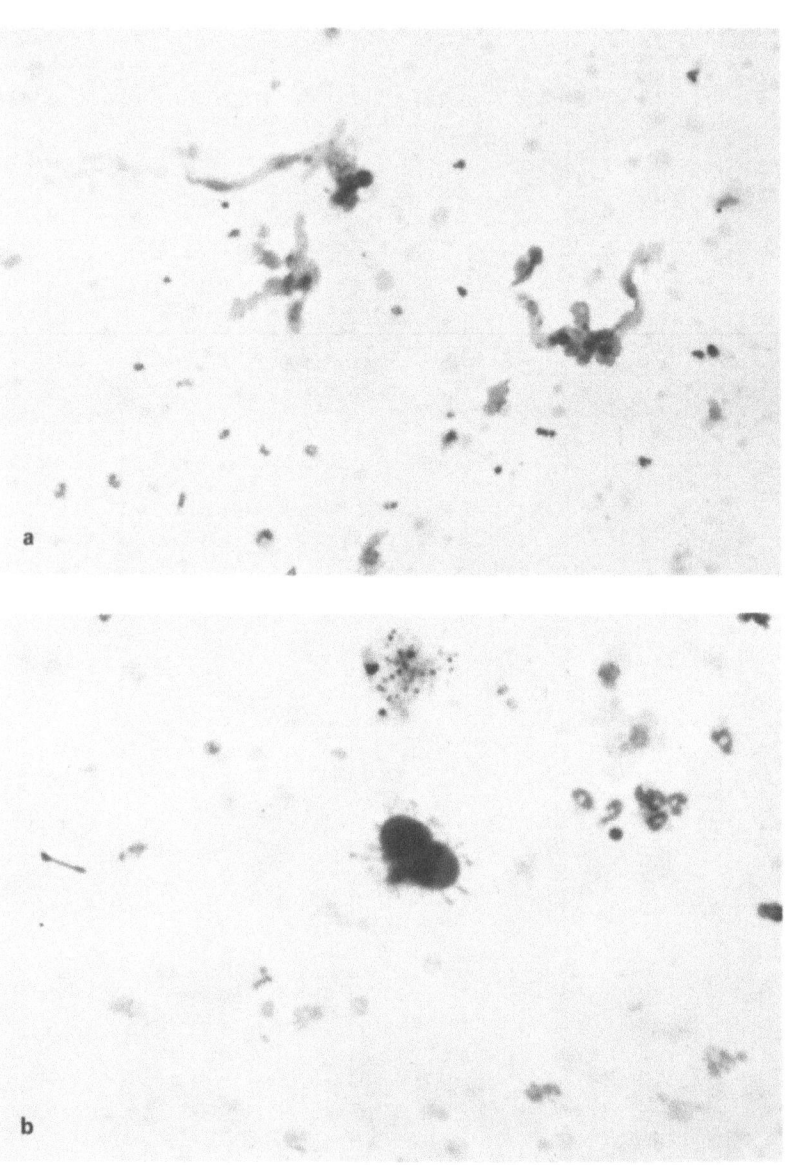

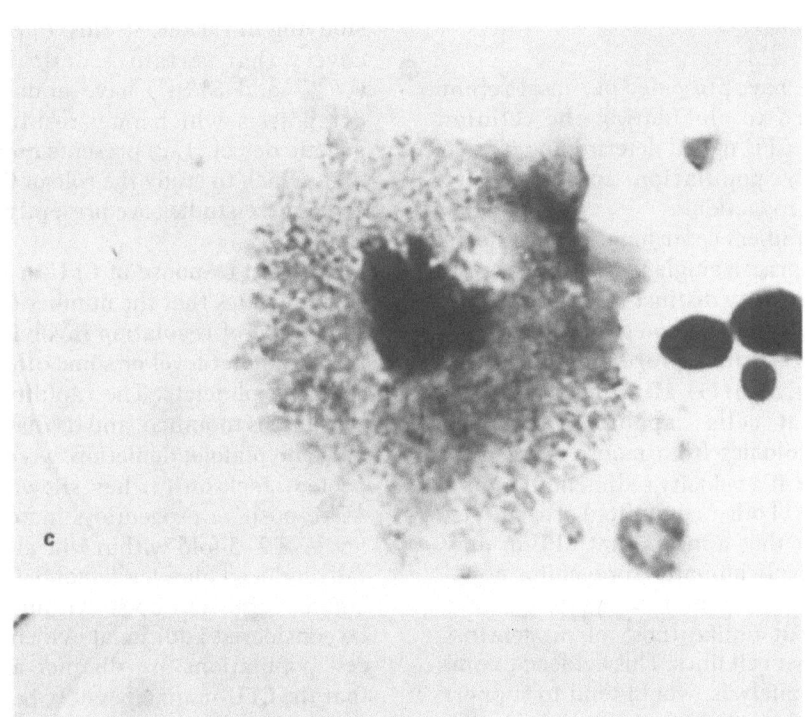

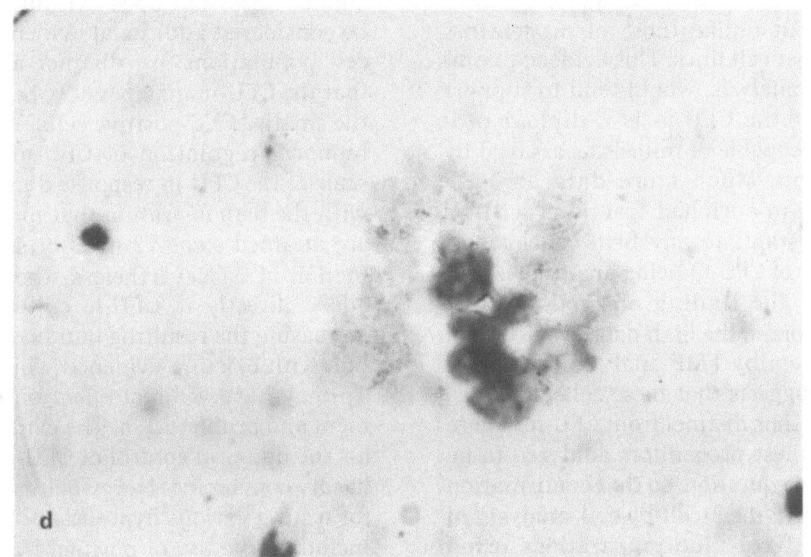

FIGURE 4. Photomicrographs of megakaryocytes in various stages of cytoplasmic maturation. "Pseudopodial" formation in: (a) mature, AChE-positive megakaryocytes (120×); and (b) immature, AChE-positive megakaryocytes (200×); (c) megakaryocyte showing evidence of cytoplasmic formation and release of platelets from the cell periphery (900×); (d) megakaryocyte at end stage of platelet release showing a thin boundary of cytoplasm and lobulated nucleus (700×).

DISCUSSION

In this report, we have presented our most recent data with regard to elucidating the cellular physiology of the CFU-m and determining its role as a progenitor population committed to megakaryocyte production.

Density gradient centrifugation has shown that CFU-m comprise a single population of cells with a peak cell density distinct from either morphologically recognizable megakaryocytes or their small, AChE-positive progenitors (1.075 vs. 1.10 g/ml, respectively) (11). Metcalf et al. (7) have also shown that cells capable of forming megakaryocyte colonies from mouse marrow in agar culture exhibit a velocity sedimentation profile characteristic of other committed progenitors, such as CFU-c. It thus appears that CFU-m may comprise a relatively immature progenitor population with cell density and, possibly, cell size characteristics not unlike those of progenitors committed to other cell lines. This evidence, combined with FMF analysis, would tend to support the concept that the CFU-m is a diploid progenitor which is capable of mitosis as assayed by colony formation. Much more data on FMF analysis of CFU-m-enriched fractions will be necessary to substantiate any firm conclusions, since the number of CFU-m being analyzed is still extremely low. The finding of small, AChE-positive progenitors in the high-density portion of the gradient shown by FMF analysis to contain polyploid cells suggests that these cells may also be polyploid and thus distinct from CFU-m. There were too few of these progenitors analyzed to be able to answer this question, so that confirmation will have to await their cell-by-cell analysis of DNA content in fixed-slide preparations using scanning microspectrophotometry. These studies are presently in progress.

The finding of different numbers of CFU-m in the femurs of different strains of mice was surprising and its meaning remains unclear. It has recently been shown that the fraction of pluripotent stem cells assayed by the spleen colony technique (CFU-s) that are in cell cycle can be different in different strains of mice (17); C3H mice have about 20% of their CFU-s in DNA synthesis, while C57Bl mice have essentially no CFU-s in the S-phase. No information exists as to whether this difference is an exclusive property of the CFU-s or whether the more differentiated progenitor cells such as the CFU-m also reflect this proliferative difference. The implication of the proliferative kinetics of CFU-m for the kinetics of megakaryocyte production may be profound and worthwhile

studying in various strains. In particular, the discovery that certain genetically anemic mice (W/W^v and Sl/Sl^d) have abnormal megakaryocytopoiesis which may result from a specific genetic defect (1, 2) presents an intriguing model with which to study the role of CFU-m in the process. These studies are presently under investigation.

The response of CFU-m to platelet depletion indicates that the number of CFU-m is under some kind of regulation involving the peripheral blood platelet level or some other function monitored by platelets. The rapidity with which the response is mounted, and its dissipation following an acute platelet depletion, were somewhat unexpected. Jackson (3) has shown that the small, AChE-positive progenitors increase to maximum levels of 2–3-fold within 6 hr after an experimentally induced platelet demand. CFU-m increase to similar levels but do so within 2–4 hr. This may be considered additional evidence that these two cell populations are distinct and, additionally, that the CFU-m may possibly be the progenitor of the small, AChE-positive cells. With regard to the humoral regulation of CFU-m, the short time scale of the CFU-m response does not fit very well with the demonstration that maximum Tp levels are attained some 12 hr after severe platelet depletion (13). Nevertheless, the finding that Tp added directly to CFU-m cultures is capable of increasing the resulting numbers of CFU-m not in line with existing evidence which indicates that Tp may have a direct effect on CFU-m commitment and proliferation. The evidence at this point for the humoral control of CFU-m proliferation is hardly convincing. Nevertheless, techniques exist for testing various hypotheses directly in culture, including the use of purified fractions of Tp prepared as described by McDonald et al. (5), which we are in the process of testing.

Recently, during a routine analysis of sera from various sources, we discovered that the substitution of horse serum for fetal calf serum supported not only the growth of megakaryocyte colonies, but also their differentiation, as shown in Fig. 3. Megakaryocytes progressed through the same maturation sequence which result in recognizable megakaryocytes by conventional cytology. This leads ultimately to platelet formation and release into the surrounding medium. McLeod et al. (6) have reported the induction, documented by electron microscopy, of platelet formation in the plasma culture system by erythropoietin. Although the exact significance of these findings is not clear at this time, the potential has been demonstrated in the plasma culture system for the possible study *in vitro* of the entire sequence of cellular events through to the formation and release of platelets in culture. Additional evidence has been communicated by Levine (4), who has demonstrated platelet formation in liquid cultures of guinea pig megakaryocytes having morphologies and pseudopodial structures identical to those which we have observed. He has been unable to observe platelets released from these cells, however, and has postulated that a "platelet-release" substance is essential to permit completion of the final step which is not present in his culture system. The possibility of fixation or culture artifacts is always present in all these studies, and may be difficult to disprove.

As shown in Fig. 5, we believe that our present plasma culture system, with the addition of horse serum, can be used to quantitate not only CFU-m, but also the number of individual megakaryocytes derived possibly from polyploid progenitors. This enables us as well as to measure their resulting maturation *in vitro*. These findings, though preliminary, open up new vistas for the study of megakaryocytopoiesis. They provide a challenge to show that what is observed *in vitro*

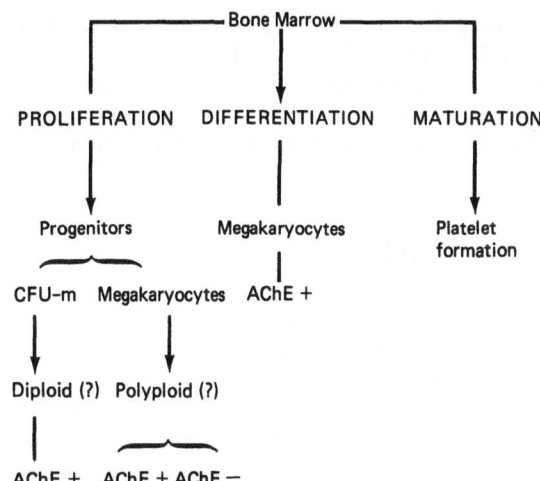

FIGURE 5. Plasma culture of colony-forming unit megakaryocyte (CFU-m) and identification by acetylcholinesterase.

is not artifactual, but rather reflects the normal process of megakaryocyte maturation which has been defined in detail *in vivo* (14).

In summary, Table 5 lists the properties and culture characteristics of megakaryocytes

TABLE 5 Cell and Colony Characteristics of Cultured Megakaryocytes[a]

PROPERTY	ASSAYED METHOD	REFERENCE
Morphology	Giemsa, orcein, electron microscopy	(7,9,10)
Maturation (immature)	Electron microscopy	(10)
	Orcein	(7)
Platelet antigen	FITC-RAMPS	(9)
Cytochemistry	AChE-positive	(7,8)
DNA content	Polyploid	(7)
Phagocytosis	Nonphagocytic	(14,15)
Adherence	Nonadherent	(14,15)
Culture dynamics	Time in cultures	
	Agar, day 6	(10)
	Plasma, day 4	(8)
	Low P.E. = 1–10/10⁵; 2–20 megakaryocytes/colony	(6,7,8)
	Small size: 2–20 megakaryocytes/colony	(6,7,8)
	CM	
	2-ME + spleen	(7)
	L-CM, PHA+ and PWM+ spleen	(8)
	M.E.F.	(9)
	Erythropoietin	(6)
	Specific stimulus	(10)
	Platelet formation	(6)
Progenitor cells		
Size	STAPUT	(7)
Density	DADGC	(11)

[a]Abbreviations: FITC-RAMPS, fluorescein isothiocyanate-labeled rabbit anti-mouse platelet serum; AChE, acetylcholinesterase; 2-ME, 2-mercaptoethanol; L-CM, L-cell-conditioned medium; PHA; phytohemagglutinin; PWM, pokeweed mitogen; M.E.F., mouse embryo fibroblasts; STAPUT, velocity sedimentaiton; DADGC, discontinuous albumin density gradient centrifugation.

and megakaryocyte colonies grown *in vitro* to date. It is clear that the clonogenic assay for CFU-m is being rapidly exploited. Our studies are using the CFU-m assay as an experimental method to elucidate the previously unknown megakaryocyte progenitor compartment and to study the interrelationship between mitotic and endoreduplicative cell processes in megakaryocyte proliferation.

SUMMARY

The development of cell culture techniques to support the proliferation and differentiation of hematopoietic cells is important to our understanding of the regulation and control of these processes. We recently reported the establishment of a clonogenic cell assay in plasma clot culture for a class of megakaryocyte precursors in mouse bone marrow entitled the CFU-m. Elucidation of its physiology and·determination of its role as a stem cell committed to megakaryocyte production have been done. This has been accomplished by using techniques which we have reported for separating megakaryocyte populations on the basis of cell size, unit velocity sedimentation and cell density and discontinuous albumin gradient centrifugation. CFU-m display both a sedimentation and a density profile similar to those for other committed stem cells (e.g., CFU-c) but show, in the latter, a profile that is clearly different from morphologically unidentifiable megakaryocytic precursors that are acetylcholinesterase (AChE)-positive. Flow microfluorimetry (FMF) analysis of the above enriched megakaryocyte fractions has been used to relate the ploidy state of CFU-m to ploidy levels in the AChE-positive precursor and maturing megakaryocytic compartments, in an attempt to understand the interplay between competing mitotic and endomitotic processes. Various classes of chemotherapeutic agents have also been used as proliferative cell probes to determine the cell-kinetic properties of CFU-m.

The humoral regulation of CFU-m has been studied in an immune-induced thrombocytopenic mouse model in which we have reason to believe that megakaryocytic responses are mediated by thrombopoietin. The number of CFU-m in femoral marrow has been determined (1) as a function of the degree of platelet demand created following increasing doses of antiplatelet serum, as well as (2) as a function of time. In order to determine whether these responses are mediated humorally, these data have been correlated with the cell-kinetic responses in the megakaryocytic system of donor mice injected with thrombopoietic mouse serum. In addition, conditioned medium from human embryo kidney cell cultures shown to contain high levels of thrombopoietin has been tested for its effect on CFU-m proliferation and differentiation. These data have been interpreted as indicating that CFU-m are regulated by a humoral factor(s) possibly produced in the kidney, and that alterations in their kinetic response are important determinants of subsequent cellular events observed in the megakaryocyte and platelet compartments.

ACKNOWLEDGMENT

Sincere appreciation is extended to James Bryan for his excellent technical assistance.

REFERENCES

(1) Ebbe, S., Phalen, E., and Stohlman, F., Jr. Abnormalities of megakaryocytes in W/W^v mice. *Blood, 42*:857, 1973.

(2) Ebbe, S., Phalen, E., and Stohlman, F., Jr. Abnormalities of megakaryocytes in S1/S1^d mice. *Blood, 42*:865, 1973.

(3) Jackson, C. W. Cholinesterase as a possible marker for early cells of the megakaryocytic series. *Blood, 42*:413, 1973.

(4) Levine, R. Personal communication.

(5) McDonald, T. P., Clift, R., Lange, R. D., Nolan, C., Tibby, I. I. E., and Barlow, G. H. Thrombopoietin production by human embryonic kidney cells in culture. *J. Lab. Clin. Med., 85*:59, 1975.

(6) McLeod, D. L., Shreeve, M. M., and Axelrad, A. A. Induction of megakaryocyte colonies with platelet formation *in vitro. Nature (London), 261*:492, 1976.

(7) Metcalf, D., MacDonald, H. R., Odartchenko, N., and Sordat, B. Growth of mouse megakaryocyte colonies *in vitro. Proc. Nat. Acad. Sci., U.S.A., 72*:1744, 1975.

(8) Nakeff, A., and Daniels-McQueen, S. *In vitro* colony assay for a new class of megakaryocyte precursor: Colony-forming unit, megakaryocyte (CFU-M). *Proc. Soc. Exp. Biol. Med., 151*:587, 1976.

(9) Nakeff, A., and Dicke, K. A. Stem cell differentiation into megakaryocytes from mouse bone marrow cultured with the thin layer technique. *Exp. Hematol., 22*:58, 1972.

(10) Nakeff, A., Dicke, K. A., and van Noord, M. J. Megakaryocytes in agar cultures of mouse marrow. *Ser. Hematol., 8*:4, 1975.

(11) Nakeff, A., and Floeh, D. P. Separation of megakaryocytes from mouse bone marrow by density gradient centrifugation. *Blood, 48*:133, 1976.

(12) Nakeff, A., and Ingram, M. Platelet count: Volume relationships in four mammalian species. *J. Appl. Physiol., 28*:530, 1970.

(13) Nakeff, A., and Roozendaal, R. J. Thrombopoietin activity in mice following an immune-induced thrombocytopenia. *Acta Haematol., 56*:340, 1975.

(14) Odell, T. T., Jr. Megakaryocytopoiesis and its response to stimulation and suppression. In Baldini, M. G., and Ebbe, S., eds., *Platelets: Production, Function, Transfusion and Storage.* New York: Grune and Stratton, p. 11, 1974.

(15) Porter, R. P., and Gengozian, N. *In vitro* proliferation and differentiation of hemic precursor cells from marrow and blood of naturally chimeric marmosets. *J. Cell. Physiol., 79*:27, 1972.

(16) Tobey, R. A., and Crissman, H. A. Unique techniques for cell cycle analysis utilizing mithramycin and flow microfluorometry. *Exp. Cell Res., 93*:235, 1975.

(17) Vassort, F., Winterholen, M., Frindel, E., and Tubiana, M. Kinetic parameters of bone marrow stem cells using *in vivo* suicide by tritiated thymidine or by hydroxyurea. *Blood, 41*:789, 1973.

PART IV

Physiology of Committed Stem Cells
(CFU-c)

Siegmund J. Baum, Ph.D.

Negative and Positive Feedback Control of the Committed Granulocytic Stem Cell Compartment

Siegmund J. Baum

INTRODUCTION

In general the hematopoietic system is comprised of three main compartments: (1) the pluripotential stem cell; (2) the committed stem cell; and (3) the differentiated compartment (Fig. 1). The pluripotent stem cells as represented by the colony-forming unit-spleen (CFU-s) have the physiologic capabilities of self-renewal and control of compartment size, which are normal requirements of precursor cells (30). Regulation of this compartment is assumed to be, at least in part, by short-range, inhibitory, cell-to-cell interaction, and it is possible that modification of cell membranes may affect recognition and initiate increases in compartment size (23). In addition, there exists evidence that the CFU-s are also influenced by long-range humoral agents (2, 20, 32). However, at this time we have no precise knowledge of the conditions responsible for the initiation of the differentiation of CFU-s into committed stem cells, such as the CFU-c. It has been hypothesized that the known migratory behavior of CFU-s, which carries them into the circulation and from there to sites of **14** hematologic significance, eventually places them within a favorable hematopoietic microenvironment for their conversion into CFU-c. It is plausible that long-range stimulatory factors, which may or may not be identical with the colony-stimulating activities (CSA), initiate the migration of CFU-s. How and by what means they eventually select the suitable ecologic niches is an enigma at present.

PHYSIOLOGIC CONTROL OF THE CFU-c

The topic of the present discussion is the physiologic control of the CFU-c which have arrived in a favorable microenvironment. As indicated schematically in Fig. 1, the cells in the CFU-c compartment are not self-sustaining. As they proliferate and differentiate into granulocytes and monocytes, their number must be replenished by the CFU-s compartment. Normally, this is achieved by the release of a small number of CFU-s, and the majority of these cells remain in the resting or G_0 state. If in response to injury the CFU-c compartment becomes depleted, CFU-s go into cyle and a larger number of cells are released. The CFU-s as well as the CFU-c must achieve a critical size before cells are released into the next compartment. Once the cells have become CFU-c there must be amplification and differentiation prior to efflux into the differentiated compart-

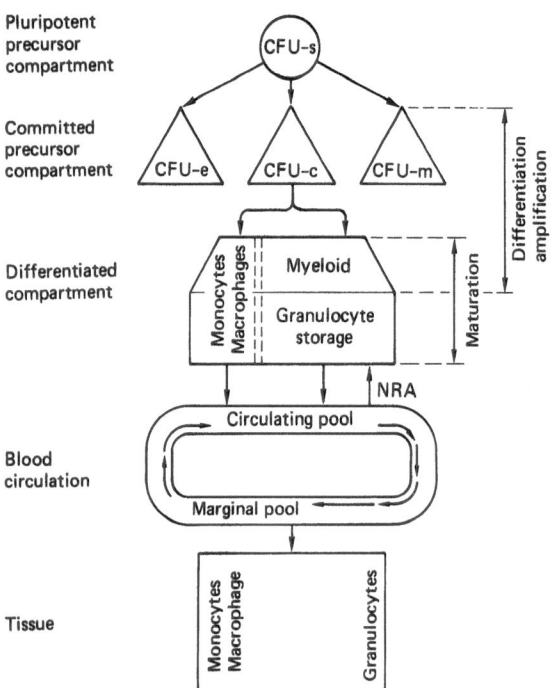

FIGURE 1. Schematic model of normal leukopoiesis.

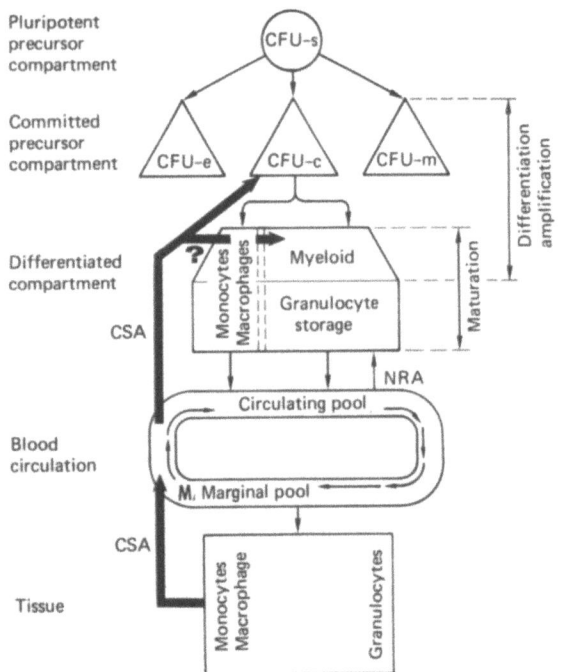

FIGURE 2. Schematic model of the positive feedback system for leukopoiesis.

ment. The CFU-c differentiate into recognizable cells, the myeloblasts; further amplification and differentiation occurs until the cells are stored in the bone marrow as metamyelocytes, bands, and mature polymorphonuclear leukocytes. This bone marrow storage pool contains about 4–6 times the number of cells present in the circulating blood. Cells leave the marrow storage pool in a sequential manner, the older cells being released first (13). Recent findings indicate that aging of granulocytes is associated with deformability and decreased stickiness (17). This obligatory maturation would, of course, favor a sequential release from the storage pools.

Upon stimulation by antigens such as endotoxin, a great number of the stored cells are released into the periphery. As may be seen from Fig. 1, a protein originally discovered by Gordon and his associates (14) may be responsible for this leukocytosis by causing a discharge of granulocytes from the storage pool. This factor was originally called leukocytosis-inducing factor; recently, however, it was more appropriately renamed neutrophil-releasing activity, or NRA (31). The transit time of the released leukocytes in the circulation is relatively short—a matter of several hr—and those cells which do not remain in the marginal pool migrate to the tissues and become the primary defense against invading antigens.

POSITIVE FEEDBACK SYSTEM FOR CFU-c REGULATION

As may be seen in Fig. 2, all the leukocytic compartments discussed above have been implicated from time to time as possible sources of stimulatory factors or leukopoietins involved with the entry, amplification, differentiation, and release of the CFU-c. The development of an assay which permitted the cloning of marrow cells, or more specifically CFU-c, in a semisolid medium (4, 25) finally demonstrated the existence of a specific stimulating factor (22). This factor was originally named colony-stimulating factor (CSF), and more recently was renamed colony-stimulating activity (CSA) (16). For the present discussion we are making the assumption that CSA is identical or related to the humoral factor responsible for *in vivo* CFU-c turnover. However, it is understood that while this concept is accepted by some workers (26), it is questioned by others (24). The prime candidate for the origin of CSA is the monocyte/macrophage system (31). In our own laboratory, we have demonstrated that the 24 hr cellular concentration in a model of an inflammatory exudate, which consisted of an acrylic cup filled with

128

Hanks's balanced salt solution and placed in a subcutaneous pouch in rodents, contained primarily monocytes (34). Cells or supernatant removed from this cup, or plasma from mice with implanted cups, stimulated increased formation of CFU-c on agar plates, as seen in Table 1. However, they also stimulated increased CFU-s formation (2).

In summary, Fig. 2 represents a conceptional model of the stimulatory feedback system for the CFU-c compartment. Presumable humoral substances produced by monocytic cells in inflammatory tissues stimulate CFU-c into cell cycle. It is quite possible that the same substances also initiate CFU-s turnover and the release of these cells into the CFU-c compartment. The relationship of CSA to NRA is not understood, at present. However, evidence was presented for a separate identity of the two (31). Although increased NRA and CSA are produced in response to endotoxin administration, the former, which mobilizes granulocytes from the storage pool, may be produced by leukocytes in the blood, whereas the latter is released by monocytes/macrophages in the tissues in response to endotoxin, and stimulates CFU-c into cycle. It is doubtful that CSA is produced in the circulation. Table 2 summarizes the known physical properties of CSA.

NEGATIVE FEEDBACK SYSTEM FOR CFU-c CONTROL

Although one could easily accept a model of CFU-c control under the influences of CSA alone, evidence from other studies clearly indicates a more complex physiologic system. Craddock et al. (11) demonstrated that effective withdrawal of large numbers of granulocytes from the blood leads to an accelerated release of cells from the bone marrow in dogs. In contrast, infusion of autologous mature granulocytes into the circulation of dogs reduced cell release from the marrow, and consequently inhibited granulocyte production (12). Destruction of leukemic cells by extracorporeal radiation of the blood appears to stimulate proliferative activity of leukemic blast cells in the bone marrow (7). Transfusion of fresh blood apparently causes a transient fall in the white blood cells of leukemic patients (35). Several of these authors suggested the operation of a negative feedback system under the control of inhibitory agents or chalones (6, 10, 27).

In earlier studies designed to determine the existence and biologic specificity of the granulocyte chalone in *in vitro* bone marrow cul-

TABLE 1 Number of CFU-c per 10^5 Bone Marrow Cells in Mice

INJECTION	MEAN NUMBER OF COLONIES PER PLATE
Exudate plasma[a]	149 ± 33.6[c]
Exudate supernatant	141 ± 36.0
Exudate cells	115 ± 16.8
Hanks's solution	102 ± 16.2
Normal plasma[b]	100 ± 12.7

[a]Plasma from mice with implanted acrylic cup.

[b]Acrylic cup not implanted.

[c]Standard error.

tures, radioautograph labeling with ^{3}H-thymidine (^{3}H-Tdr) showed that extract of granulocytes significantly decreased granulocytic precursors (28). No changes in labeling index were detected for other bone marrow cells. Although some doubts were expressed about the validity of these findings (27), recently, using a more refined technique of analysis, the strict specificity of action of partially purified extracts from granulocytes on their own precursors was confirmed (1). The effects of extracts from mature granulocytes can be determined by measuring the structuredness of the cytoplasmic matrix of the receiving cells (19). Such studies have clearly demonstrated that granulocyte extracts affect only granulocytic precursor cells. Incidentally, these fractions were tested at concentrations of 33 μg/ml.

Studies utilizing diffusion chambers in an *in vivo* closed culture system have given a better understanding of the specific action of granulocytic chalone. In an early experiment, extracts of granulocytes and of liver were injected into mice implanted with two different culture chambers (3). It was shown that granulocyte extracts inhibited DNA synthesis in proliferating granulocytes

TABLE 2 Characterization of CSA

Chemical composition	Glycoprotein
Temperature of inactivation	Above 60°C
Sedimentation coefficient	4.5–7.0
Electrophoretical migration	α-Globulin-post albumin range
Molecular weight by gel filtration	70,000
Sucrose gradients	45,000
Human CSA by gel filtration	45,000
Inactivated by	α-Chymotrypsin, subtilisin

but not in immunoblasts and macrophages. It appears, then, that the granulocyte chalone, like other chalones, is a cell line-specific but not species-specific regulator substance. It inhibits cell proliferation in the granulocyte system in a reversible manner. The latter was established in an interesting experiment by Laerum and Maurer (15).

In our own laboratory, Drs. MacVittie and McCarthy (21) used the *in vivo* diffusion chamber technique for mouse marrow culture to determine the proliferative responses of CFU-s and CFU-c after exposure to granulocyte inhibitor. Presented in Fig. 3 is their experimental design. They prepared a granulocyte-conditioned media (GCM) from granulocytes obtained from rat peritoneal cavities. These cells were induced to accumulate in the peritoneal cavity by injection of 10 ml of 2.5% sterile oyster glycogen. The GCM was derived from media in which white cells were incubated for 20 hr. Based on previous tests by the authors this material contained inhibitors or chalones for granulocytes. Mice (Swiss-Webster) were then radiated with 900 rad ^{60}Co γ-radiation at 154 rad/min. Four hr later they were implanted with diffusion chambers made from 0.22 μm Millipore filters. Each chamber contained separated cell marrow suspensions which were prepared from the femurs of several normal mice in tissue

culture media CMRL 10 with 10% fetal calf serum. Two diffusion chambers were implanted into each mouse. The GCM was administered during the initial 48 hr of chamber culture. The diffusion chambers were removed at selected intervals and cells were harvested. Nucleated cell counts were performed prior to pooling the cell suspensions for assay of CFU-s and CFU-c. As is seen in Fig. 4, GCM administration effectively reduced the total nucleated cell production within the diffusion chambers. The specificity of the GCM is attested to by the fact that inhibition of growth did not occur in cultures of mouse fibroblasts (L-929 cells) grown in identical conditions as the marrow cells.

In Fig. 5 it was shown that the GCM containing the chalone has no effect on CFU-s. On the other hand, as can be seen in Fig. 6, the GCM significantly reduced the number of granulocyte progeny formed within the diffusion chamber, in part by reducing cell turnover at the level of the committed granulocyte progenitor cell (CFU-c).

To date it has been difficult to demonstrate chalone activity *in vivo*. However, Schütt and Langen (29) inhibited granulopoiesis in rats by injections of granulocyte extracts. These authors labelled bone marrow cells with ^{3}H-Tdr, and then produced an aseptic inflammation in the peritoneal cavity of rats. The chalone reduced the

FIGURE 3. Flow diagram of the *in vivo* diffusion chamber technique for mouse bone marrow culture.

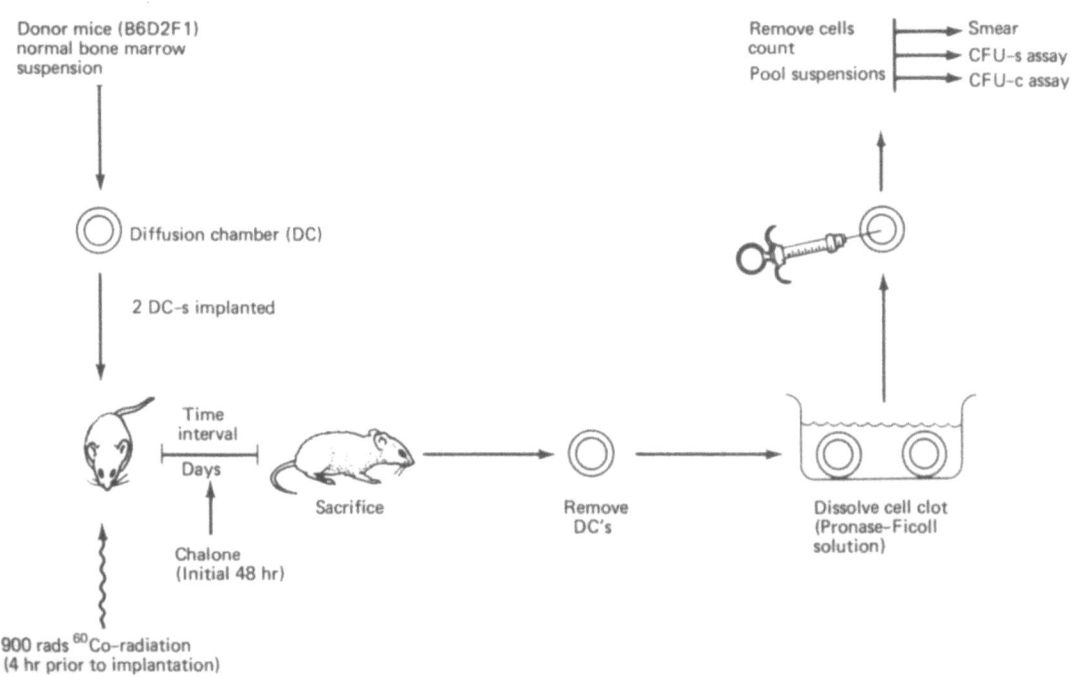

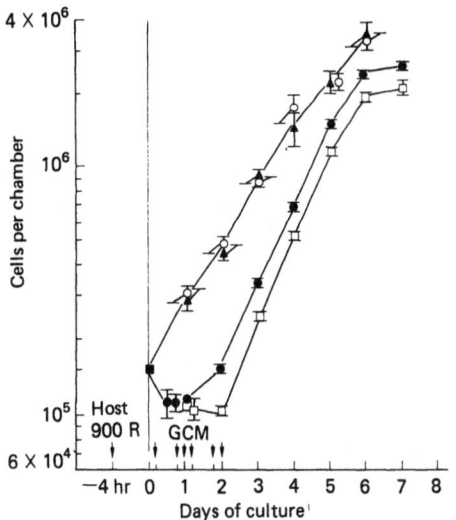

FIGURE 4. The effect of granulocyte-conditioned media (GCM) on the growth of mouse fibroblasts (L-929) inoculated at 10^5 cells per chamber and on nucleated cells (granulocytes and macrophages) inoculated at concentrations of 1.5×10^5 normal bone marrow cells per chamber. Mean values ($\pm$ SEM) are results of six replicate experiments, and observed differences are significant at $p<0.001$ level. Fibroblasts: ▲, control; ○, GCM. Bone marrow: ●, control; □, GCM. (By permission of the author.)

DNA-specific activity of the granulocytes recovered from the inflammatory exudate by 50%.

All the reports on granulocyte chalone indicate that it inhibits granulopoiesis in a tissue-specific manner. Most likely, cells are arrested in the G_1 phase (15). These studies support the contention that granulocyte chalone is a polypeptide

FIGURE 5. The effect of GCM on the growth of CFU-s in diffusion chambers. ●, Control; □, GCM. Chambers were inoculated with 1.5×10^5 normal bone marrow cells. Values ($\pm$ SEM) are results of at least four replicate experiments in which chamber cells were pooled and used to inject eight to 10 assay mice per point. (By permission of the author.)

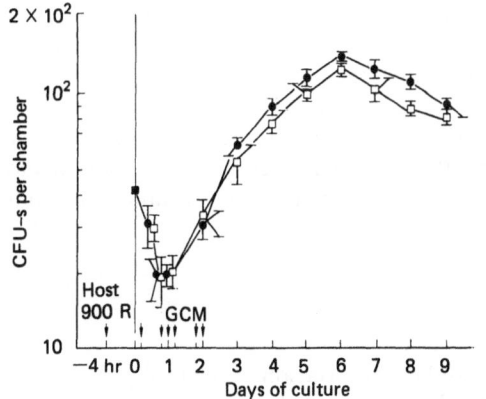

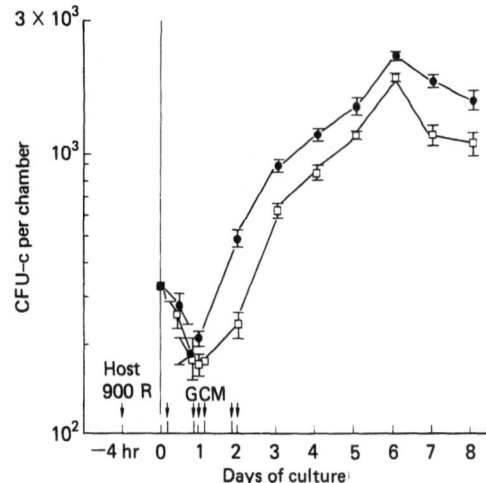

FIGURE 6. The effect of GCM on the growth of CFU-c in diffusion chambers inoculated with 1.5×10^5 normal bone marrow cells. ●, Control; □, GCM. Values ($\pm$ SEM) are results of at least four replicate experiments in which chamber cells were pooled and used to inoculate four agar plates per point. (By permission of the author.)

with a molecular weight of 4000 daltons (33). However, in addition a nondialyzable inhibitor of granulocytic colony growth has been found in the sera of humans and mice. It is assumed to be a lipoprotein of possibly much higher molecular weight (9).

Figure 7 enables us to summarize the suggested anatomic areas of chalone production and its possible physiologic functions. Nearly all reports seem to agree that granulocytic chalone is produced by adult functional granulocytes (33). These granulocytes were obtained from the tissues (21), the circulation (18), and the bone marrow (26). This, of course, fits precisely the description of a chalone as postulated by Bullough (5). Each cell line or tissue produces specific inhibitors which regulate its mitotic rate. Injury or cell death reduces the concentration of the inhibitor and the inhibition, with a net result of increased mitotic activity and cellular production. Evidence has been presented that the action of granulocyte chalone is on the CFU-c compartment (21).

POSSIBLE INTERACTION OF THE TWO FEEDBACK SYSTEMS IN THE REGULATION OF THE CFU-c COMPARTMENT

A positive feedback system of CFU-c regulation with CSA operating as the controlling agent was described. Evidence was presented that CSA might be produced by monocytes and/or mac-

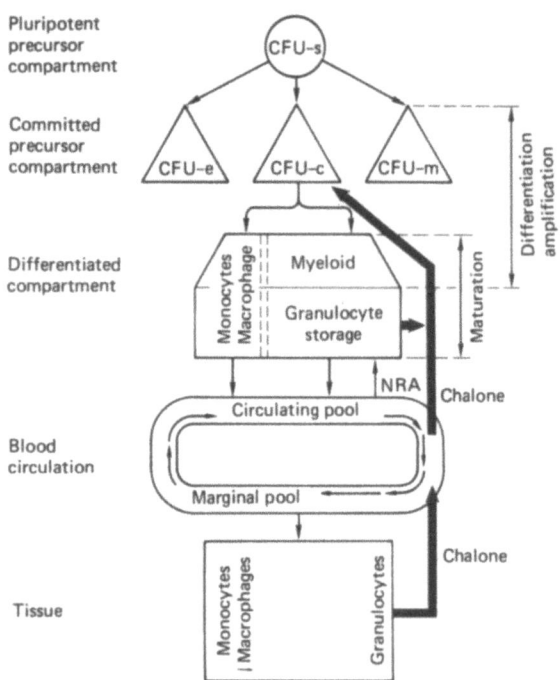

FIGURE 7. Schematic model of the negative feedback system for leukopoiesis.

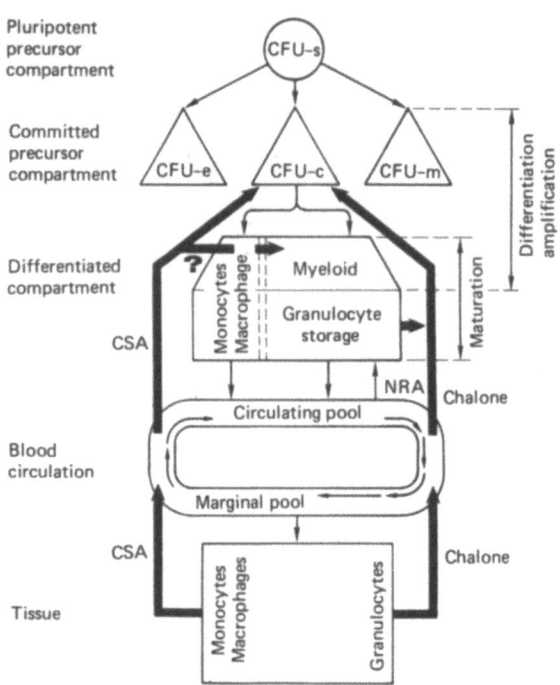

FIGURE 8. Model for leukopoiesis featuring suggested actions of the dual system controlled by both positive and negative feedback systems.

rophages in the tissues, probably in inflammatory loci or exudates. It probably does not act on the bone marrow storage compartment and is not related to NRA, which does release neutrophils from it (31). Although cellular proliferation and differentiation of the CFU-c could easily be accomplished under the stimulatory action of CSA alone, strong evidence has also been produced for the existence of a negative feedback system regulated by the granulocytic chalone (27, 33). The chalone probably is produced by adult functional granulocytes, and its action again might very well be on the CFU-c (21).

Figure 8 presents the possible model for leukocytopoiesis and indicates suggested actions of the dual system controlled by both positive and negative feedback systems. All that remains to be done now is to explain to the satisfaction of the workers in this field how the two regulatory systems may interact. Obviously, if both systems operated simultaneously but independently, responses in times of stress might be chaotic and disastrous. For example, if the chalone titer were high, CFU-c could not rapidly respond to sudden

stimulation by CSA. If CFU-c were under strong control of CSA the result would be increased granulocyte production, which in turn would increase the chalone concentration. The net effect could be decreased granulocyte production, regardless of whether or not the stressful condition which initially induced increased CSA production still existed. This situation could obviously not be tolerated. A clue to the possible interaction of the two feedback systems comes from the work of Chan and Metcalf, who reported that wholebody radiation caused an acute, dose-dependent rise in serum CSA levels and concomitant fall in inhibitor level (8). This might well lend support to the idea that under stressful conditions demanding increased granulocyte production, the positive feedback system is under control and chalone production is suppressed. It is possible that the negative feedback system controls the normal concentration of CFU-c and consequently the production of adult functional granulocytes. Under stressful conditions CSA or a granulopoietin alone is responsible for maximum granulocyte production.

SUMMARY

Although there exists some evidence that proliferation and differentiation of the multipotential stem cell (CFU-s) is at least partly controlled by cell-to-cell interaction, it

132

appears that the committed granulocytic colony forming cell (CFU-c) is under the influence of stimulatory or inhibitory humoral agents. The latter is supported by the fact that the addition of material containing colony stimulating activities (CSA) to the semisolid medium of *in vitro* bone marrow cultures permits the formation of granulocyte and macrophage colonies. CSA has to date only been partially purified and appears to be a glycoprotein with a molecular weight of 45,000 daltons. At present there exists no final proof that CSA stimulates granulopoiesis *in vivo*, however, indirect results from several studies appear to make it a good candidate. Along these lines our studies utilizing either (1) the diffusion chamber methodology and cytoxan treatment of mice or (2) murine models of inflammatory exudates support the hypothesis that CSA represent long distance humoral agents possibly produced by mononuclear cells that migrated into inflammatory exudates.

Shortly after the discovery of CSA, inhibitors of granulocyte-macrophage colony formation were discovered. These inhibitors were separated by gel filtration on Sephadex G-150 into two distinct areas of activity, one being a lipoprotein while the other was not. From diffusion chamber studies it appears that the inhibitors or chalones were specific for CFU-c and have no affect on CFU-s. Furthermore, granulocytic chalones are produced by mature granulocytes and inhibit only granulocytic precursors.

It appears that positive and negative feedback loops regulate CFU-c. CSA as well as granulocyte chalone are candidate regulators of the committed granulocyte colony forming cells. Possibilities of interactions are discussed.

REFERENCES

(1) Bateman, A. E. Cell specificity of chalone type inhibitors of DNA synthesis released by blood leukocytes and erythrocytes. *Cell Tissue Kinet.*, 7:451, 1974.

(2) Baum, S. J. MacVittie, T. J., Brandenburg, R. T., and Levin, S. G. Stimulation of stem cell release by humoral agents produced in inflammatory exudates. Exp. Hemats., 4 (Suppl.):55, 1976.

(3) Benestad, H. B., Rytömaa, T., and Kiviniemi, K. The cell specific effect of the granulocyte chalone demonstrated with the diffusion chamber technique. *Cell Tissue Kinet.*, 6:147, 1973.

(4) Bradley, T. R. and Metcalf, D. The growth of mouse bone marrow cells *in vitro*. Aust J. Exp. Biol. Med. Sci., 44:287, 1966.

(5) Bullough, W. S. Mitotic and functional hemeostasis: A speculative review. *Cancer Res.*, 25:1683, 1965.

(6) Chan, B. W. B., and Hahyoe, F. G. J. Changes in proliferative activity of marrow leukemic cells during and after extracorporeal irradiation of blood. *Blood*, 37:657, 1971.

(7) Chan, B. W. B., Hahyoe, F. G. J., and Bullimore, J. A. Effect of extracorporeal irradiation of the blood on bone marrow activity in acute leukaemia. *Nature (London)*, 221:972, 1969.

(8) Chen, S. H., and Metcalf, D. Local and systemic control of granulocytic and macrophage progenitor cell regeneration after irradiation. *Cell Tissue Kinet.*, 6:185, 1973.

(9) Chen, S. H., Metcalf, D., and Stanley, E. R. Stimulation and inhibition by normal human serum of colony forming *in vitro* by bone marrow cells. *Br. J. Haematol.*, 20:329, 1971.

(10) Craddock, C. G. Granulocyte kinetics. In Williams, W. J., Beutler, E., Erslev, A. J., and Rundles, R. W., eds., *Hematology*. New York: McGraw-Hill, p. 593, 1972.

(11) Craddock, C. G., Perry, S., and Lawrence, J. S. The dynamics of leukopoiesis and leukocytosis, as studied by leukopheresis and isotope techniques. *J. Clin. Invest.*, 35:285, 1956.

(12) Craddock, C. G., Perry, S., Lawrence, J. S., Buxbaum, L., and Pieper, G. Production and distribution of granulocytes and the control of granulocyte release. In Wolstonholme, G. E. W., and O'Connor, M., eds., *Ciba Foundation Symposium on Haemopoiesis*. London: Churchill, p. 237, 1960.

(13) Cronkite, E. P., and Vincent, P. C. In Stohlman, Jr., F., ed., Granulocytopoiesis. Symposium of Hemopoietic Cellular Proliferation. New York: Grune and Stratton, p. 211, 1970.

(14) Gordon, A. S., Handler, E. T., Siegel, C. D., Dornfest, B. S., and LoBue, J. Factors influencing leukocyte release in rats. *Ann. N.Y. Acad. Sci.*, 113:766, 1964.

(15) Laerum, O. D., and Maurer, H. R. Proliferation kinetics of myelopoietic cells and macrophages in diffusion chambers after treatment with granulocyte extracts (chalone). *Virchows Archiv. Pathol. Anat. Physiol.* (b) Zellpath., 14:293, 1973.

(16) Lajtha, L. G. Hemopoietic stem cells. *Br. J. Haematol.* 29:529, 1975.

(17) Lichtman, M. A., and Weed, R. I. Alteration of the cell periphery during granulocyte maturation: Relationship to cell function. *Blood*, 39:301, 1972.

(18) Lord, B. I. Modification of granulopoietic cell proliferation by granulocyte extracts. *Boll. Ist. Sieroter. Milan.*, *54*:187, 1975.

(19) Lord, B. I., Cercek, L., Cercek, B., Shan, G. P., Dexter, T. M., and Lajtha, L. G. Inhibitors of haemopoietic cell proliferation: Reversibility of action. *Br. J. Cancer*, *29*:407, 1974.

(20) MacVittie, T., and McCarthy, K. Increased proliferation of hematopoietic stem cells within diffusion chambers implanted into irradiated mice pretreated with cyclophosphamide. *Radiat. Res.*, *59*:291, 1974.

(21) MacVittie, T., and McCarthy, K. F. The influence of a granulocytic inhibitor(s) on hematopoiesis in an *in vivo* culture system. *Cell Tissue Kinet.*, *8*:553, 1975.

(22) Metcalf, D., Bradley, T. R., and Robinson, W. Analysis of colonies developing *in vitro* from mouse bone marrow cells stimulated by kidney feeder layers or leukemic serum. *J. Cell. Physiol.*, *69*:93, 1967.

(23) Monette, F., Morse, B., Howard, D., Nishamen, E., and Stohlman, F., Jr. Hemopoietic stem cell proliferation and migration following *Bordetella pertussis* vaccine. *Cell Tissue Kinet.*, *5*:121, 1972.

(24) Paukovits, W. R., and Paukovits, J. B. Separation, identification and mechanism of action of the granulocytic chalone. *Boll. Ist. Sieroter. Milan.*, *54*:177, 1975.

(25) Pluznik, D. H., and Sachs, L. The cloning of normal "mast" cells in tissue culture. *J. Cell. Comp. Physiol.*, *66*:319, 1965.

(26) Robinson, W. A., and Mangalik, M. The kinetics and regulation of granulopoiesis. *Semin. Hematol.*, *12*:7, 1975.

(27) Rytömaa, T. Biology of the granulocyte chalone. *Boll. Ist. Sieroter. Milan.*, *54*:195, 1975.

(28) Rytömaa, T., and Kiviniemi, K. Control of granulocyte production. I. Chalone and antichalone, two specific humoral regulators. *Cell Tissue Kinet.*, *1*:329, 1968.

(29) Schütt, M., and Langen, P. Comments on granulocyte chalone action. Proc. Symp. on Active Control of Nucleic Acid Metabolism. *Studio Biophys. 31/32*:311, 1972.

(30) Stohlman, F., Jr. Stem cell regulation. In Baldini, M. G., and Ebbe, S. E., eds., *Platelets: Production, Function, Transfusion and Storage.* New York: Grune and Stratton, p. 1, 1974.

(31) Stohlman, F., Jr., Quesenberry, P. J., and Tyler, W. S. The regulation of myelopoiesis as approached with *in vivo* and *in vitro* techniques. In Brown, E. B. ed., *Progress in Hematology.* New York: Grune and Stratton, Vol. 8, p. 259, 1973.

(32) Tyler, W., Niskamen, E., Stohlman, F., Jr., Keane, J., and Howard, D. The effect of neutropenia on myeloid growth and the stem cell in an *in vivo* culture system. *Blood*, *40*:634, 1972.

(33) Vogler, W. R., and Winton, E. F. Humoral granulopoietic inhibitors: A review. *Exp. Hematol.*, *3*:337, 1975.

(34) Wyant, D. E., and Baum, S. J. An *in vivo* model for the definitive analysis of cellular inflammation in normal and irradiated rats. *J. Reticuloendothelial Soc.*, *21*:53–59, 1977.

(35) Wetherley-Mein, G., and Cotton, D. G. Fresh blood transfusion in leukaemia. *Br. J. Haematol.*, *2*:25, 1956.

INTRODUCTION

Colony-stimulating factor (GM-CSF), which specifically stimulates mouse bone marrow cells to proliferate and form neutrophil and/or macrophage colonies in semisolid agar cultures (1, 22), has been detected in many biologic fluids (8, 23–26, 30). Most of the sources of GM-CSF are unsuitable for the isolation of this factor and only human urine (30), mouse L-cell-conditioned medium (8, 28), and mouse lung-conditioned medium (MLCM) (23–26) appear to contain sufficient GM-CSF to allow the molecule to be isolated. Although some progress has been made in the purification of GM-CSF (8, 24, 29), little is known about the mechanism by which GM-CSF acts on bone marrow cells (29).

In this report the purification and biologic activity of GM-CSF derived from mouse lung-conditioned medium will be described. GM-CSF retained biologic activity after radioiodination, and the serum half-life of ^{125}I-GM-CSF was measured directly. Finally, studies were performed on the early effects of GM-CSF on RNA, protein, and DNA synthesis by mouse bone marrow cells.

MATERIALS AND METHODS
PURIFICATION OF GM-CSF$_{MLCM}$

All procedures performed during the purification of GM-CSF$_{MLCM}$ were at 8° C or less. Buffers contained 0.02% sodium azide and 0.01% Triton X-100 (Packard Instruments Company, Chicago Ill.).

Step I. MLCM was prepared from the lungs of mice previously injected with endotoxin (5 μg). After incubation in serum-free Dulbecco's Modified Eagle's Medium (Gibco, Grand Island, New York) for 48 hr at 37°C in a fully humidified atmosphere of 10% CO_2 in air, the medium was filtered through gauze and heated for 30 min at 56°C. Four-liter batches of MLCM were dialyzed against 3 changes of 3 volumes of distilled water, and centrifuged at 12,000 g for 15 min to remove the precipitated proteins (23, 24).

Step II. Calcium phosphate gel (11) (5 mg gel/mg MLCM protein) was stirred into the solution of GM-CSF$_{MLCM}$ from Step I. After 2 hr, the supernatant fluid was discarded and the gel collected by centrifugation. The gel was washed initially with 2 volumes of 0.01 M sodium phosphate buffer, pH 6.8, and with 1.5 gel volumes of

Colony-Stimulating Factor and the Differentiation of Granulocytes and Macrophages

Antony W. Burgess and Donald Metcalf

15

0.05 M phosphate buffer, pH 6.8, to elute the GM-CSF.

Step III. GM-CSF$_{MLCM}$ from Step II was dialyzed against 3 changes of 0.06 M Tris-HCl buffer, pH 7.4, and applied to a column of DEAE cellulose (Whatman, Pty. Ltd., London, England) equilibrated with the same buffer. When all of the GM-CSF$_{MLCM}$ solution had been applied to the column as well as an additional 100 ml of starting buffer, a linear gradient of NaCl was obtained (0–0.2 M NaCl in 1 liter). Fractions containing CSF activity were pooled and adjusted to 0.2 M sodium acetate, pH 5.0, with sodium acetate buffer (1 M). Ca^{2+}, Mg^{2+}, and Mn^{2+} ions were added as their chloride salts, to yield a final concentration of 1 mM.

Step IV. GM-CSF$_{MLCM}$ from Step III was applied to a column (1.5 × 25 cm) of concanavalin A-Sepharose (Pharmacia, Stockholm, Sweden) equilibrated with 0.2 M sodium acetate buffer, pH 5.0, containing Ca^{2+}, Mn^{2+}, and Mg^{2+} ions (1 mM). After the GM-CSF solution and a further 130 ml of starting buffer had been applied, the column was eluted with a solution of methyl-α-D-glucopyranoside (0.05 M) dissolved in the starting buffer. The fractions eluted with the methyl-α-D-glucopyranoside, and which contained GM-CSF, were pooled and concentrated to 20 ml by ultrafiltration (Amicon, TCF-10 apparatus) on a PM-10 membrane.

Step V. CSF$_{MLCM}$ from Step IV was applied to a column (1.7 × 200 cm) of Ultrogel AcA44 (LKB, Stockholm, Sweden) previously equilibrated with 0.1 M Tris-HCl buffer, pH 7.4. Fractions from the column which contained GM-CSF were pooled and concentrated to 7 ml using ultrafiltration with a UM 05 membrane.

Step VI. Sucrose (10% w/v) and bromophenol blue (10 μl of a 1% w/v solution in distilled water) were added to the GM-CSF$_{MLCM}$ solution from Step V. Preparative polyacrylamide gel electrophoresis was performed with a stacking gel system (3, 21) in a cylindrical (1.7 × 6 cm) gel tube. The running gel was made with 15% acrylamide monomer and 0.8% bis acrylamide cross-linking reagent (upper reservoir buffer: Tris-glycine pH 8.3, 0.025 M; spacer gel buffer: Tris-HCl, pH 6.8, 0.06 M; running buffer: Tris-HCl, pH 8.8, 0.035 M). The current was maintained at 6 mA until the bromophenol blue had stacked onto the top of the running gel. The current was then increased to 9 mA and the electrophoresis continued for 9 hr. The gel was sliced

into 2.5 mm sections and CSF eluted by crushing the slices in 0.03 M Tris-HCl, pH 7.4 (2 ml per slice). The eluates containing GM-CSF$_{MLCM}$ were pooled and concentrated by ultrafiltration using a UM 05 membrane.

RADIOIODINATION

Biologically active ^{125}I-GM-CSF$_{MLCM}$ was prepared by a modification of the chloramine-T method (4, 9). Dimethylsulphoxide (20 μl), sodium phosphate buffer 0.5 M, pH 7.0 (2 μl), tracer-free sodium ^{125}I- (5 μl), 78.5 μCi/ml, and Chloramine-T (1 μl of a 10 μg/μl solution in water) were added to 20 μl of GM-CSF$_{MLCM}$ solution (3.5 μg in Tris-HCl buffer pH 7.4, 0.01% Triton X-100, and 0.02% sodium azide). The reaction was stopped after 10 min with 1 μl of sodium metabisulfite (25 μg/μl) and 5μl of potassium iodide. Utilizing gel filtratrion on Biogel P6 and affinity chromatography with concanavalin A-Sepharose, ^{125}I-GM-CSF$_{MLCM}$ was separated from the reaction mixture.

GM-CSF ASSAYS

All assays for GM-CSF were performed in 1 ml cultures in 35 mm petri dishes containing 75,000 pooled C$_{57}$BL marrow cells. The general technique has been described elsewhere (15), but in the present experiments Dulbecco's Modified Eagle's Medium (containing 15% fetal calf serum and 5% horse serum) was used in 0.3% agar.

Because of the peculiar titration characteristics of GM-CSF$_{MLCM}$ (25), all assays were performed using serial 5-fold dilutions. For each dilution 0.1 ml was added to duplicate culture dishes before the addition of the bone marrow cells in agar medium. The GM-CSF$_{MLCM}$ was mixed with the agar medium and, after gelling, the cultures were incubated at 37°C for 7 days in a fully humidified atmosphere of 10% CO$_2$ in air.

Colony counts were performed at ×35 using an Olympus dissection microscope, scoring as colonies all discrete aggregates of 50 or more cells. The colony-stimulating activity of a sample was calculated from the linear portion of the dose-response curve.

LIQUID CULTURES OF C$_{57}$BL BONE MARROW CELLS

RNA protein and DNA synthesis were studied (2) *in vitro* using suspension cultures (19) of mouse bone marrow cells in the presence or

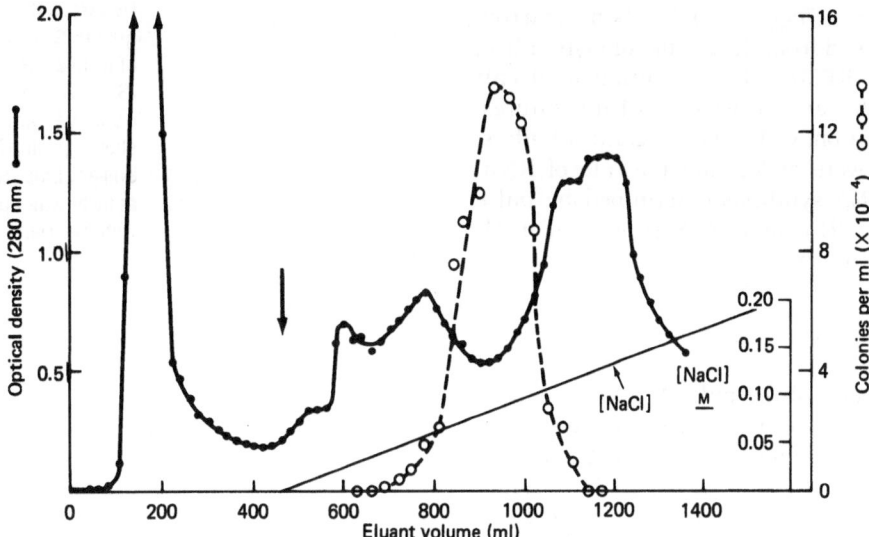

FIGURE 1. DEAE-cellulose column elution profile for CSF$_{MLCM}$ (Step III). Protein concentration (●——●), GM-CSF activity (○-----○), and the sodium chloride gradient are shown. The arrow indicates the start of the salt gradient.

FIGURE 2. Ultrogel AcA44 column elution profile. Protein concentration (●——●) and GM-CSF activity (○-----○) are shown. Blue dextran was used to determine the void volume, and the elution volumes of several marker proteins are indicated. The elution volume of GM-CSF$_{MLCM}$ corresponds to a molecular weight of 29,000 daltons.

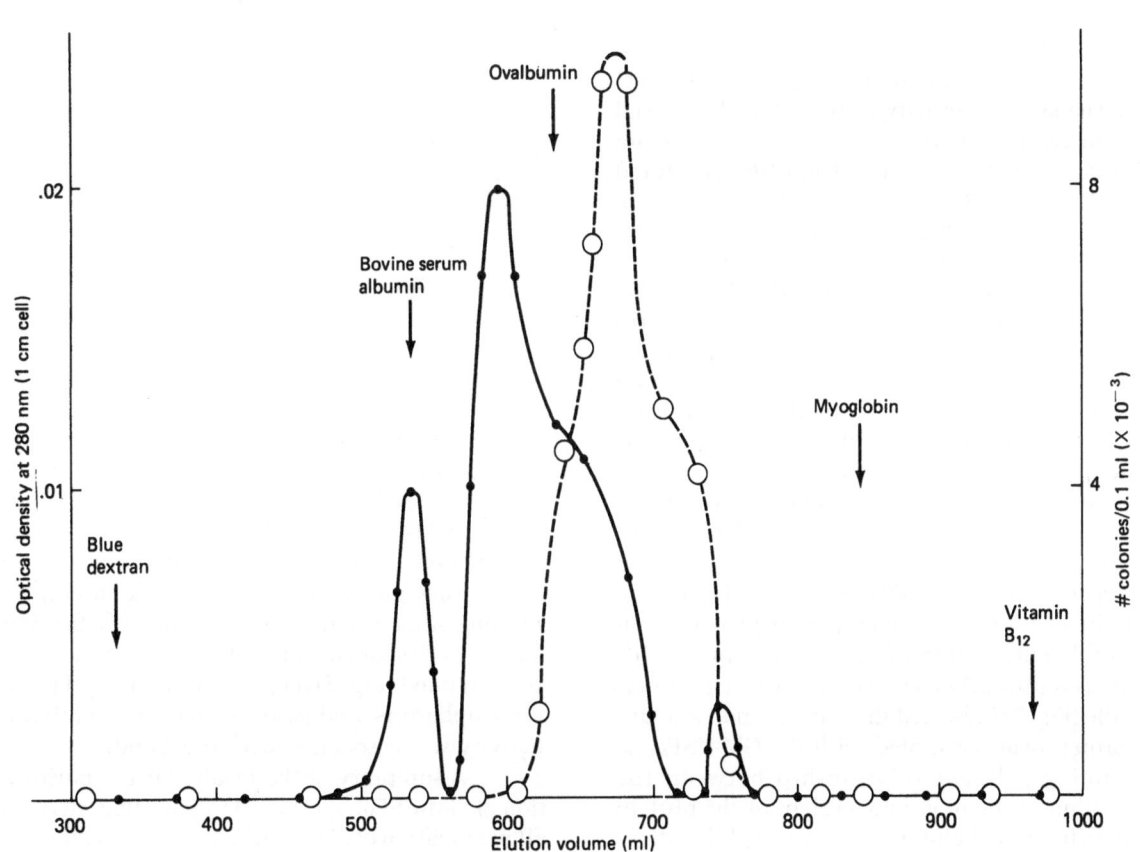

absence of GM-CSF$_{MLCM}$. Briefly, bone marrow cells were flushed from the femurs of C$_{57}$BL mice, suspended in RPMI-1640 containing fetal calf serum (10% v/v), and dispensed as 1 ml cultures. After equilibration, GM-CSF$_{MLCM}$, control buffers, or inhibitors were added and the rate of RNA, protein, or DNA synthesis determined by pulse labeling with 5-^{3}H-uridine, 4, 5-^{3}H-leucine, or ^{3}H-thymidine, respectively.

RADIOAUTOGRAPHY

At the end of the pulse labeling the cells were washed twice with normal saline, suspended in fetal calf serum, and spread on gelatin-coated glass slides. The slides were placed in absolute methanol for 30 min, washed with deionized water, air dried at room temperature, dipped in photographic nuclear track emulsion NTB2 (Eastman Kodak, Rochester, N.Y.) exposed, developed, and stained with May Grunewald and Giemsa.

RESULTS

PURIFICATION OF GM-CSF$_{MLCM}$

After dialysis against distilled water, the specific activity of Step I GM-CSF$_{MLCM}$ was 6×10^4 colonies/mg protein (24). GM-CSF$_{MLCM}$ absorbed to calcium phosphate gel and was eluted with 0.05 M and 0.1 M sodium phosphate buffer, pH 6.8. The specific activity of this Step II material was increased 4-fold as a result of the calcium phosphate gel absorption. GM-CSF$_{MLCM}$ eluted from DEAE cellulose when the NaCl was between 0.1 M and 0.16 M (Fig. 1), and the specific activity of this Step III material was 10-fold higher than Step II GM-CSF$_{MLCM}$. When Step III GM-CSF$_{MLCM}$ was chromatographed on concanavalin A-Sepharose, colony-stimulating activity was associated both with the protein failing to bind (20% of the total colony-stimulating activity) and the protein eluted with 0.05 M methyl-α-D-glucopyranoside (80% of the total colony-stimulating activity). The amount of GM-CSF$_{MLCM}$ which eluted from concanavalin A-Sepharose with methyl-α-D-glucopyranoside represented an overall recovery of 25% of GM-CSF, and had a specific activity of 2.6×10^6 colonies per mg protein. Considerable purification was achieved using gel filtration on Ultrogel AcA44. The gel filtration elution profile (Fig. 2) indicated that only a small amount of protein was associated with the GM-CSF$_{MLCM}$. Assuming a linear relationship between the logarithm of the molecular weight of the protein and its fractional elution volume on gel filtration,

FIGURE 3. Polyacrylamide gel (15%) electropherogram of purified GM-CSF$_{MLCM}$ (Step VI). Only one protein band was apparent after staining with Coomassie brilliant blue, and the biologic activity was coincident with this band.

the apparent molecular weight of GM-CSF$_{MLCM}$ was calculated to be 29,000 daltons. Although analytical polyacrylamide gel electrophoresis Step V GM-CSF eluted from Ultrogel AcA44 detected more than 12 protein bands, the specific activity had increased to 3.0×10^7 colonies per mg protein. After the final stage of the purification, Step VI GM-CSF$_{MLCM}$ appeared to run as a single protein band (Fig. 3) on polyacrylamide gel (15%) electrophoresis, and *all* of the colony-stimulating activity was associated with this band.

A summary of the results for the purification of GM-CSF$_{MLCM}$ (Table 1) shows that an overall purification of 3500-fold was achieved after the

TABLE 1   Purification of GM-CSF$_{MLCM}$

STEP	PROCEDURE	TOTAL PROTEIN (MG)	SPECIFIC ACTIVITY[a] (COLONIES × 10^{-7}/MG)	YIELD (%)
I	Conditioned medium	5,300	0.002	100
II	Calcium phosphate gel	620	0.009	50
III	DEAE-cellulose	60	0.10	50
IV	Concanavalin A-Sepharose	13	0.26	30
V	Ultrogel Ac 44	0.75	3.0	20
VI	Electrophoresis	0.10	70	7

[a]Cultures contained 75,000 C$_{57}$BL bone marrow cells. Colonies were counted after 7 days of incubation and the specific activity calculated as total number of colonies per milligram protein.

preparative gel electrophoresis. The molecular weight of purified GM-CSF$_{MLCM}$ as determined by SDS-polyacrylamide gel electrophoresis (13) was 23,000 daltons, which agreed with earlier studies which had used zone sedimentation under non-dissociating conditions (24).

BIOLOGICAL ACTIVITY OF PURIFIED GM-CSF$_{MLCM}$

Titration of Step VI GM-CSF$_{MLCM}$ in cultures of 75,000 C$_{57}$BL marrow cells gave the sigmoid dose-response curve shown in Fig. 4. It was previously noted that unpurified GM-CSF$_{MLCM}$ gave a curiously flat dose-response curve (25) compared with other forms of GM-CSF. This dose-response relationship was also observed using purified GM-CSF$_{MLCM}$ where, in the linear portion of the dose-response curve, an 8-fold reduction of GM-CSF levels was required to reduce colony numbers 2-fold.

Morphologic analysis of 7-day colonies stimulated by GM-CSF$_{MLCM}$ at the various steps of purification (Table 2) showed that during purification the material retained its capacity to stimulate both granulocytic and macrophage colony

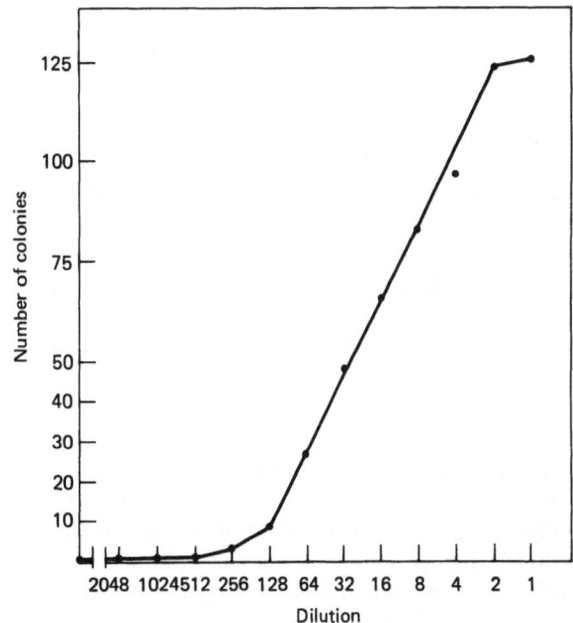

FIGURE 4.   Effect of purified GM-CSF$_{MLCM}$ (Step VI) on colony formation in cultures of 75,000 C$_{57}$BL mouse bone marrow cells. Note the fall in colony numbers with decreasing concentrations of GM-CSF$_{MLCM}$. Mean data from duplicate cultures.

TABLE 2   Morphology of Colonies Stimulated to Develop in Cultures of Mouse Bone Marrow Cells by Material from the Various Fractionation Steps in GM-CSF$_{MLCM}$ Purification

STEP	MEAN NUMBER OF COLONIES	PERCENT COLONIES		
		Granulocytic	Mixed Granulocytes and Macrophage	Macrophage
I	115	72	28	0
II	129	60	27	13
III	162	69	27	4
IV	69	40	28	32
V	164	62	34	4
VI	92	31	28	41

Cultures contained 75,000 C$_{57}$BL marrow cells, Colonies typed on day 7 of incubation.

TABLE 3 Effect of Varying the Concentration of Purified GM-CSF$_{MLCM}$ on the Frequency and Morphologic Type of Colonies Developing

GM-CSF$_{MLCM}$ DILUTION	MEAN NUMBER OF COLONIES	PERCENT COLONIES		
		Granulocytic	Mixed Granulocytic and Macrophage	Macrophage
1:1	127	51	35	14
1:2	124	50	38	12
1:4	97	45	23	32
1:8	83	33	26	41
1:16	66	42	21	37
1:32	49	19	28	53
1:64	27	10	21	69
1:128	9	0	15	85
1:256	4	0	0	100

Cultures contained 75,000 C$_{57}$BL marrow cells. Colonies analyzed on day 7 of incubation. Forty-five sequential colonies examined from cultures containing each GM-CSF$_{MLCM}$ dilution.

formation. In confirmation of the relationship seen with unpurified GM-CSF$_{MLCM}$, progressive reduction in the concentration of purified GM-CSF$_{MLCM}$ caused a concomitant fall in the proportion of pure granulocytic colonies and a progressive rise in the proportion of macrophage colonies (Table 3). Paralleling this change was a progressive reduction in colony size in cultures containing lower concentrations of the purified material.

The morphologic appearance of the colonies stimulated by purified GM-CSF$_{MLCM}$ was identical with that of colonies stimulated by unpurified material. With concentrations of Step VI GM-CSF$_{MLCM}$ 10-fold or more in excess of those stimulating plateau numbers of colonies to develop, the special multicentric granulocytic colonies were observed and comprised 5–10% of total colonies. Previous studies had shown that these multicentric granulocytic colonies only developed in the presence of high concentrations of

GM-CSF. The remaining colonies developing in cultures stimulated by Step VI material were either tight aggregates of granulocytic cells, loose colonies of macrophages, or loose colonies of macrophages with a tight central region of granulocytes (Fig. 5). As with unpurified GM-CSF material, there was an asynchronous onset of proliferation in the colony-forming cells and a marked variation in colony size. As in cultures stimulated by unpurified material, clusters (3–50 cells) outnumbered colonies by a factor of 3–5:1.

Careful inspection of the cultures failed to detect eosinophil colonies or clusters, and no megakaryocyte colonies developed regardless of the concentration of purified GM-CSF$_{MLCM}$. Control cultures with lymphocyte-conditioned medium developed the expected frequency of eosinophil and megakaryocyte colonies. No inhibition of eosinophil or megakaryocyte colony-formation was observed in cultures containing a

FIGURE 5. Orcein-stained colonies from a culture of C$_{57}$BL bone marrow cells, stimulated by purified GM-CSF$_{MLCM}$ (Step VI). Note multiple granulocytic colony (left), mixed granulocyte and macrophage colony (center), and macrophage colony (right).

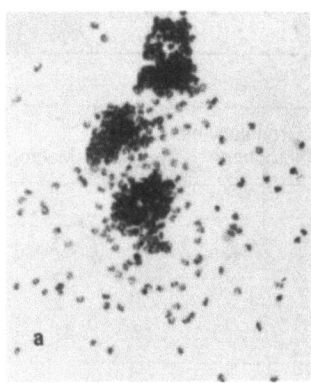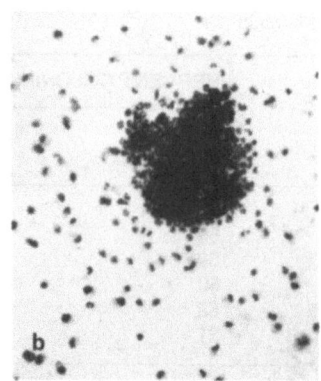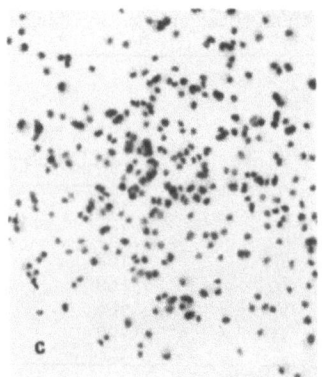

TABLE 4 Iodination of GM-CSF Purified from MLCM

| | % RECOVERY OF: | | | |
| | Biologic Activity[a] | | ^{125}I | |
DIMETHYLSULPHOXIDE (%, V/V)	Biogel P6	Con A-Sepharose	Biogel P6	Con A-Sepharose
0	0	0	25	0.2
15	5	0.3	14	0.5
50	57	20	30	5.0

[a]GM-CSF$_{MLCM}$ (7.5 μg in 50 μl of 0.03 m Tris-HCl, pH 7.4) was reacted with carrier free ^{125}I. The recoveries were calculated from the initial number of colony-stimulating units (3.7 × 10^5) and the initial amount of ^{125}I (600 μCi) in the reaction mixture.

mixture of GM-CSF from Step VI of the purification scheme and lymphocyte-conditioned medium.

In the absence of mercaptoethanol, Stage VI material failed to stimulate B-lymphocyte colony formation in cultures of C$_{57}$BL spleen cells, even in the presence of 2-mercaptoethanol; the addition of Stage VI material did not increase the number of colonies over those occurring in the presence of 2-mercaptoethanol alone.

RADIOIODINATION OF GM-CSF$_{MLCM}$

Iodination of GM-CSF$_{MLCM}$ (Step VI) using chloramine-T (4, 9) usually led to the complete loss of biologic activity (Table 4), unless sufficient dimethyl sulphoxide was present (27, 29) during the iodination reaction, whence the recovery of biologic activity increased to 60%. Only 25% of the biologic activity and 15% of the ^{125}I activity could be eluted from concanavalin A-Sepharose using 0.05 M methyl-α-D-glucopyranoside. The remaining ^{125}I bound to concanavalin A-Sepharose could be eluted with ethylene diamine tetracetic acid (1 mM) or citrate buffer (0.01 M, pH 6.0). However, only ^{125}I-GM-CSF$_{MLCM}$, specifically eluted with the sugar, was used in the serum half-life experiments.

The mean serum half-life for ^{125}I-GM-CSF$_{MLCM}$ was 6.4 ± 1.3 hr in syngeneic C$_{57}$BL mice, and 4.6 ± 1.1 hr in allogeneic BALB/c mice. Under similar experimental conditions the serum half-life of ^{125}I-mouse serum albumin was 5.0 ± 0.9 hr. The distribution of ^{125}I-GM-CSF$_{MLCM}$ throughout the organs of C$_{57}$BL mice was similar to the distribution of ^{125}I-mouse serum albumin. There was no apparent accumulation of ^{125}I-GM-CSF$_{MLCM}$ in the bone marrow or lungs at 1 or 6 hr. After 6 hr the level of ^{125}I-GM-CSF$_{MLCM}$ in the kidney was 3 times higher than the corresponding level of ^{125}I-mouse serum albumin. This was not due to an increased level of ^{125}I in the urine, which was lower in the mice injected with ^{125}I-GM-CSF$_{MLCM}$ than in the mice given ^{125}I-mouse serum albumin.

EFFECT OF GM-CSF$_{MLCM}$ ON MACROMOLECULAR SYNTHESIS IN BONE MARROW CELLS

In vitro cultures of bone marrow cells indicated that GM-CSF$_{MLCM}$ stimulated RNA synthesis within 10 min. When GM-CSF was added at the same time as the ^{3}H-uridine and the incubation continued for a further 20 min, the rate of RNA synthesis appeared to increase as much as 1.3-fold [e.g., acid insoluble ^{3}Hcpm/10^6 cells equalled 4150 ± 300 (mean ± S.D.) in control cultures, whereas this value increased to 5300 ± 100 in cultures containing GM-CSF$_{MLCM}$ (2)]. RNA synthesis was stimulated in the presence of GM-CSF$_{MLCM}$ at all times between 10 min and 24 hr, but the maximum stimulation usually occurred between 9 and 11 hr. In one such experiment the acid insoluble ^{3}Hcpm/10^6 cells was 6000 ± 300 in bone marrow cultures stimulated by GM-CSF$_{MLCM}$ for 9 hr and pulse labeled for 20 min with ^{3}H-uridine (5 μCi). The corresponding value for the control cultures was only 2300 ± 100. Thus, in this experiment GM-CSF$_{MLCM}$ stimulated RNA synthesis by 2.6-fold.

GM-CSF$_{MLCM}$ did not alter the overall viability of the bone marrow cells in these cultures. In both control and GM-CSF-stimulated cultures the viability of cells was more than 90% even after 8 hr, and by 20 hr the cell viability was still greater than 80%. In the presence of GM-CSF$_{MLCM}$, the rate of RNA synthesis in the bone marrow cultures remained constant during the first 10 hr. In the control cultures the rate of RNA synthesis decreased, with an apparent half-life of 6.3 hr. The magnitude of the stimulation of RNA synthesis by GM-CSF$_{MLCM}$ was dependent on the concentration of GM-CSF in the liquid culture (2). Stimulation of RNA synthesis was maximal when the concentration of GM-CSF$_{MLCM}$ was equal to or greater than 10^4 units per ml, but statistically significant stimulation was detectable even with concentrations of GM-CSF$_{MLCM}$ as low as 800 units per ml.

Several experiments were performed in which bovine serum albumin (2–10 mg per ml) was substituted for fetal calf serum in the culture medium. The results were similar to the experiments in which fetal calf serum was used: after 4 hr in the presence of GM-CSF$_{MLCM}$ the acid insoluble cpm/10^6 cells was 5040 ± 200, and only 3800 ± 150 in the corresponding control cultures.

Actinomycin D (0.5 μg/ml) inhibited RNA synthesis (greater than 90%) in bone marrow cultures (2), and consequently no significant stimulation was observed even in the presence of GM-CSF. Even though puromycin (100 μg/ml) inhibited protein synthesis in bone marrow cultures by more than 85%, the rate of RNA synthesis in the control cultures was stimulated by 20% (acid insoluble ^{3}Hcpm/10^6 cells in the presence of puromycin was 1300, but without puromycin was only 1100). In the presence of GM-CSF$_{MLCM}$, the effect of puromycin on RNA synthesis was similar, but less pronounced (< 5%). Cycloheximide (100 μg/ml) inhibited protein synthesis in bone marrow cultures by more than 70%, but appeared to stimulate RNA synthesis in the absence of GM-CSF by almost 70% (2).

GM-CSF$_{MLCM}$ also appeared to stimulate protein and DNA synthesis (Table 5), although neither of these effects was measurable as early as the stimulation of RNA synthesis (2). ^{3}H-leucine and ^{57}Se-methionine were used to monitor protein synthesis, and both isotopes yielded identical results. There was no significant increase in the incorporation of labeled ^{3}H-leucine into protein before 6 hr, but after 9 hr in the presence of GM-CSF, protein synthesis in bone marrow cultures had increased by 60% (Table 5). The magnitude of the stimulation of protein synthesis by GM-CSF$_{MLCM}$ increased during the first 20 hr, by which time the rate of protein synthesis in cultures containing GM-CSF was more than double the rate in the control cultures. Although there is evidence to indicate that GM-CSF stimulates GM-CFC's into DNA synthesis within 6 hr (19), it was not possible to detect a significant stimulation by GM-CSF of the rate of DNA synthesis in bone marrow cells until 16 hr. After 19 hr in the presence of GM-CSF the rate of DNA synthesis (as determined by pulse labeling with ^{3}H-thymidine) was 50% greater than the rate in the corresponding control cultures (Table 5). It was difficult to ascertain whether this was a direct stimulation of the original bone marrow cells, or whether, after 20 hr of *in vitro* culture, different cells were responding to the pulse label. The morphology of the cells after 20 hr of *in vitro* culture was quite different in the presence and in the absence of GM-CSF.

SPECIFICITY OF GM-CSF STIMULATION OF RNA SYNTHESIS

GM-CSF$_{MLCM}$ only stimulated RNA synthesis in cell suspensions which contained a significant proportion of granulocytes, monocytes, or their precursors (2). Although bone marrow cells were the most responsive to the action of GM-CSF$_{MLCM}$, RNA synthesis in peripheral blood leucocytes was also stimulated (10–20%) in the presence of GM-CSF. Peritoneal cells stimulated by GM-CSF$_{MLCM}$ for 3 hr and pulse labeled with ^{3}H-uridine (5 μCi) for 20 min yielded a value for the acid insoluble cpm/10^6 cells of 3300 ± 500,

TABLE 5 Effect of GM-CSF$_{MLCM}$ on Protein and Deoxyribonucleic Acid Synthesis[a]

ISOTOPE	TIME OF[b] STIMULATION (HR)	ACID INSOLUBLE ^{3}HCPM/10^6 CELLS		INCREASE IN RATE OF SYNTHESIS (-FOLD)[c]
		GM-CSF$_{MLCM}$	Control	
^{3}H-Leucine	3	6400 ± 300	6100 ± 400	N.S.
	9	3300 ± 200	2100 ± 100	1.6
	19	3100 ± 300	1400 ± 100	2.2
^{3}H-Thymidine	2	7000 ± 700	7800 ± 600	N.S.
	9	7200 ± 800	7500 ± 800	N.S.
	19	640 ± 40	430 ± 40	1.5

[a]Incubations were performed in triplicate with 5×10^6 bone marrow cells in 1 ml of RPMI 1640 medium containing fetal calf serum (10%, v/v).

[b]Cultures were preincubated for at least 3 hr before adding GM-CSF or control buffer. Either ^{3}H-leucine or ^{3}H-thymidine (5 μCi) was added after the time indicated and the incubation continued for 30 min.

[c]N.S. = not significant.

whereas in control cultures the corresponding measurement was only 2200 ± 400. RNA synthesis in spleen, thymus, and subcutaneous and mesenteric lymph node cells was not stimulated by GM-CSF$^{(2)}_{MLCM}$, and in several experiments GM-CSF appeared to inhibit (10%) RNA synthesis in cultures of cells from these tissues.

Radioautographic studies were performed on $C_{57}BL$ bone marrow cells to determine the morphology of the cells exhibiting increased RNA synthesis following incubation with GM-CSF$_{MLCM}$. The cells were classified into five groups: blast cells, myelocytes, metamyelocytes and polymorphs combined, lymphocyte-like cells, and nucleated erythroid cells. The most obvious effect of the incubation of bone marrow cells with GM-CSF$_{MLCM}$ was an increase in the percentage of labeled metamyelocytes and polymorphs, and an increase in their average grain count (Table 6). Metamyelocytes showed higher grain counts than polymorphs, but the effect of GM-CSF was observed for both cell types. Most blast cells and myelocytes were labeled even in the control cultures, but the average grain counts for both cell types was increased in the presence of GM-CSF$_{MLCM}$. Neither the lymphocyte-like cells nor the nucleated erythroid cells showed an increase in percentage labeling or average grain count following incubation with GM-CSF$_{MLCM}$.

DISCUSSION

The procedure described above for the purification of GM-CSF$_{MLCM}$ from mouse lung-conditioned medium led to a protein with a specific activity of 7×10^7 colonies/mg protein. Purified GM-CSF$_{MLCM}$ bound to concanavalin A-Sepharose, which confirmed that it is a glycoprotein (26). Only one band was observed on polyacrylamide gel (15% electrophoresis).

The purification scheme described in this report produces a 3500-fold purification compared with the starting conditioned medium. The level of specific activity of purified GM-CSF$_{MLCM}$ corresponds to an 80,000-fold increase over the specific activity of GM-CSF in endotoxin serum (16) and a 900,000-fold increase over the specific activity of GM-CSF in unfractionated human urine (29). It was important to include a nonionic detergent in the buffers to prevent the loss of GM-CSF during the purification.

The capacity of highly purified GM-CSF to stimulate the proliferation of both granulocytic and macrophage populations needs emphasis. Both populations appear to be generated from the same pool of progenitor cells, but these cells have been shown to be heterogeneous (18). The present results indicate that this form of CSF can stimulate both types of colony-forming cell; alternatively, it is possible that there are two or more different types of GM-CSF present even in the apparently homogeneous GM-CSF$_{MLCM}$ from Step VI of the purification.

Purified GM-CSF$_{MLCM}$ was radioiodinated in the presence of dimethyl sulphoxide (27, 29) without a significant loss of biologic activity. Only 50% of the radioiodinated GM-CSF retained its ability to bind to concanavalin A-Sepharose, which suggested that some of the carbohydrate moiety was damaged during the iodination pro-

TABLE 6 Effect on GM-CSF on ^{3}H-Uridine Labeling of Bone Marrow Cells

CELL TYPE[a]	% LABELED CELLS		MEAN GRAIN COUNT[b]		PROPORTIONAL[c] STIMULATION BY GM-CSF (-FOLD)
	GM-CSF	Control	GM-CSF	Control	
Blasts	96	100	56 ± 38	45 ± 34	1.24
Myelocytes	100	96	49 ± 30	30 ± 26	1.63
Metamyelocytes and polymorphs	84	57	21 ± 18	12 ± 6	1.75
Lymphocyte-like cells	36	39	17 ± 13	19 ± 15	0.89
Nucleated erythroid	7	4	15 ± 12	19 ± 17	0.79

[a]Cells were preincubated for 1 hr before adding GM-CSF or control buffer. After stimulation with GM-CSF for 6 hr the cultures were pulse labeled with ^{3}H-uridine (5 μCi) for 20 min. The incubation was stopped by washing the cells three times with cold normal saline (10 ml) and the cells spread on gelatine-coated slides for autoradiography. Five replicate cultures from the GM-CSF-stimulated cultures and control cultures were processed to determine the amount of ^{3}H-uridine incorporated into RNA (stimulated cultures the acid insoluble cpm/10^6 cells = 7700 ± 300 and in control cultures the acid insoluble cpm/10^6 cells = 5100 ± 200).

[b]Mean grain counts of labeled cells (5 or more grains) ± standard deviations from 100 cells.

[c]Proportional increase in grain counts following incubation with GM-CSF.

cedure. The half-life of ^{125}I-GM-CSF$_{MLCM}$ in the serum of C$_{57}$BL mice was approximately 6 hr, which was longer than that found for human urinary GM-CSF (17). This relatively short half-life would require, however, injections every 5–6 hr to maintain high levels of GM-CSF in the serum. The levels of ^{125}I-GM-CSF in the kidney indicated some selective localization in this organ, although the rate of appearance of ^{125}I in the urine did not indicate that the rate of clearance of ^{125}I-GM-CSF into the urine was higher than that of ^{125}I mouse serum albumin. The appearance of ^{125}I in the urine after injection of ^{125}I-GM-CSF was not associated with colony-stimulating activity. This finding was in agreement with the hypothesis of Sheridan and Metcalf (25) that the colony-stimulating activity associated with human urine is probably produced by the kidney rather than results from filtration of the serum CSF through the kidney.

GM-CSF$_{MLCM}$ effectively stimulated RNA synthesis in suspension cultures of mouse bone marrow cells. The kinetics of this stimulation were similar to the action of erythropoietin on rat bone marrow (6, 7, 12) and on fetal liver (14, 20). Stimulation of RNA synthesis in both systems was detected as early as 10 min, and reached apparent maximum levels after 10 hr. The larger stimulation of RNA synthesis by GM-CSF$_{MLCM}$ probably reflects the proportion of granulocytic cells in bone marrow. When similar experiments

were performed with fetal liver, erythropoietin stimulated RNA synthesis by 200–300% (14, 20), whereas GM-CSF only increased RNA synthesis by 20% (Burgess and Johnson, unpublished data).

Puromycin and cycloheximide did not inhibit RNA synthesis in cultures containing GM-CSF$_{MLCM}$. Bone marrow cultures without GM-CSF appeared to be stimulated in the presence of these inhibitors, an effect observed in several *in vitro* culture systems (10). This effect was not observed when cycloheximide was present in suspension cultures of rat bone marrow cells (7). Thus the stimulation of RNA synthesis by GM-CSF did not appear to be dependent on the prior induction of protein synthesis.

Radioautographic studies indicated that granulocytic and monocytic cells were specifically stimulated by GM-CSF (2). It was interesting that the bone marrow cells contributing most to the increased levels of RNA synthesis as a result of the action of GM-CSF were polymorphs. Polymorphs are postmitotic cells and the increased RNA synthesis observed in these cells suggests that GM-CSF, like erythropoietin (5), may be able to stimulate specific biosynthetic processes unrelated to cell division. Recent findings of Stanley et al. (28) proving that macrophage growth factor and purified GM-CSF are identical also suggest that GM-CSF is able to stimulate postmitotic cells at the end of a differentiative pathway.

SUMMARY

The factor stimulating neutrophil and macrophage colony-formation by mouse bone marrow cells (GM-CSF) was purified 3500-fold from C$_{57}$BL mouse lung-conditioned medium. GM-CSF from the final stage of the purification appeared to be homogeneous on polyacrylamide gel electrophoresis. Almost 7% of the initial GM-CSF was recovered, and purified GM-CSF had a specific activity of 7×10^7 colonies per mg protein. At low concentrations (less than 10 μg/ml) GM-CSF activity was lost from solution unless stabilizing agents such as Triton X-100 (0.01%, v/v) were included in all of the buffers. Purified GM-CSF stimulated the proliferation of all subtypes of granulocytic and macrophage colonies and exhibited concentration-dependent effects on colony numbers and morphology. In the presence of dimethylsulphoxide, GM-CSF was iodinated without loss of biologic activity. ^{125}I-GM-CSF was eluted from concanavalin A-Sepharose and used to determine its serum half-life in C$_{57}$BL mice (6.2 ± 0.8 hr). Distribution studies using ^{125}I-GM-CSF showed a preferential localization in the kidney but not in the bone marrow. GM-CSF stimulated RNA synthesis in bone marrow cells but not in spleen, thymus, or lymph node cells. RNA synthesis in bone marrow was stimulated within 10 min of adding GM-CSF, but the maximal stimulation (270%) occurred after 10 hr. Radioautographic studies indicated that GM-CSF-stimulated RNA synthesis in blast cells, myelocytes, metamyelocytes, and polymorphs, but not in small lymphocytes or cells of the erythroid series. Protein synthesis in bone marrow cells was stimulated by GM-CSF, but this effect could be detected only after several hr. No significant stimulation of DNA synthesis was observed after 9 hr, but after 16 hr the presence of GM-CSF stimulated DNA synthesis in bone marrow cells by 15%.

ACKNOWLEDGMENTS

The authors are indebted to Ms. Sue Hollins for skillful technical assistance.

This work was supported by the Queen Elizabeth II Fellowship Fund, the Anti-Cancer Council of Victoria, and the National Cancer Institute, Washington, D.C. (U.S.A.), Contract No. NOI-CB-33854.

REFERENCES

(1) Bradley, T. R., and Metcalf, D. The growth of mouse bone marrow cells *in vitro. Aust. J. Exp. Biol. Med. Sci., 44*:287, 1966.

(2) Burgess, A. W., and Metcalf, D. The effect of colony stimulating factor on the synthesis of ribonucleic acid by mouse bone marrow cells *in vitro. J. Cell. Physiol.* (in press).

(3) Davis, B. J. Disc gel electrophoresis. III. Method and application to human serum proteins. *Ann. N.Y. Acad. Sci., 121*:404, 1964.

(4) Greenwood, F. C., Hunter, W. M., and Glover, J. S. The preparation of ^{131}I-labelled human growth hormones of high specific radioactivity. *Biochem. J., 89*:114, 1963.

(5) Goldwasser, E. Erythropoietin and the differentiation of red blood cells. *Fed. Proc., 34*:2285, 1975.

(6) Gross, M., and Goldwasser, E. On the mechanism of action of erythropoietin-induced differentiation. V. Characterization of the ribonucleic acid formed as a result of erythropoietin action. *Biochemistry, 8*:1795, 1969.

(7) Gross, M., and Goldwasser, E. On the mechanism of action of erythropoietin-induced differentiation. XI. Stimulated RNA synthesis independent of protein synthesis. *Biochim. Biophys. Acta, 287*:514, 1972.

(8) Guez, M., and Sachs, L. Purification of the protein that induces cell differentiation to macrophages and granulocytes. *Fed. Eur. Bioch. Soc. Lett., 37*:149, 1973.

(9) Hunter, W. M., and Greenwood, F. C. Preparation of *i*odine-131 labelled human growth hormones of high specific activity. *Nature (London), 194*:495, 1962.

(10) Kay, J. E., and Korner, A. Effect of cycloheximide on protein synthesis and ribonucleic acid synthesis in cultures of human lymphocytes. *Biochem. J., 100*:815, 1966.

(11) Keilin, D., and Hartree, E. F. On the mechanism of the decomposition of hydrogen peroxide by catalase. *Proc. R. Soc. London (Biol.)*, B 124:397, 1938.

(12) Krantz, S. B., and Goldwasser, E. On the mechanism of erythropoietin induced differentiation. II. The effect on RNA synthesis. *Biochim. Biophys. Acta, 103*:325, 1965.

(13) Laemmli, U. K. Cleavage of structural proteins during the assembly of the head of bacterophage T4. *Nature (London), 227*:680, 1970.

(14) Maniatis, G. M., Rifkind, R. A., Bank, A., and Marks, P. A. Early stimulation of RNA synthesis by erythropoietin in cultures of erythroid precursors. *Proc. Nat. Acad. Sci., U.S.A., 70*:3189, 1973.

(15) Metcalf, D. Studies on colony formation *in vitro* by mouse bone marrow cells. II. Action of colony stimulating factor. *J. Cell. Physiol., 76*:89, 1970.

(16) Metcalf, D. Antigen-induced proliferation *in vitro* of bone marrow precursors of granulocytes and macrophages. *Immunology, 20*:727, 1971.

(17) Metcalf, D., and Stanley, E. R. Serum half life in mice of colony stimulating factor prepared from human urine. *Br. J. Haematol., 20*:549, 1971.

(18) Metcalf, D., and MacDonald, H. R. Heterogeneity of *in vitro* colony- and cluster-forming cells in the mouse marrow: Segregation by velocity sedimentation. *J. Cell. Physiol., 85*:643, 1975.

(19) Moore, M. A. S., and Williams, N. Functional, morphologic and kinetic analysis of the granulocyte-macrophage progenitor cell. In Robinson, W. A., ed., *Hemopoiesis in Culture, Second International Workshop.* Washington, D.C.: U.S. Department of Health, Education, and Welfare, publication number (NIH) 74-205, p. 17, 1973.

(20) Nicol, A. G., Conkie, G. D., Tanyon, W. G. Drewienkiewicz, C. E., Williamson, R., and Paul, J. Characteristics of erythropoietin-induced RNA synthesis from fetal mouse liver erythropoietic cultures and the effects of 5-fluorodeoxyuridine. *Biochim. Biphys. Acta, 277*:342, 1972.

(21) Ornstein, L. Disc electrophoresis. I. Background and Theory. *Ann. N.Y. Acad. Sci., 121*:321, 1964.

(22) Pluznik, D. H., and Sachs, L. The induction of clones of normal mast cells by a substance from conditioned medium. *Exp. Cell Res., 43*:553, 1966.

(23) Sheridan, J. W., and Metcalf, D. CSF production and release following endotoxin. In Robinson, W. A., Ed., *Hemopoiesis in Culture, Second International Workshop.* Washington, D.C.: U.S. Department of Health, Education, and Welfare, publication number (NIH) 74–205, p. 135, 1973.

(24) Sheridan, J. W., and Metcalf, D. Purification of mouse lung-conditioned medium colony stimulating factor (CSF). *Proc. Soc. Exp. Biol. Med., 146*:218, 1974.

(25) Sheridan, J. W., and Metcalf, D. A low molecular weight factor in lung conditioned medium stimulating granulocyte and macrophage colony formation *in vitro. J. Cell. Physiol., 81*:11, 1973.

(26) Sheridan, J. W., and Stanley, E. R. Tissue sources of the bone marrow colony stimulating factor. *J. Cell. Physiol., 78*:451, 1971.

(27) Stagg, B. H., Temperley, J. M., Rochman, H., and Morley, J. S. Iodination and the biological activity of gastrin. *Nature (London), 228*:58, 1970.

(28) Stanley, E. R., Cifone, M., Heard, P. M., and Defendi, V. Factors regulating macrophage production and growth: Identity of colony-stimulating factor and macrophage growth factor. *J. Exp. Med., 143*:631, 1976.

(29) Stanley, E. R., Hansen, G., Woodcock, J., and Metcalf, D. Colony stimulating factor and the regulation of granulopoiesis and macrophage production. *Fed. Proc., 34*:2272, 1975.

(30) Stanley, E. R., and Metcalf, D. Purification and properties of human urinary colony stimulating factor (CSF). In Harris, R., Allin, P., and Viza, D., eds., *Cell Differentiation.* Copenhagen: Munksgaard, p. 272, 1972.

Characteristics of the in Vitro Monocyte-Macrophage Colony-Forming Cells Detected within Mouse Thymus and Lymph Nodes

T. J. MacVittie and T. L. Weatherly

16

INTRODUCTION

The population of *in vitro* colony-forming cells (CFU-c) has been found to be heterogeneous with respect to several characteristic parameters such as buoyant density (10, 21, 32), sedimentation rate (29), adherence to glass beads (21), cell cycle (19, 20), and response to various colony-stimulating activities (CSA) as well as colony-enhancing factors (28, 29, 32), In addition, the degree of heterogeneity has varied with the species and age of animal (20, 23) as well as the tissue source from which the CFU-c are derived. Recently, several subpopulations or different classes of colony-forming cells have been described (5, 13, 14, 16, 32). Lin and coworkers (5, 12–15) have reported on a class of monocyte-macrophage colony-forming cell (CFC) derived from thioglycollate-induced peritoneal exudate (PE-CFC) and pleural effusion (PL-CFC) as well as from free alveolar cells. van den Engh and Bol (29) and Williams and van den Engh (32) have also reported the detection of an additional subpopulation(s) of bone marrow-derived CFU-c that responded in the presence of CSA to the addition of specific enhancing factors. These subpopulations were identified by their differential buoyant density (32) and response to the enhancing factor (28, 29). In addition, we have recently reported the detection of monocyte-macrophage CFC within the thymus (T-CFC) and lymph nodes (LN-CFC) of the mouse (16). It was found that the T-derived and LN-derived CFC shared two characteristics with those CFC derived from the peritoneal exudate, pleural effusion, and alveolar space which differed significantly from the same parameters measured for bone marrow-derived CFU-c. They were: the markedly long lag period (6–10 days) prior to initiation of colony formation, and the apparent unidirectional monocyte-macrophage line of differentiation. Two additional parameters characteristic for T-derived and LN-derived CFC were at variance with those reported for BM-derived CFU-c as well as PE-derived, PL-derived, and alveolar cell-derived CFC. Colonies derived from thymus and LN-CFC exhibited a markedly slower rate of appearance once colony formation was initiated, and colony formation was apparently dependent on the presence of specific factor(s) in pregnant mouse uterus extract (PMUE) alone (16).

The present experiments were designed to determine additional properties of this class of CFC derived from thymus (T) and lymph node (LN) in an effort to examine their relationship to those CFC derived from peritoneal exudate (PE),

147

pleural effusion, and alveolar cells as well as those CFU-c derived from bone marrow (BM), spleen (SPL), and peripheral blood leukocytes (PBL). Properties were studied such as cluster-to-colony ratio, sensitivity to the absence of PMUE in culture, radiation and drug sensitivity, and the fraction in cell cycle.

MATERIALS AND METHODS
TISSUE SOURCE

Thymuses, cervical and mesenteric lymph nodes, spleens, peripheral blood leukocytes, peritoneal exudate cells, and femoral bone marrow cells were obtained from 8- to 12-week-old male and female mice (Cumberland View Farms, Clinton, Tenn.) of strains AKR/Cum-BR and BALB/C Cum-BR, B6D2F1/Cum-BR, C57BL/6 Cum-BR, and random-bred Ha/ICR-BR. The animals were maintained on a 6:00 AM to 6:00 PM (light–dark) cycle. Wayne Lab-Blox and acidified (pH 2.5) water were available *ad libitum*. All mice were acclimated to laboratory conditions for 2 weeks. During this time they were certified free of lesions of murine pneumonia complex, and of oropharyngeal *Pseudomonas* spp. Approximate numbers of thymus, lymph node, spleen, and bone marrow cells were prepared and suspended in McCoy's 5A medium with 25 mM Hepes buffer and 15% fetal calf serum (McCoy's 5A/15% FCS). Peripheral blood leukocytes were obtained by dextran sedimentation of heparinized blood obtained by cardiac puncture of mice under ether anesthesia. Exudate cells from the peritoneal cavity were stimulated to migrate there by an injection of 2.5% thioglycollate medium (TM) 3 days prior to lavage with 5 ml of Spinner modified Hanks' balanced salt solution (HBSS) containing 5 μg heparin/ml (15). Harvested exudate cells were pooled, centrifuged, and washed two times and resuspended in McCoy's 5A/15% FCS prior to counting and dilution for culture.

IN VITRO CULTURE

The culture technique used was similar to that described by Bradley et al. (2) using the extract of pooled mouse placentae, membranes, and gravid uteri (PMUE) as the source of CSA. PMUE was prepared in the same manner, using water extraction, ammonium sulfate fractionation, dialysis, and heating. Each pooled extract was made up to a final volume so that 1 ml of extract was obtained per gram of wet weight tissue used.

Also used were L-cell conditioned medium (LCM) as prepared by Austin et al. (1), and mouse sera collected after stimulation with bacterial endotoxin (*Salmonella typhosa*, Difco, Detroit, Mich.) as described by Quesenberry et al. (24). The serum was diluted 1:6 with McCoy's 5A prior to use.

The ability of the blood leukocytes, marrow, spleen, peritoneal exudate, lymph node, and thymus cells to form colonies in culture was determined by using each of the various stimulating factors mixed (v/v) with culture medium (CMRL 1066/10% FCS + 10% trypticase soy broth + 5% horse serum and supplemented with 30 μg/ml of L-asparagine) plus 0.5% agar. The cell types were each suspended in various concentrations in an upper layer of culture medium and 0.3% agar. Cultures were incubated at 37°C in a humidified CO_2 in air atmosphere. Colonies of more than 50 cells counted after 7 to 10 days of incubation were considered derived from CFU-c, while those colonies counted after 22 to 25 days of culture were considered to be derived from CFC. In order to determine the survival of CFU-c and CFC in the absence of PMUE, a group of marrow and thymus cell cultures were prepared without PMUE. At selected days after incubation, PMUE was added to the culture dish in a 0.5 ml aliquot of 0.33% agar:media. Control cultures were initiated immediately after they had jelled. Respective marrow-derived and thymus cell-derived cultures were then incubated for an additional 10 to 24 days after addition of PMUE.

COLONY MORPHOLOGY

Morphology was determined by removing individual colonies from the agar by means of a pasteur pipette and placing in a small volume of McCoy's 5A/15% FCS. Glass slides were prepared through use of a cytospin centrifuge (Shandon Southern Instruments Limited, Sewickley, Penn.), air-dried, and stained with a Wright-Giemsa solution.

CELL CYCLE STATUS OF COLONY-FORMING CELLS

In vitro cell killing with tritiated thymidine (^{3}H-TdR). BM-,T-, and LN-derived cell suspensions were incubated at a concentration of 4–5×10^6 cells/ml in HBSS at 37°C for 10 min. Two hundred microcuries of ^{3}H-TdR (specific activity 30 Ci/mM, New England Nuclear, Boston, Mass.) were added per ml of culture volume and incubated for an additional 20 min. Duplicate cell suspensions were incubated in the presence of 20 μg/ml of unlabeled thymidine. After incubation all culture tubes were diluted with 10 ml of ice-cold HBSS containing 20 μg/ml of thymidine and centrifuged at 1000 g for 10 min. Cell pellets were washed three times with ice-cold HBSS, counted, and cultured in agar at several concentrations.

In vivo killing with hydroxyurea (HU). Mice were injected i.p. with hydroxyurea (Aldrich Chemical Co., Milwaukee, Wis.) (900 mg/kg) 2 hr prior to sacrifice and removal of the T, LN, SPL, and BM for preparation of cell suspensions for *in vitro* culture. Peritoneal cells were collected by lavage with Spinner modified HBSS (5 μg heparin/ml) 2 hr after injection with HU. All cell suspensions were cultured at several concentrations.

IRRADIATION AND DRUG SENSITIVITY

Whole body radiation (WBI) of mice was performed by midline bilateral exposure from the Armed Forces Radiobiology Research Institute ^{60}Co source at a dose rate of 42rad/min. Cyclophosphamide (CY) (Cytoxan, Mead Johnson, Evansville, Ind.) and Vinblastine (VLB) ("Velbe", Lilly and Company, Indianapolis, Ind.) were administered i.v. as a single injection. Animals were

euthanatized 24 hr after radiation or exposure to cytotoxic drugs; tissues were removed and prepared for culture.

RESULTS

RELATIVE INCIDENCE OF CFC, CFU-c AND CLUSTER-TO-COLONY RATIOS

T-CFC and LN-CFC had significantly lower concentrations relative to both TM-stimulated PE-CFC as well as BM-derived and SPL-derived CFU-c. In addition, T-derived and LN-derived cultures had markedly lower cluster-to-colony ratios (0.8 to 1.4) than those observed for PE-CFC cultures (5.8) and those CFU-c cultures derived from BM (6.4), SPL (3.6), and PBL (10.8) (Table 1).

Significant differences were observed in the number of T-CFC in several strains of mice (Table 2). The AKR strain had a greater concentration ($p < 0.001$) while BALB/C and C57BL/6 were lower ($p < 0.001$) than Ha/ICR and B6D2F$_1$ strains. On an organ basis, only the T-CFC in BALB/C and C$_{57}$BL/6 remained less than the other strains tested.

SURVIVAL OF T-CFC AND BM CFU-c IN THE ABSENCE OF PMUE

T-CFC were markedly resistant to the absence of PMUE in culture (Fig. 1). The more sensitive BM CFU-c were reduced to 60% of control

TABLE 1 The Frequency and Cluster-to-Colony Ratios of Colony-Forming Cells (CFC, CFU-c) from Various Hematolymphopoietic Sites in the Mouse[a]

	COLONIES PER 10^6 CELLS	COLONIES PER ORGAN	CLUSTER-TO-COLONY RATIO
CFC			
Peritoneal exudate, TM	27.3 ± 4.7	2,317 ± 369[c]	5.8 (4)[b]
Peritoneal exudate, normal	8 ± 3	18 ± 4	(4)
Thymus	31 ± 4	2,506 ± 295	0.8 (20)
Cervical lymph node	22 ± 4	915 ± 267	1.3 (18)
Mesenteric lymph node	20 ± 6	329 ± 93	1.4 (12)
CFU-c			
Bone marrow (femur)	1,802 ± 122	39,284 ± 3,902	6.4 (15)
Spleen	62 ± 9	6,363 ± 1,168	3.6 (16)
Peripheral blood	10 ± 5	74 ± 17[d]	10.8 (8)

[a]PMUE used at 3.0% (final v/v) of culture medium-agar mix. Mean values ± SEM. Peritoneal exudate values × 10^3. B6D2F1 male and female mice.

[b]Number of experiments.

[c]Expressed as total CFC harvested from peritoneal cavity.

[d]Expressed as CFU-c per milliliter of blood.

TABLE 2 The Incidence of Colony-Forming Cells (CFC) and Cell Clusters in Thymus Cell Suspensions from Various Strains of Mice[a]

MOUSE STRAIN	CFC PER 10⁶ CELLS ± SEM	CFC PER ORGAN ± SEM	CLUSTER-TO-COLONY RATIO
AKR/Cum	41.4 ± 2.2	3,188 ± 237	0.9 (7)[b]
Ha/ICR	31.6 ± 4.2	2,647 ± 346	1.1 (14)
B6D2F1/Cum	31.9 ± 3.6	2,506 ± 295	0.8 (20)
B6CBF1/Cum	26.7 ± 5.1	2,136 ± 320	0.8 (10)
BALB/c Cum	12.2 ± 3.3	1,013 ± 137	1.4 (12)
C57BL/6 Cum	19.8 ± 6.0	1,531 ± 125	1.1 (12)

[a]PMUE used at 3.0% (final v/v) of culture medium-agar mixture.

[b]Number of experiments.

within 24 hr of culture without PMUE, whereas it took 6 days without PMUE for an equivalant effect with T-CFC.

APPEARANCE OF COLONIES IN CULTURE

The marked delay (6–10 days) in appearance of colonies derived from T-CFC and LN-CFC was reported earlier (16) and is similar to that previously observed for PE-CFC (15) and confirmed here (Fig. 2). Once colony formation is initiated, the rate of appearance of colonies derived from T-CFC and LN-CFC is much slower than that observed for colonies derived from PE-CFC. The rate of appearance of colonies from PE-CFC is not significantly different from that observed for BM, SPL, and PBL colonies derived from CFU-c. The number of T-CFC-derived and LN-CFC-derived colonies doubled every 58 and 42 hr, respectively. The number of PE-CFC and colonies derived from BM, SPL, and PBL CSU-c doubled within a range of every 17 to 23 hr.

FIGURE 1. The surviving fraction of bone marrow- (o) and thymus- (●) derived CFU-c versus time in culture in the absence of PMUE. Values are individual means of four culture dishes from each experiment.

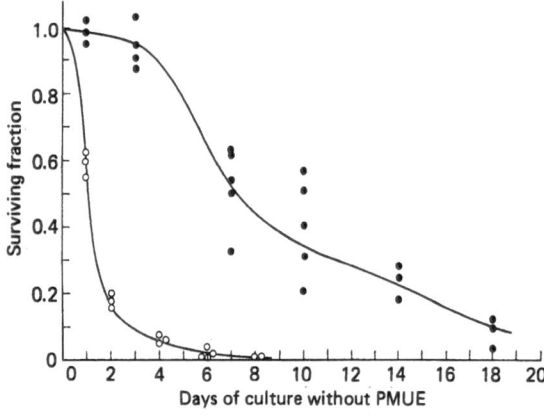

FRACTION OF COLONY-FORMING CELLS IN S-PHASE OF THE CELL CYCLE

The fraction of T-CFC in S-phase of the cell cycle did not differ significantly from that fraction measured for BM or SPL CFU-c when measured either by the ³H-TdR or the HU techniques (Table 3). Mean values of the percentage reduction using ³H-TdR ranged from 37.0 to 43.8%, while mean values using HU ranged from 37.0 to 40.1%. The percentage reduction in colony formation of PE-CFC was slightly higher, at 48.9% of control values.

COMPARATIVE RADIOSENSITIVITY OF CFU-c DERIVED FROM BM AND SPL AND CFC DERIVED FROM PE, T, AND LN CELLS

The radiosensitivity of T-CFC, LN-CFC, and PE-CFC did not differ significantly from that observed from BM CFU-c (Fig. 3). Thymus-CFC and LN-CFC, however, were found to be significantly more radiosensitive than PE-CFC and PL-CFC (5). The T-CFC and LN-CFC had respective D_0 values of 85 ± 5 rad (SEM) and 80 ± 10 rad, while PE-CFC and PL-CFC had reported D_0 values of 117 rad and 116 rad, respectively. The D_0 value for PE-CFC observed in our laboratory was 100 ± 7 rad. Spleen CFU-c were consistently more radiosensitive than BM-derived CFU-c.

SENSITIVITY OF T-CFC, BM CFU-c, AND SPL CFU-c TO CYCLOPHOSPHAMIDE (CY) AND VINBLASTINE (VLB)

CY is an agent classified as proliferation-dependent in its action on target cells (4). The dose-survival curves are exponential in form through 300 mg/kg of body weight of CY (Fig. 4). No significant difference in sensitivity was ob-

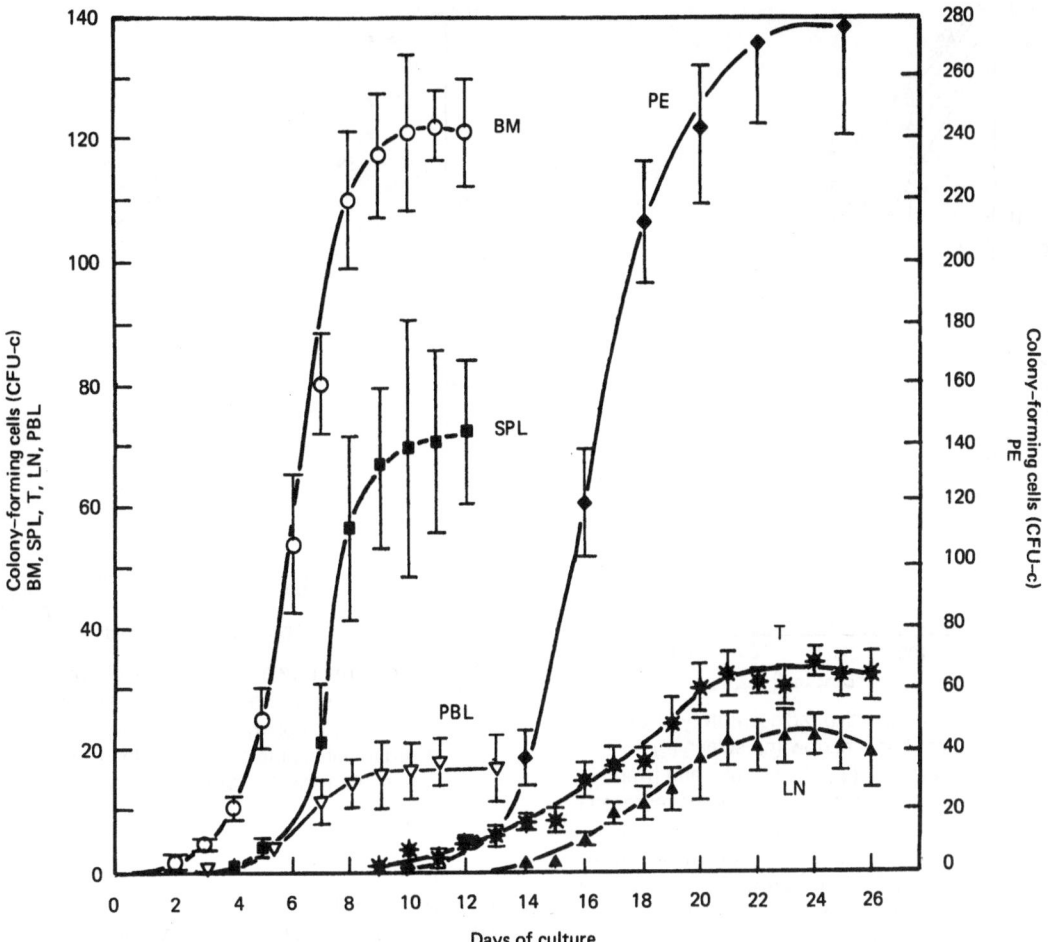

FIGURE 2. The appearance of CFU-c derived from 5 × 10⁴ bone marrow (○), 1 × 10⁶ spleen (■), 1 × 10⁶ peripheral blood leukocytes (▽), 1 × 10⁴ peritoneal exudate stimulated with thioglycollate (◆), 1 × 10⁶ thymus (*), and 1 × 10⁶ cervical lymph node (▲) cell suspensions. Values are means (± SEM) from four to six replicate experiments.

TABLE 3 Tritiated Thymidine (*in Vivo*) and Hydroxyurea (HU) (*in Vitro*) Killing of Colony-Forming Cells Derived from Bone Marrow, Spleen, Thymus, and Peritoneal Exudate Cultures[a]

BONE MARROW		SPLEEN		THYMUS		PERITONEAL EXUDATE[b]
HU	³H-TdR	HU	³H-TdR	HU	³H-TdR	HU
		Percent Depression in Colony Formation				
37.3	25.0	38.6	45.8	39.1	48.1	55.5
38.6	38.7	35.8	40.2	39.3	43.1	44.2
54.5	48.2	41.7	30.6	40.6	33.3	46.9
42.0	47.3	39.1	31.3	39.0	34.3	—
46.4	46.2	—	—	—	—	—
		Mean Values (+ SEM)				
43.8 ± 4.2	40.1 ± 4.6	38.8 ± 1.4	37.0 ± 3.7	37.0 ± 2.8	39.5 ± 3.6	48.9 ± 3.9

[a]HU values (± SEM) are percent depression per organ (femur, spleen, and thymus) with the exception of peritoneal exudate (percent depression per 10⁴ CFC). ³H-TdR values are percent depression per concentration of CFU-c, CFU in culture (BM, 10⁵ cells; SPL, 10⁶ cells; T, 10⁶ cells).

[b]Peritoneal CFC stimulated to peritoneal site by injection of thioglycollate medium 3 days prior to harvest.

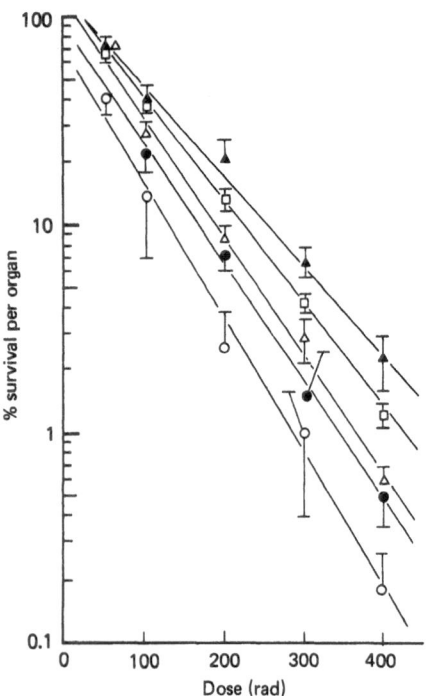

FIGURE 3. The comparative radiosensitivity of CFU-c derived from bone marrow (□), spleen (○), peritoneal exudate (▲), thymus (△), and lymph node (●) cell suspensions. Each point represents the mean percentage survival of CFU-c per organ (± SEM) of four to 10 replicate experiments. Doses in rads were 90 for bone marrow, 70 for spleen, 85 for thymus, 80 for lymph node, and 100 for peritoneal exudate.

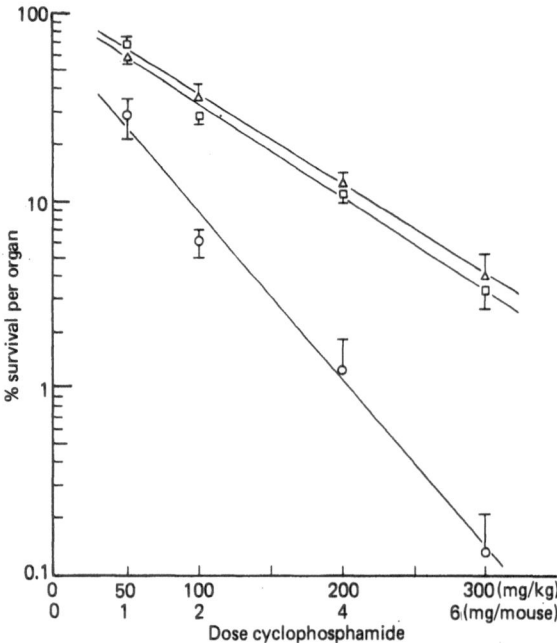

FIGURE 4. The percentage survival of bone marrow-derived (□), spleen-derived (○), and thymus-derived (△) colony-forming cells (CFU-c, CFC) per organ versus the dose of cyclophosphamide. Values are means (± SEM) of three replicate experiments.

served between BM-derived CFU-c and T-derived CFC, while the SPL CFU-c showed a significantly greater sensitivity to the drug. An LD_{10} dose of 200 mg/kg resulted in 10–12% survival for BM CFU-c and T-CFC, while SPL CFU-c survival was approximately 1%.

VLB is phase specific in its mode of action (4). The survival curves decreased exponentially with low doses of VLB, then approached constant values of survival with dose at a saturation concentration of VLB (Fig. 5). The percentage survival of colony-forming cells per organ at maximum drug concentration was 12% for BM CFU-c, 2% for T-CFC, and 0.3% for SPL CFU-c.

DISCUSSION

In these experiments, we compared the cluster-to-colony ratios, survival in the absence of CSA, radiosensitivity, sensitivity to the proliferation and cell cycle, phase-specific drugs, CY and VLB, respectively, and the fraction in cell cycle of T-derived and LN-derived CFC with the same

parameters for PE-CFC and CFU-c derived from BM, SPL, and PBL (Table 4). We recently reported (16) their low incidence, singular direction toward monocyte-macrophage differentiation, apparent sole specificity of PMUE to initiate colony formation, the long delay prior to colony formation, and the more gradual appearance of colonies in culture compared to BM-derived CFU-c. In addition the present data indicate that the T-CFC and LN-CFC were significantly different from CFU-c derived from BM, SPL, and PBL with respect to their survival in the absence of CSA and their cluster-to-colony ratios. Similarities with the CFU-c were observed in radiosensitivity, response to cyclophosphamide, and the fraction in cell cycle. In comparison with the CFC derived from peritoneal exudate, two additional similarities were observed: namely, the marked survival in the absence of CSA and the same fraction in cell cycle. Thymus and LN-CFC were, however, significantly lower in cluster-to-colony ratio and more sensitive to ^{60}Co γ radiation than PE-CFC.

The similarities between T-CFC, LN-CFC and PE-CFC, PL-CFC and alveolar CFC with respect to (a) the long lag time prior to colony formation, (b) singular production of monocyte-macrophage colonies, (c) survival in the absence

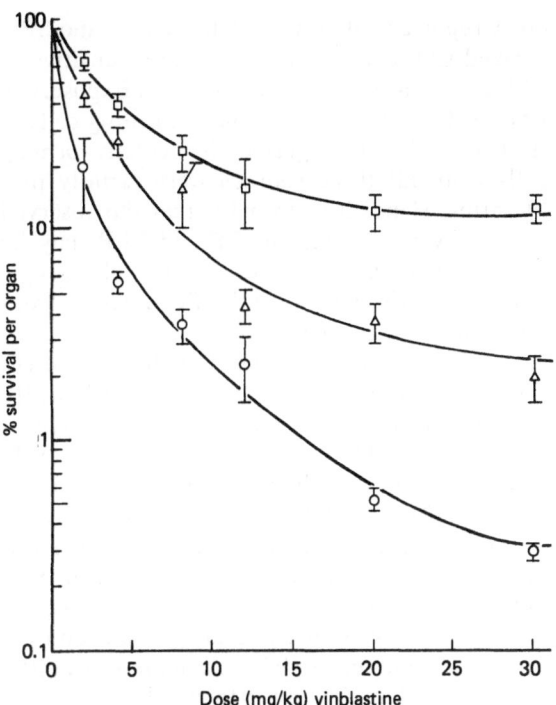

FIGURE 5. The percentage survival of bone marrow-derived (□), spleen-derived (○), and thymus-derived (Δ) colony-forming cells (CFU-c, CFC) per organ, versus the dose of vinblastine. Values are means (± SEM) of three replicate experiments.

of CSA, and (d) the fraction in cell cycle support the contention that these CFC are probably of the same class or population of progenitor cells (5, 16). A number of investigators have shown that monocyte-macrophage cells detected within inflammatory exudate, peritoneal exudate, and

lung alveoli have their origin in a bone marrow precursor (7, 8, 30, 31). It is most likely that the CFC have a similar origin in the heterogeneous bone marrow CFU-s and/or CFU-c population(s). The granulocyte-to-macrophage transition is a characteristic feature of the CFU-c culture system (20). If we assume the existence of a common progenitor cell for both granulocyte and macrophage cell lines, the persistence of monocyte-macrophage morphology in CFC cultures may be a product of the inductive influence of lymphoid organ specific microenvironment and/or locally produced factors to commit the heterogeneous marrow-derived CFU-c to a single line of differentiation. The same influence might be expected within inflammatory tissue, the serous cavities, and the alveolar tissue.

The relatively low incidence of T-CFC and LN-CFC as well as the low cluster-to-colony ratio of approximately 1.0, would be indicative of a resident population of progenitor cells with a limited monocytopoietic capacity within the lymphoid microenvironment. This is based on the assumption that cell clusters are the near progeny of the CFC, a relationship similar to that ascribed to clusters forming in CFU-c cultures (17). Cultures derived from CFU-c within BM, SPL, and PBL as well as those derived from PE-CFC were characterized by larger cluster-to-colony ratios ranging from 3.6 to 10.8 (Table 1). The higher ratios were characteristic of cell populations with extensive granulo- and/or monocytopoiesis. A similar situation may be found within certain fetal tissues. CFU-c are found within the fetal yolk sac and liver. Those organs, however, are characteristi-

TABLE 4 Characteristic Parameters of Colony-Forming Cells[a]

	CFC			CFU-c		
	T	LN$_c$	PE	PBL	SPL	BM
Incidence per 10^6	31	22	27,300	10	63	1,800
Morphology[b]	M	M	M	G-M	G-M	G-M
CSA[c]	A	A	A,B,C,+	A,B,C,+	A,B,C,+	A,B,C,+
Cluster-to-colony ratio	1.0	1.0	5.8	10.8	3.6	6.4
Lag time (days)	9–12	13–15	10–12	3	3	2
Rate of colony appearance[d]	58	43	22	28	17.5	17.5
Survival minus CSA[e]	60%	—	40%	—	—	3%
Radiosensitivity, D$_0$ (rad)	85	80	100	—	70	90
Cell cycle status	Rapid	Rapid	Rapid	—	Rapid	Rapid

[a]B6D2F1, adult, male and female mice. Source: T = thymus, LN$_c$ = cervical lymph node, PE = thioglycollate-stimulated peritoneal exudate, PBL = peripheral blood leukocytes, SPL = spleen, BM = bone marrow.

[b]Morphology: M = monocyte-macrophage, G = granulocyte.

[c]CSA, A = PMUE, B = L-cell conditioned media, C = endotoxin-stimulated mouse sera, + = other CSA's active.

[d]Rate of colony appearance in culture as doubling time (hours) of colonies per culture.

[e]Percentage survival of CFC and CFU-c after 6 days in absence of CSA PE = CFC value (ref. 13).

cally nongranulopoietic and subsequently have low cluster-to-colony ratios of approximately 2.0 (20, 23).

Thymus CFC survival in the absence of PMUE was remarkably similar to that observed for the peritoneal exudate CFC in the absence of L-cell conditioned medium (15). Both showed relatively little sensitivity to the absence of stimulating factor during the first 3 days of incubation, during which time marrow CFU-c had decreased to less than 20% of control values. The marked sensitivity of marrow CFU-c is in agreement with that previously shown by Metcalf (18). It took approximately five times as long (15 days) in the absence of stimulating factor to reduce colony formation in thymus and peritoneal exudate cultures to values 20% of their control.

The inordinate lag period prior to colony formation, marked survival in the absence of PMUE, and slower rate of appearance in culture indicated that perhaps the CFC population detected within the T and LN had proliferative characteristics different from those observed for BM- and SPL-derived CFU-c. Bruce and coworkers (4) have classified cyclophosphamide and vinblastine as agents which are proliferation-dependent and phase-specific (mitosis) in their respective cytotoxic effects on target cells. We used these agents in addition to the S-phase specific, high specific activity tritiated thymidine and hydroxyurea to determine the proliferative state of the colony-forming cell population detected within the thymus relative to PE-CFC and CFU-c within Femoral BM and SPL.

The marked depression of colony-forming cells by *in vitro* ^{3}H-TdR and *in vivo* HU incorporation suggested that the T-CFC and PE-CFC are in a state of rapid cycle similar to that observed for BM and SPL CFU-c. Spleen-derived CFU-c, however, have been reported to be less sensitive to ^{3}H-TdR incorporation (19) and therefore have a lesser percentage of the CFU-c in active cycle. This is in contrast to our results using the B6D2F$_1$ strain of mouse which showed no significant difference between the reduction of CFU-c derived from BM and SPL.

The response of T-CFC to cyclophosphamide did not differ from that of the BM-derived CFU-c, while SPL CFU-c showed a marked sensitivity to the action of the drug. Marrow CFU-c and T-CFC were reduced to 10–12% of control values by a dose of 200 mg/kg. This falls within the response observed for BM CFU-c of 7–30% survival noted by Brown and Carbone (3) and Millard et al. (22), respectively. Although the sensitivity of BM CFU-c to VLB was within previ-

ously reported values (3, 22), the T-CFC and SPL-derived CFU-c were markedly more sensitive to VLB at the saturation concentration. Exposure of BM-, SPL- and T-derived colony-forming cells to ^{3}H-TdR and HU suggested that colony-forming cells from all three sources were rapidly proliferating. Hence it is probable that the observed variability in response to CY and VLB may be accounted for by differences in the phenotypic makeup of the CFU-c and CFC derived from different tissue sources.

The radiosensitivities (D$_0$ values) of T-CFC, LN-CFC, and PE-CFC were not significantly different from that of marrow CFU-c. Thymus CFC and LN-CFC were, however, significantly more radiosensitive than PE-derived CFC (Table 4). LIN (12) and CHU and LIN (5) recently reported a similar radiosensitivity between BM CFU-c and PE- and PL-derived CFC. Only the SPL CFU-c proved to be consistently more radiosensitive in terms of both D$_0$ value (70·rad) and extrapolation number (<0.7). Guzman and Lajtha (9) observed a similar relationship between pluripotent stem cells (CFU-s) derived from marrow and spleen. They reported D$_0$ values of 69 rad and 82.5 rad with extrapolation numbers of 0.8 and 1.5 for SPL- and BM- derived CFU-s, respectively. The SPL-derived CFU-c may retain the characteristic sensitivity to radiation and drugs through a parent-progeny relationship, and/or the marked sensitivity may be solely a function of the splenic microenvironment which altered the phenotype of its resident stem and progenitor cells. This heterogeneity within a class of progenitor or stem cells was also observed for the T-CFC and LN-CFC. They differed from those CFC derived from peritoneal exudate, pleural effusion, and alveolar space in terms of (a) the sole specificity of PMUE to initiate colony formation (16), (b) their low cluster-to-colony ratio, (c) their greater radiosensitivity, and (d) their more gradual appearance in culture (16).

These differences observed for progenitor cells assumed to be within the same population emphasize the fact that the colony detected in the culture plate—just as the colony observed in the *in vivo* spleen assay—is not homogeneous with respect to several characteristic properties. The pluripotent stem cell (6, 9, 11, 25, 26), the granulocyte-monocyte progenitor cell (10, 19–21, 27, 32), and monocyte-macrophage progenitor cells are heterogeneous populations. These differences may in part reflect the influence of tissue specific microenvironment on determining the phenotypic expression of resident stem and/or progenitor cells.

154

SUMMARY

The enigmatic presence of *in vitro* colony-forming cells (CFC) within the thymus (T) and lymph node (LN) organs prompted us to determine additional characteristics of this cell population in an effort to examine their relationship to other colony-forming cells (CFC, CFU-c) derived from several hematopoietic sites. Their cluster-to-colony ratio, survival in the absence of colony-stimulating activity (CSA), radiosensitivity, drug sensitivity, and fraction in cell cycle were compared to the same parameters for CFU-c derived from bone marrow (BM), spleen (SPL), and peripheral blood leukocytes (PBL) and those CFC derived from thioglycollate-stimulated peritoneal exudate (PE). When compared with the parameters characteristic of CFU-c derived from BM, SPL, and PBL, the T-CFC and LN-CFC differed markedly in cluster-to-colony ratio, sensitivity to absence of pregnant mouse uterus extract (PMUE) in culture, and rate of appearance of colonies in culture. Similarities were observed in sensitivity to radiation (D_0 value) and cytotoxic drugs as well as the fraction in cell cycle. In comparison with the CFC derived from the peritoneal exudate, two similarities were observed; namely, the marked survival in the absence of CSA in culture and the same fraction in cell cycle. However, T-CFC and LN-CFC were significantly lower in cluster-to-colony ratio and more sensitive to ^{60}Co γ radiation than PE-CFC.

Data are provided on the nature of a subpopulation of the ubiquitous monocyte-macrophage CFC located in the thymus and lymph nodes of the mouse. The results indicated that the CFC is a heterogenous population in several respects and support the implication that the tissue microenvironment may alter the phenotypic expression of the resident CFC population.

ACKNOWLEDGMENTS

The authors wish to acknowledge the excellent technical assistance of Ms. E. McCarthy, Mr. R. Brandenburg, Mr. J. Atkinson, and HMC R. Mitchell.

REFERENCES

(1) Austin, P. E., McCulloch, E. A., and Till, J. E. Characterization of the factor in L-cell conditioned medium capable of stimulating colony formation by mouse marrow cells in culture. *J. Cell. Physiol.*, 77:121, 1971.

(2) Bradley, T. R., Stanley, E. R., and Sumner, M. A. Factors from mouse tissues stimulating colony growth of mouse bone marrow cells *in vitro*. *Aust. J. Exp. Biol. Med. Sci.*, 49:595, 1971.

(3) Brown, C. H., and Carbone, P. P. Effects of chemotherapeutic agents on normal mouse bone marrow growth *in vitro*. *Cancer Res.*, 31:185, 1971.

(4) Bruce, W. R., Meeker, B. E., and Valeriote, F. A. Comparison of the sensitivity of normal hematopoietic and transplanted lymphoma colony-forming cells to chemotherapeutic agents administered *in vivo*. *J. Natl. Cancer Inst.*, 37:233, 1966.

(5) Chu, J.-Y., and Lin, H. Induction of macrophage colony-forming cells in the pleural cavity. *J. Reticuloendothel. Soc.*, 20:299, 1976.

(6) Gidali, J., Feher, I., and Antal, S. Some properties of the circulating hemopoietic stem cells. *Blood*, 43:573, 1974.

(7) Godleski, J. J., and Brain, J. D. The origin of alveolar macrophages in mouse radiation chimeras. *J. Exp. Med.*, 136:630, 1972.

(8) Goodman, J. W. On the origin of peritoneal fluid cells. *Blood*, 23:18, 1964.

(9) Guzman, E. and Lajtha, L. G. Some comparisons of the kinetic properties of femoral and splenic hemopoietic stem cells. *Cell Tissue Kinet.*, 3:91, 1970.

(10) Haskill, J. S., McKnight, R. D., and Galbraith, P. R. Cell-cell interaction *in vitro*: Studied by density separation of colony-forming, stimulating, and inhibiting cells from human bone marrow. *Blood*, 40:394, 1972.

(11) Kretchmar, A. L., and Conover, W. R. A difference between spleen-derived and bone marrow-derived colony-forming units in ability to protect lethally irradiated mice. *Blood*, 26:772, 1970.

(12) Lin, H. S. Peritoneal exudate cells. III. Effect of gamma-irradiation of mouse peritoneal colony-forming cells. *Radiat. Res., 63*:560, 1975.

(13) Lin, H. S., Kuhn, D., and Kuo, T. T. Clonal growth of hamster free alveolar cells in soft agar. *J. Exp. Med., 142*:877, 1975.

(14) Lin, H. S., and Stewart, C. C. Colony formation by mouse peritoneal exudate cells *in vitro. Nature (London), 243*:176, 1973.

(15) Lin, H. S. Peritoneal exudate cells. I. Growth requirement of cells capable of forming colonies in soft agar. *J. Cell. Physiol., 83*:369, 1974.

(16) MacVittie, T. J., and McCarthy, K. The detection of *in vitro* monocyte-macrophage colony forming cells in mouse thymus and lymph nodes. J. Cell. Physiol., In Press, 1977.

(17) Metcalf, D. Studies on colony formation *in vitro* by mouse bone marrow cells. I. Continuous cluster formation and relation of clusters to colonies. *J. Cell. Physiol., 74*:323, 1969.

(18) Metcalf, D. Studies on colony formation *in vitro* by mouse bone marrow cells. II. Action of colony stimulating factor. *J. Cell. Physiol., 76*:89, 1970.

(19) Metcalf, D. Effect of thymidine suiciding on colony formation *in vitro* by mouse hematopoietic cells. *Proc. Soc. Exp. Biol. Med., 139*:511, 1971.

(20) Metcalf, D., and Moore, M. A. S. In Neuberger, A. and Tatum, E. L., eds, *Hemopoietic Stem Cells.* Amsterdam, North Holland Publishing Co., Chap. 2, pp. 33–40, Chap. 3, pp. 123–140, 1971.

(21) Metcalf, D., and Shortman, K. Adherence column and buoyant density separation of bone marrow stem cells and more differentiated cells. *J. Cell. Physiol., 78*:441, 1971.

(22) Millard, R. E., Blackett, N. M., and Okell, S. F. A comparison of the effect of cytotoxic agents on agar colony forming cells, spleen colony forming cells and the erythrocyte repopulating ability of mouse bone marrow. *J. Cell. Physiol., 82*:309, 1973.

(23) Moore, M. A. S., and Williams, N. Analysis of proliferation and differentiation of foetal granulocyte-macrophage progenitor cells in hemopoietic tissue. *Cell Tissue Kinet., 6*:461, 1973.

(24) Quesenberry, P., Morley, A., Stohlman, F. Jr., Richard, K., Howard, D., and Smith, M. Effect of endotoxin on granulopoiesis and colony stimulating factor. *N. Engl. J. Med., 286*:227, 1972.

(25) Schofield, R. A comparative study of the repopulating potential grafts from various hemopoietic sources: CFU repopulation. *Cell Tissue Kinet., 3*:119, 1970.

(26) Siminovitch, L., Till, J. E., and McCulloch, E. A. Radiation responses of hemopoietic colony-forming cells derived from different sources. *Radiat. Res., 24*:482, 1965.

(27) Sutherland, D., Till, J. E., and McCulloch, E. A. Short-term cultures of mouse marrow cells separated by velocity sedimentation. *Cell Tissue Kinet., 4*:479, 1971.

(28) van den Engh, G. J. Quantitative *in vitro* studies on stimulation of murine haemopoietic cells by colony stimulating factor. *Cell Tissue Kinet., 7*:537, 1974.

(29) van den Engh, G. J., and Bol, S. The presence of a CSF enhancing activity in the serum of endotoxin-treated mice. *Cell Tissue Kinet., 8*:479, 1975.

(30) Virolaisen, M. Hematopoietic origin of macrophages as studied by chromosome markers in mice. *J. Exp. Med., 127*:943, 1968.

(31) Volkman, A., and Gowans, J. L. The production of macrophages in the rat. *Br. J. Exp. Pathol., 46*:50, 1965.

(32) Williams, N., and van den Engh, G. J. Separation of subpopulations of *in vitro* colony forming cells from mouse marrow by equilibrium density centrifugation. *J. Cell. Physiol., 86*:237, 1975.

INTRODUCTION

The hemopoietic tissues of mice and men contain a cell population which gives rise to clones of granulocytes and monocytes in culture when exposed to a proper proliferation stimulus (3, 16). These cells, which occur in low numbers in the bone marrow, are designated as colony-forming units culture (CFU-c). The inducer of colony formation has been shown to be present in many tissue extracts, and is designated as colony-stimulating factor or CSF.

In vivo, the incidence and the kinetics of CFU-c are related to the size of the pool of pluripotent stem cells (26). The latter cell type can be detected by means of an *in vivo* spleen colony assay, and is therefore referred to as colony-forming unit spleen (CFU-s). The coherent behavior of CFU-s and CFU-c indicates that there is a close relationship between them.

When CFU-s and CFU-c are compared with respect to physiologic and physical characteristics, some differences in their properties become apparent. The two cell types have been shown to differ in their density distributions and in their sedimentation rate distributions. The sensitivities of these cells to cell cycle-specific cytotoxic agents show that a large proportion of CFU-c (~ 50%) are engaged in DNA synthesis, whereas the majority of CFU-s (~ 90%) are in a noncycling state under normal conditions (9, 10). More recently, an antigenic membrane marker which is expressed on CFU-s but which cannot be demonstrated to be present on CFU-c has been described (7). These experiments indicate that CFU-c, although closely related to CFU-s, represent more differentiated cell types.

It is possible that the capacity to react to CSF marks the first stage at which differentiating blood cells become restricted to maturation into granulocytes and/or monocytes. This stage may be indicated as the committed progenitor cell of the granulocyte/monocyte series.

To answer the question as to whether CFU-c represent this stage of differentiation, it is of importance to determine if CFU-c correspond to a homogeneous cell population of a single cell type. Since CFU-c form colonies of mature cells via a process of proliferation and differentiation, it is possible that the culture system also allows the proliferation of intermediate maturation stages. Consequently, CFU-c can be expected to be a heterogeneous cell population. Reports in the literature in which the properties of CFU-c are

Physical Characterization of a Subpopulation of Granulocyte/Monocyte Progenitor Cells (CFU-c)

Ger van den Engh, Dries Mulder, Neil Williams, and Simon Bol

17

analyzed show, indeed, that a heterogeneity may exist among CFU-c (11, 13).

In our laboratory, a number of preparations with colony-stimulating activity have been compared with respect to the shape of their dose-response curves (5, 6). From these studies, it has been concluded that many CSF-containing preparations contain other factors which modify the action of CSF, in the sense that the number of colonies is increased. CFU-c induced to colony formation by these enhancing factors differ in their density properties from the CFU-c which are stimulated in the presence of CSF alone (2, 23). In these studies, semipurified CSF prepared from pregnant mouse uteri (CSF-pmue) served as a reference preparation. At plateau levels, this material induces low colony numbers, as compared to other sources of stimulating material. It seems possible, therefore, that CSF-pmue selects for a specific subpopulation of CFU-c. In this report, the density and sedimentation rate profiles of CFU-c cultured in the presence of CSF-pmue are described. These characteristics are compared to the density and sedimentation rate distributions of a hypothetical cell population of a single cell type.

MATERIALS AND METHODS
PREPARATION OF CSF-pmue

Colony-stimulating factor from pregnant mouse uterus extract (CSF-pmue) was prepared according to the methods described by Bradley et al. (4) and Stanley et al. (20). Whole pregnant mouse uteri were homogenized in 3 volumes of cold glass-distilled water and, after centrifugation (15,000 g, 30 min), the supernatant was saturated to 50% with ammonium sulfate. Following centrifugation (20,000 g, 20 min), the precipitate was discarded and the supernatant was fully saturated with ammonium sulfate. The precipitate was spun down (20,000 g, 30 min) and resuspended in one-fourth of the original volume of cold glass-distilled water. Following dialysis against glass-distilled water (3 days at 4°C), heat treatment (56°C, 30 min), and centrifugation, the solution was adsorbed on calcium phosphate gel and eluted by 0.09 M phosphate buffer, pH, 7.4.

The biologic activity of the resulting preparation was determined by comparing its titration curve with that of a standard preparation (5). If necessary, the heat inactivation and dialysis were repeated until all traces of enhancing or inhibiting substances were removed. The ultimate

preparation was sterilized by millipore filtration and 20 ml aliquots were stored at −20°C.

The activity of the preparation is expressed in an arbitrary unit system proportional to the CSF concentration. One unit CSF/ml represents the lowest concentration at which cells are stimulated; 5 units/ml represent the threshold for colony formation; and a plateau is reached at 40 units/ml (5).

CFU-c ASSAY

The method is described in detail by Metcalf and Moore (12). Bone marrow cells from the femurs were suspended in phosphate buffered saline or in Hanks's balanced salt solution, buffered with Hepes buffer, and kept at 0°C until use. Desired quantities of this cell suspension were added to nutrient agar consisting of 0.3% agar (Bacto agar, Difco) in Dulbecco's medium (prepared from 10 × stock solution, Biocult), supplemented with 20% of a mixture of equal volumes of horse serum, fetal calf serum (Flow), and Trypticase Soy Broth (3%) (BBL). One ml aliquots of this nutrient agar containing 5×10^4 nucleated cells were distributed to 30 mm Petri dishes (Greiner) which contained 0.1 ml CSF in different dilutions. All cultures were incubated at 37°C in an atmosphere of 10% CO_2 in air and 100% humidity (National incubator). The number of colonies (cell aggregates containing more than 50 cells) was counted at day 6 or day 7. Hybrid mice of CBA × C57BL/Rij of 6–9 weeks of age were used in all experiments.

EQUILIBRIUM DENSITY CENTRIFUGATION

Bone marrow cells were separated on the basis of their buoyant density in continuous albumin gradients using the technique of Shortman (17, 18). Preparation of bovine serum albumin (BSA) solutions, pH 5.1, for cell separation studies has been described elsewhere (19). Cells were dispersed directly into the dense medium. After generating a linear gradient, cells were centrifuged to equilibrium at 4000 g for 30 min. Approximately 20 fractions were collected over a density range of 1.06–1.09 g/cm³. The cells in each fraction were recovered by dilution and centrifugation. The cell pellet was resuspended in balanced salt solution. The mean yields of cells were 80% for CFU-c and 90% for all nucleated cells.

VELOCITY SEDIMENTATION AT UNIT GRAVITY

This cell separation technique was carried out using the procedure described by Miller and Phillips (15). Further information can be found in the report of Visser et al. (Chapter 3).

³H-THYMIDINE CYTOCIDE

The proportion of S-phase cells in a cell population was determined by incubating cells in suspension with 15 μCi ³H-TdR/ml (specific activity, 22 Ci/mM) for 30 min at 37°C. The treatment was terminated by washing the cells with cold Thymidine solution.

RESULTS

In previous studies, it has been shown that semipurified CSF prepared from pregnant mouse uteri (CSF-pmue) is a convenient reference material for comparing different preparations with colony-stimulating activity. Similarly, the use of this stimulator appears to be advantageous for the analysis of subpopulations of CFU-c. Figure 1 shows the density distribution of CFU-c cultured at various levels of CSF-pmue. The profile is remarkably restricted in its density range. It reaches a peak value at approximately 1.074 g/cm³. The band width at half the peak height is approximately 0.005 g/cm³. This width is of the same order of magnitude as the resolution of the density separation technique. From the rebanding experiment of Williams and Shortman (24), a value of ± 0.002 g/cm³ can be derived for the resolution of the method.

The density profile of CFU-c is independent of the amount of CSF-pmue which has been added to the plates. The response to two doses of CSF situated on the upward slope of the titration curve (10 and 20 units CSF), as well as the response to an optimal dose (160 units), are shown. The concentration of the stimulus does not markedly influence the density distribution of responding CFU-c; the variation observed is less than the error of counting. These observations show that the CFU-c which are cultured with CSF-pmue have a homogeneous density distribution.

The homogeneity of this particular population of CFU-c is also expressed in the sedimentation rate distribution. Since the cells are of the same density, they are distributed relative to their size when submitted to the velocity sedimenta-

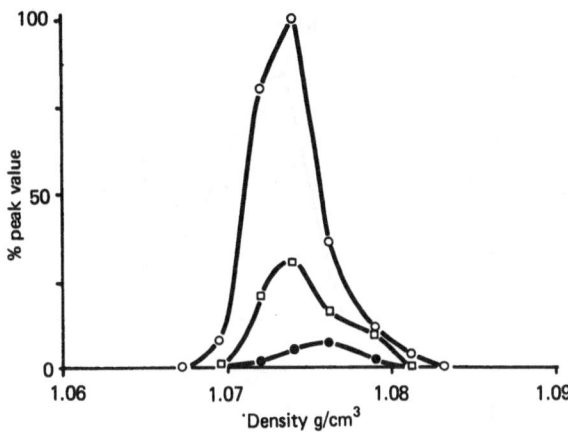

FIGURE 1. The effect of CSF-pmue concentration on the density distribution of CFU-c from mouse bone marrow. Cells from each fraction were cultured with 10 units (●), 20 units (□), or 160 units (○) CSF-pmue/ml. Each point is the mean value of three cultures and is expressed as a percentage of the peak fraction stimulated with 160 units CFS/ml. The profiles are expressed as cells per fraction per density increment.

tion procedure. Thus, the sedimentation rate profiles can be analyzed according to the criteria which are derived in the paper of Visser et al. (Chapter 3).

Figure 2 shows the distribution of sedimentation rates of CFU-c cultured with CSF-pmue. Two peaks of activity are found at sedimentation rates of 5.0 mm/hr and 8.2 mm/hr, respectively, the latter peak being approximately one-third the height of the peak of the more slowly sedimenting CFU-c. The ratio of the two rates is 1.60, which is close to the theoretical sedimentation rate ratio of 1.59 of G_2 and G_1 cells. This information is combined in a semi-logarithmic graph in the bottom panel of Fig. 2. The observed profile (black dots) is compared to a theoretical distribution of an exponentially growing population with a G_1 phase sedimenting at a rate of 5.0 mm/hr. There is a good correlation between the measured values and the distribution of an ideal homogeneous cell population.

The sedimentation rate distribution may be further analyzed by determining the distribution of DNA synthesizing cells. Figure 3 shows an experiment in which all fractions of a velocity sedimentation separation experiment were incubated with 15 μCi of heavily labeled ³H-TdR. The open dots represent the CFU-c numbers which were observed in fractions treated with cold TdR. The solid dots represent CFU-c numbers after ³H-TdR cytociding. The area under the distribu-

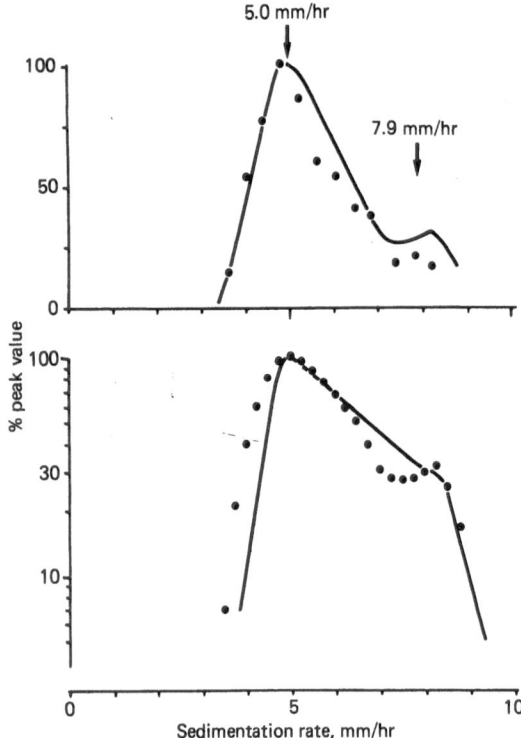

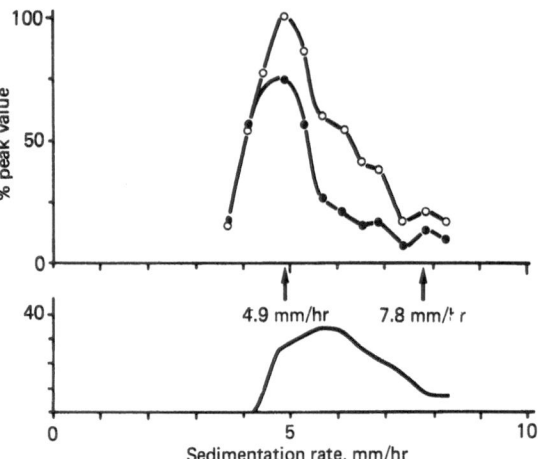

FIGURE 2. Sedimentation rate distribution of CFU-c from mouse bone marrow cultured with 160 units CSF-pmue/ml. Top panel: the drawn line represents the average profile from four separate experiments. The points represent measurements obtained in a typical single experiment. Each point is the mean value of three cultures. The values represent total cells per fraction expressed as a percentage of peak value. Bottom panel: the drawn line represents the theoretical distribution of an exponentially growing homogeneous cell population in which the G_1 cells sediment at a rate of 5.0 mm/hr. The dots are values from the drawn line in the top panel.

FIGURE 3. Sedimentation rate distribution of CFU-c from mouse bone marrow before and after thymidine cytociding. The results of a typical experiment are shown: ○, CFU-c cultured with 160 units CSF-pmue/ml; ●, CFU-c content of the fractions after exposure to 15 μCi of ³H-TdR/ml, 22 Ci/mMol. The bottom graph shows the difference between the two distributions. The values are expressed as a percentage of the maximum value of the untreated controls. The area under the graph of the ³H-TdR-treated cells represents 60% of the control graph.

tion which is obtained after ³H-TdR-incorporation is 60% of the area of the curve representing the untreated cells. Therefore, the ³H-TdR treatment has resulted in a 40% reduction in CFU-c numbers. This is in agreement with the reduction which is observed when the CFU-c in total bone marrow are subjected to ³H-TdR cytociding. A relatively small reduction in CFU-c numbers is observed at the two peaks in the profile, indicating that the majority of DNA-synthesizing cells is located in the area between the peaks.

DISCUSSION

In previous studies, CFU-c have been described as being a heterogeneous cell population which cov-

ers a broad density range (8, 13, 25). In contrast to these findings, the CFU-c population which was studied in these experiments is homogeneous in its density properties, and may consist of a single cell type with a modal density of 1.074 g/cm³.

This observation is confirmed by the sedimentation rate distribution of these cells. The sedimentation profile of CFU-c cultured with CSF-pmue coincides with a theoretical distribution of a homogeneous cell population, with the cells in G_1 phase sedimenting at a rate of 5.0 mm/hr. The distribution of the CFU-c which are killed by high doses of ³H-TdR is in agreement with the assumed G_1 and G_2 peaks in the profile.

Since these CFU-c are restricted in their densities, the sedimentation profile is a measure of the size distribution. When the measured values are substituted in the equations derived by Miller (14), the CFU-c in G_1 phase measures 7.8 μm in diameter, whereas CFU-c in G_2 have a cross-section of 9.7 μm. These values are slightly different from the size estimates of cycling CFU-s (7.3–9.2 μm) as calculated by Visser et al. (21), which indicates once more that these two cell types may represent closely related stages of development.

The discrepancy between the conclusions about CFU-c reported here as a single cell type and the conclusions of previously mentioned

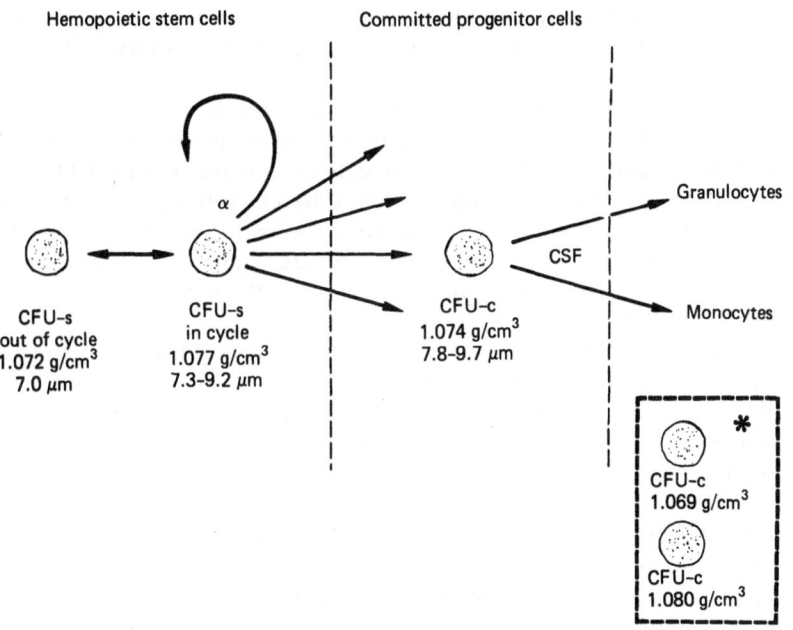

FIGURE 4. Schematic representation of the early steps in hemopoietic differentiation. Three cell types with uniform physical characteristics can be recognized: the resting pluripotent stem cell: the proliferating pluripotent stem cell; and the committed progenitor cell of the granulocyte/monocyte series. *: Density separation experiments show that there are two additional types of progenitor cells with colony-forming potential *in vitro*: a cell type requiring enhancing activity of the erythrocyte lysate type (modal density, 1.080 g/cm³) (23), and a cell type requiring enhancing activity as is present in post-endotoxin serum (modal density, 1.069 g/cm³) (2). These cells represent as-yet unknown stages of differentiation.

studies which show CFU-c to be heterogeneous, can be ascribed to the material which is employed to stimulate the cultures. As compared to stimulating material used by most investigators, semipurified CSF-pmue may be considered to be a poor stimulus, both with respect to absolute colony numbers as well as to the slope of its dose-response curve (5, 6). If "richer" stimulators are used under identical conditions as described in this report, broad density distributions are indeed observed (22, 23). The data reported here lead, therefore, to the conclusion that semipurified CSF from pmue (and possibly equivalent stimulating molecules from other sources) acts on a single step in the hemopoietic cell differentiation series. Bol (1) has shown that the CFU-c at this stage is still capable of giving rise to both granulocytes and monocytes; it is possible, therefore, that this type of CFU-c represents the first cell which is committed to differentiate along the granulocyte/monocyte pathway. This information and the conclusions of the paper of Visser et al. (21) are combined in the hypothetical scheme of Fig. 4. This figure represents the first steps in hemopoietic cell differentiation. The exact place of the light density CFU-c (modal density, 1.069 g/cm³) and the high density CFU-c (modal density, 1.080 g/cm³) (2, 23) which are mentioned in the introduction is as yet unclear. Elucidation of the differentiation stage of these cell types may very well link the functionally defined cell types of experimental hematology to the first morphologically recognizable cells of the granulocyte/monocyte differentiation series.

SUMMARY

Culture systems in which a population of hemopoietic progenitor cells (CFU-c) proliferates and gives rise to colonies of granulocytes and monocytes have become standard techniques in experimental hematology. The *in vitro* growth of CFU-c from mouse bone marrow is dependent on the presence of a humoral factor termed colony-stimulating factor or CSF. Techniques for *in vitro* colony formation may therefore be used to enumerate CFU-c or to measure levels of stimulating activity under various experimental conditions.

Although these methods have been frequently employed, the nature of the CFU-c is still unclear. From a number of reports in the literature in which the physical properties of these cells were measured, it can be concluded that CFU-c are a heterogeneous group of cells. This indicates that different cell types, possibly representing different stages of maturation, may form colonies *in vitro*. In contrast, it was found that CFU-c stimulated by semipurified CSF from pregnant mouse uteri are restricted in their density distribution and that this subpopulation of CFU-c may consist of a single cell type.

In the present studies, it is investigated whether these CFU-c also behave like a single cell population when submitted to sedimentation rate separation. By this method cells of a uniform density are separated according to their size and, therefore, a cell population of a single species is divided into fractions containing cells at different stages of the cell cycle. On the basis of these considerations the sedimentation rate distribution of CFU-c is compared to a theoretical distribution of a homogeneous cell population. The distribution of CFU-c is further analysed by incubating the cell fractions with high doses of ^{3}H-TdR. In this manner the sedimentation rate distribution of CFU-c in S-phase is determined. According to both criteria, the CFU-c subpopulation studied coincides with a uniform population of a single cell type. It is concluded that CFU-c which are stimulated by semipurified CSF have distinct physical properties. This cell type has a modal density of 1.074 g/cm^3. Its diameter varies from 7.8 μm (G$_1$ cells) to 9.7 μm (G$_2$ cells).

ACKNOWLEDGMENTS

This investigation is part of a study on the regulation of hemopoiesis which is supported by a program grant of The Netherlands Foundation for Medical Research (FUNGO), which is subsidized by The Netherlands Organization for the Advancement of Pure Research (ZWO). Special thanks are due to Mrs. M. G. C. Platenburg and Mrs. H. Jackson for their expert technical assistance.

REFERENCES

(1) Bol, S., and Williams, N. T. Physical separation of mouse myeloid progenitor cells with different differentiation kinetics. *Exp. Hematol.* (Suppl.), *4*: 136, 1976.

(2) Bol, S., Williams, N., and van den Engh, G. J. Unpublished observations, 1975.

(3) Bradley, T. R., and Metcalf, D. The growth of mouse bone marrow cells *in vitro. Aust. J. Exp. Biol. Med. Sci., 44*: 287, 1966.

(4) Bol, S., Stanley, E. R., and Sumner, M. A. Factors from mouse tissues stimulating colony growth of mouse marrow cells *in vitro. Aust. J. Exp. Biol. Med. Sci., 49*: 595, 1971.

(5) van den Engh, G. J. Quantitative *in vitro* studies on stimulation of murine haemopoietic cells by colony stimulating factor. *Cell Tissue Kinet., 7*: 537, 1974.

(6) van den Engh, G. J., and Bol, S. The presence of a CSF enhancing activity in the serum of endotoxin-treated mice. *Cell Tissue Kinet., 8*: 579, 1975.

(7) van den Engh, G. J., and Golub, E. S. Antigenic differences between hemopoietic stem cells and myeloid progenitors. *J. Exp. Med., 139*:1621,1974.

(8) Haskill, J. S., McNeill, T. A., and Moore, M. A. S. Density distribution analysis of *in vivo* and *in vitro* colony forming cells in bone marrow. *J. Cell. Physiol., 75*: 167, 1970.

(9) Iscove, N. N., Till, J. E., and McCulloch, E. A. The proliferative status of mouse granulopoietic progenitor cells. *Proc. Soc. Exp. Biol. Med., Sci. 134*: 33, 1970.

(10) Lajtha, L. G., Pozzi, L. V., Schofield, R., and Fox, M. Kinetic properties of haemopoietic stem cells. *Cell Tissue Kinet., 2*: 39, 1969.

(11) Messner, H., Till, J. E., and McCulloch, E. A. Density distributions of marrow cells from mouse and man. *Ser. Haematol., 5*: 22, 1972.

(12) Metcalf, D., and Moore, M. A. S. Haemopoietic cells. In Neuberger, A., and Tatum, E. L., eds., *Frontiers of Biology.* Amsterdam: North-Holland Publishing Company, vol. 24, 1971.

(13) Metcalf, D., Moore, M. A. S., and Shortman, K. Adherence column and buoyant density s3paration of bone marrow stem cells and more differentiated cells. *J. Cell. Physiol., 78:* 441, 1971.

(14) Miller, R. G. Separation of cells by velocity sedimentation. In Pain, R. H., and Smith, B. J., eds., *New Techniques in Biophysics and Cell Biology*. London: John Wiley and Sons, 1973.

(15) Miller, R. G., and Phillips, R. A. Separation of cells by velocity sedimentation. *J. Cell. Physiol., 73:* 191, 1969.

(16) Pluznik, D. H., and Sachs, L. The cloning of normal "mast" cells in tissue culture. *J. Cell. Comp. Physiol., 66:* 319, 1965.

(17) Shortman, K. The separation of different cell classes from lymphoid organs. II. The purification and analysis of lymphocyte populations by equilibrium density gradient centrifugation. *Aust. J. Exp. Biol. Med. Sci., 46:* 375, 1968.

(18) Shortman, K. Physical procedures for separation of animal cells. *Ann. Rev. of Biophys. Bioengineering, 1:* 93, 1972.

(19) Shortman, K., Williams, N., and Adams, P. The separation of different cell classes from lymphoid organs. V. Simple procedures for the removal of cell debris, damaged cells, and erythroid cells from lymphoid cells. *J. Immunol. Methods, 1:* 273, 1972.

(20) Stanley, E. R., Metcalf, D., Maritz, J. S., and Yeo, G. F. Standardized bioassay for bone marrow colony-stimulating factor in human urine: Levels in normal man. *J. Lab. Clin. Med., 79:* 657, 1972.

(21) Visser, J. W. M., van den Engh, G. J., Williams, N. T., and Mulder, A. H. Physical separation of the cycling and non-cycling compartments of murine haemopoietic stem cells. Chapter this volume.

(22) Williams, N. T. As yet unpublished observations, 1976.

(23) Williams, N. T., and van den Engh, G. J. Separation of subpopulations of *in vitro* colony forming cells from mouse bone marrow by equilibrium density centrifugation. *J. Cell. Physiol., 86:* 237, 1975.

(24) Williams, N. T., and Shortman, K. The separation of different cell classes from lymphoid organs: The effect of pH on the buoyant density of lymphocytes and erythrocytes. *Aust. J. Exp. Biol. Med. Sci., 50:* 133, 1972.

(25) Worton, R. G., McCulloch, E. A., and Till, J. E. Physical separation of haemopoietic stem cells from cells forming colonies in culture. *J. Cell. Physiol., 74:* 171, 1969.

(26) Wu, A. M., Till, J. E., Siminovitch, L., and McCulloch, E. A. Cytological evidence for a relationship between normal hematopoietic colony-forming cells and cells of the lymphoid system. *J. Exp. Med., 127:* 455, 1968.

INTRODUCTION

Functional mature cells of the hematopoietic system in adults are derived from proliferation and differentiation of their progenitor cells, the committed stem cells (Fig. 1). By definition, the committed stem cells derived from multipotent stem cells through differentiation and proliferation processes can give rise to only one type of differentiated cell. The stem cells should respond to a physiologic control for a normal function. This control is apparently mediated through some specific regulatory functions, i.e., transitions of precursor cells to differentiated cells require regulatory factors. Some regulatory factors have a long-range effect on cellular differentiation, whereas others are closely associated with a specific environment and exert a short-range effect (33). Because the normal function of the hematopoietic system is dependent on a continuous supply of mature cells, appropriate cell-factor interactions are critical for the maintenance of the normal function of blood cells. Derangement of this interaction will naturally impair the normal function of hemopoietic tissue. Leukemia may result from this derangement (12, 21, 34, 49).

Cell-Factor Interaction in Populations of Normal and Leukemic Blood Cells

18

Alan M. Wu, Francis W. Ruscetti, and Robert C. Gallo

This chapter presents some of our studies on the fractionation of heterogeneous species of regulatory factors in conditioned medium prepared from short-term culture of phytohemagglutin- (PHA)-stimulated human lymphocytes. These regulatory factors are divided into two categories based on the assay system used. Those factors that stimulate the formation of blood cell colonies in semisoft culture medium are termed "colony stimulators" (CS), whereas those that stimulate the growth of blood cells in suspension culture are termed "growth stimulators" (GS). The results suggest that some CS and GS activities are copurified at some stages of purification; others, however, are separable. We also discuss derangements of cell-factor interactions attributable either to an abnormality in the factor-responsive cells (FRC) or an abnormality in the factor-producing cells, or in the factor per se. These possibilities are used to explain most of the observed responses of leukemic cells to factors in culture. Finally, we present a model for the induction of growth and differentiation of myelogenous cells based on our observation that PHA-LyCM enhanced the growth and differentiation of blood and bone marrow cells from people with myelogenous leukemia first stimulated by phytohemagglutinin in a short-term culture as recently reported by Dicke et al. (16).

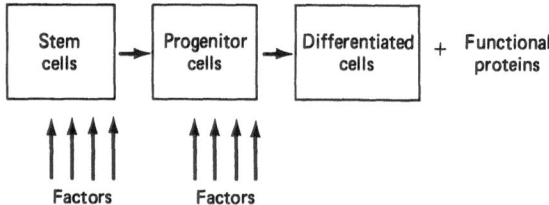

FIGURE 1. Three compartment models of maturation of hemopoietic cells. This model is designed for application to the hemopoietic system in postneotal life. The stem cell compartment and its regulatory proteins are still hypothetical, except in the case of mice the stem cells of which can be measured by the spleen-colony technique (52). The committed stem cells are generally identified by colony assay in semisolid culture medium. This compartment of stem cells can give rise to only one type of differentiated cell. The functional proteins include both those residing inside the cells and those released from the cells.

RESULTS

FACTORS AFFECTING THE DIFFERENTIATION OF GRANULOCYTIC COMMITTED STEM CELLS

Measurement of Committed Stem Cells Differentiation of the committed stem cells can be measured by clonal or nonclonal methods (see Table 1). The clonal method is based on colony formation on a solid surface (7, 42) and rests on the self-renewing ability of the stem cell. Growth is dependent on certain factors obtained either from certain conditioned media, tissue extracts, or body fluids (e.g., urine and serum). At present, by using appropriate factors, the colony assay is applicable to each cell type, at least in the mouse, i.e., granulocytic, erythrocytic (27, 51), megakaryocytic (35, 37), and lymphoid cells (17, 38). Those factors required for the colony formation in semisolid culture systems are categorically termed "colony stimulators" (CS),* and in this chapter we follow the tradition that uses CS only for those factors that affect the growth of

*At present, CS is generally used to name stimulators for granulocytic colony formation. However, for consistency of terminology in the future, it seems that CS would be more appropriately used to name all kinds of stimulators for colony formation. Therefore, it can be subdivided into granulocytic CS (G-CS) [equivalent to colony-stimulating activity (CSA)], or colony-stimulating factor (CSF); erythrocytic CS (È-CS) (equivalent to erythropoietin); megakaryocytic CS (M-CS) (which may or may not be equivalent to thrombopoietin); and lymphoid CS (L-CS) (yet to be identified).

granulocytic cells. Each type of committed stem cell appears to have some corresponding factor(s) to regulate their growth and differentiation. For example, colony-stimulating activity (CSA) induces granulocytic colony formation, partially purified erythropoietin preparations stimulate erythrocytic colony formation (27, 51), and probably megakaryocytic colony formation (35), as well as some components of PHA-stimulated lymphocyte conditioned medium cause B-lymphocytic colony formation (38). CS can also be obtained from the feeder layer in the culture system. However, those present in the cell-free preparation are superior to those present in the feeder layer for studying the mechanism of action of the CS, as the former are more reproducible and easier to measure quantitatively.

In spite of the fact that there are many advantages in using a clonal assay to study the differentiation of the committed stem cells, a major drawback of this system is the difficulty in obtaining sufficient amounts of cultured cells for detailed studies, espically biochemical and biomolecular investigations. This disadvantage may be overcome by developing mass suspension cultures in which the growth and differentiation of the committed stem cells must depend upon the presence of some regulatory factors. These factors are categorically termed "growth stimulators" (GS) (56, 57). Unlike fibroblasts or some epithelial cells, hemopoietic cells are generally difficult to grow long-term in culture. Most of the time they survive in culture for only 1 to 2 weeks. However, in some cases, long-term cell lines have been established. These established cell lines are generally ill-differentiated lymphoblastoid cells and contain EBV antigens and B-cell markers. However, in rare cases, some cell lines containing T-cell markers, such as the Molt cell lines, were established from patients with lymphoid leukemia. All these cell lines are independent of the added regulatory factors. At present, the only factor-independent myelogenous cell lines are the mouse M-1 cell line (25) and a human line derived from CML that possesses Ph' + chromosome positive cells, but which shows little evidence of differentiation (32).

When conditioned medium containing CS was added to normal marrow cells in a short-term culture, differentiation of myelocytic cells was often observed, but the cells begin to die in approximately 2 weeks. The committed stem cells in this system can probably renew themselves only to a limited degree. It was recently observed in this laboratory that some protein factors in conditioned medium derived from PHA-stimulated T-cell enriched lymphocytes (PHA-LyCM) can

TABLE 1 A Summary of the Methods for Measurement of Hemopoietic Precursor Cells

Potentiality of stem cells	Type of stem cells	Method	Stimulant
		A. CLONAL METHOD	
Multipotent	CFU-s[a]	Repopulation of radiated mouse spleen (52)	Mouse environment
	G-CFU	Semisoft agar or methylcellulose culture (7,42,48)	Conditioned medium Cell lines (human kidney (9) L cells (1), fibroblasts (30,53) Embryo cell strains Short-term culture of blood cells, PHA-stimulated lymphocyte (13,44) leukocytes (26) Urine (46) and serum (18) Tissue extracts (8,50) Cellular or tissue feeder layers (45)
		Diffusion chamber with agar medium (23)	Mouse environment
Erythrocytic committed stem cells	TE-CFU	Radiated mouse spleen (24)	Erythropoietin and mouse environment
	E-BFU	Plasma clot (2)	Erythropoietin
	E-CFU	Plasma clot (8) or methylcellulose medium (14,27)	Erythropoietin (27,50) or type-C virus infection (11,40)
Megakaryocytic committed stem cells	M-CFU	Semisolid agar (37) Plasma clot (35)	PHA-LyCM (37) Erythropoietin preparation (35)
Lymphocytic committed stem cells	B-CFU	Semi-solid agar medium (17,38)	PHA-LyCM, PWM (17)
	T-CFU	Semisolid agar medium (17)	PHA and PWM (17)
		B. NONCLONAL METHOD	
Multipotent stem cells		Diffusion chamber (6) Short-term suspension culture Suspension culture of Dexter and Testa (10,15)	Mouse humoral factor (5) Activated T-cell factor (10) Autologous conditioned medium (15)
Granulocytic committed stem cells		Short-term semisuspension culture (4) Dexter's suspension culture	Feeder layer (4) or conditioned medium containing CS (3) Autologous conditioned medium (15)
Erythrocytic committed stem cells		Short-term suspension culture Reticulocytosis 59Fe uptake/spleen	Erythropoietin (22,54) Erythropoietin and mouse environment (28) Mouse spleen environment (28)
Lymphocytic precursor cells		B-cell suspension culture T-cell suspension culture PB cell in repopulated irradiated mice (29)	EBV infection (43) T-GS (39) Immune response (33)

[a]*Abbreviations:* CFU-s: Spleen colony-forming unit; G-CFU: granulocytic CFU; TE-CFU: transient erythrocytic colony-forming unit; E-BFU: erythrocytic burst-forming unit; E-CFU: erythrocytic CFU; M-CFU: megakaryocytic CFU; CS: colony stimulator; PHA-LyCM: phytohemagglutinin-stimulated lymphocyte conditioned medium; B-CFU: B-lymphocyte CFU; T-CFU: T-lymphocyte CFU; PB: precursor of B-lymphoid cells; ASC: antigen-sensitive cells; EBV: Epstein-Barr virus; T-GS: growth stimulator for T-lymphoid cells; PWM: pokeweed mitogen.

stimulate the growth of lymphoid cells containing T-cell markers for several months, and possibly of permanent duration (39). The growth of these T-cells is strictly dependent on the added factors (39). This discovery may make it possible to study specific T-cell antigens, receptors, and cellular differentiation in a relatively pure system for the first time. Erythropoietin is another example of GS, as hemoglobin synthesis from cultured cells is induced when erythropoietin is added in culture (55).

Relationship between CS and GS Functionally, CS and GS are similar, but they are not identical. As shown in Table 2, some conditioned media contain both CS and GS activities (horizontal rows 6 and 8), whereas others contain high CS activity without detectable GS activity (rows 5, 7, and 9), and *vice versa*. Both CS and GS consist of heterogeneous species of functional proteins, some of which are present both in CS and GS, whereas some are found only either in CS or GS (57). This relationship is shown in Table 3, where C, D, and E species of proteins are common both to CS and GS activities. A and B are unique to CS, and F and G to GS. The number and the name of the protein species are, of course, hypothetical. In our laboratory, experiments to support this nonidentity–heterogeneity hypothesis of regulatory factors are based chiefly on our results with fractionation of PHA-stimulated lymphocyte-

TABLE 3 Model Showing Relationship between CSA and GSA

ACTIVITY	HYPOTHETICAL PROTEIN SPECIES
CSA	A.B.C.D.E.
GSA	C.D.E.F.G.

conditioned medium (PHA-LyCM). Lymphocytes were prepared by removing adherent cells from pools of buffy coat from normal individuals by passing them through a nylon fiber column (31). The eluted nonadherent cells are mainly T-lymphocytes. The post column fraction was then incubated at 37°C in the presence of 3% PHA and 1% homologous serum for 72 hr. We had previously shown that PHA-LyCM contains CS for human marrow cells; when PHA-LyCM is fractionated by gel filtration with Sephadex G-150 followed by DEAE-cellulose chromatography, three peak activities of CS are found, only two of which could stimulate ^{3}H-thymidine uptake in suspension culture (57). In those experiments, we assayed only for CS activity. Therefore, we were unable to detect those factors which had GS activity and not CS activity. We recently repeated the gel-filtration experiment with Sephadex G-150 and screened both CS and GS activities in eluates simultaneously. In this experiment, CS activity

TABLE 2 Some Results That Differentiate CSA and GSA

SOURCE OF FACTOR-RESPONSIVE CELLS	CONDITIONED MEDIUM	COLONY FORMATION IN SEMISOFT AGAR MEDIUM[d]	PROLIFERATION IN SUSPENSION CULTURE[e]
AML(AS)[a]	none	−	−
	LyCM	−	+++
	WHE-1 CM[c]	−	+++
AMML(JC)	none	−	+++
NA[b]	LyCM	+++	±
	WHE-1 CM	+++	+
Normal(EH)	none	+	−
NA	LyCM	+++	++
	WHE-1CM	+	−

[a]Letters in parentheses refer to patients' initials.

[b]NA: Nonadherent cells. These were prepared according to the procedure of Messner et al. (36).

[c]WHE-1 CM denotes conditioned medium prepared from whole human embryo cell strain no. 1 in our laboratory. The conditioned medium was harvested 48 hr after WHE-1 cells reached stationary phase.

[d]The procedures for the agar colony assay are performed according to published procedures (44). Briefly, the culture system consists of two layers of semisoft agar medium. The upper layer contained 10^5 nucleated human marrow cells in 0.8 ml McCoy's 5A medium containing 20% of fetal calf serum and 0.3% agar. The lower layer contained 20% each of conditioned medium in 2.5 ml McCoy's 5A medium containing 20% of fetal calf serum and 0.5% agar. Duplicate plates were prepared for each condition. The plates were incubated at 37°C with 10% CO_2. Expression of the observations: (−) No colony formation; (+) one colony per 10^4 nucleated cell; (+++) one colony per 10^3 nucleated cell.

[e]Based on results obtained from viable cell count. (−) No stimulation of cell proliferation; (+) slight stimulation of cell growth; (++) moderate stimulation of cell growth; (+++) exponential growth of the factor responsive cells.

was measured by colony formation in semisoft agar medium and GS by the cell growth or ability to stimulate the incorporation of ^{3}H-thymidine in suspension culture. The last two measurements may not assay the same biologic activities, although the latter must be more sensitive than the former as an indication of cellular proliferation. As shown in Table 4, pools III and IV contained both CS and GS activities; pool I contained GS activity, and pool II contained CS activity. (See Table 4 footnote for details.) In most cases, the ability to enhance ^{3}H-thymidine incorporation activity and the cell-growth-stimulating activities appear in the same fraction. Pool IV appears to contain T-GS, as we were able to maintain growth of lymphoid cells in suspension culture relatively long-term and more than 80% of these lymphoid cells were able to form E-rosettes at 20°C incubation. For screening purposes, we normally grow cells for only 2 to 3 weeks. In the case in which both CS and GS were cofractionated, CS and GS are sometimes separated by further fractionation. For example, when pool IV was applied to a concanavalin A (Con A) sepharose column, most CS activities that stimulate large colony formation were recovered only in the flowthrough fraction when the CSA were measured in agar medium, but large colony formation was detected both in the flowthrough fraction, borate buffer, and α-D-

methylmannoside eluates when the CSA were measured in methylcellulose medium (Table 5). This suggests that the conditions required for colony formation in agar and in methylcellulose medium are different. It is not known whether this is attributable to different factors, to different conditions favoring different cell populations, or to unfavorable agar medium. Cluster-stimulating activity (aggregates containing more than 20 cells but less than 50 cells) was also found in borate eluate and α-D-methylmannoside eluate (Table 5). At this step of fractionation, only very weak GS activity (measured by cell growth) was found in α-D-methylmannoside eluate and weak GS activity (measured by [^{3}H]thymidine incorporation activity) in the flowthrough and α-D-methylmannoside eluate. Results from this fractionation study clearly demonstate that there are heterogeneous species of proteins both in CS and GS activities. At this stage of purification, some of them still functionally overlap and some do not.

Biochemical Properties of CS and GS   CS activities obtained from PHA-LyCM have the following properties.

1. They are not dialyzable and have a molecular weight ranging from 160,000 to 5000 as determined by gel filtration.

2. They are relatively heat stable (stable

TABLE 4   Fractionation of CS and GS Activities from PHA-LyCM by Sephadex G-150 Filtration Procedure[a]

POOL NUMBER	MOLECULAR WEIGHT (× 10⁻⁴)	GROWTH-STIMULATING ACTIVITY		COLONY-STIMULATING ACTIVITY
		^{3}H-TdR Incorporation Activity	Cell Growth	
I	8.4–22.0	+	±	−
II	3.7–6.4	±	−	+
III	2.4–3.0	+	+	++
IV	0.8–1.7	+	++	++
V	0.23–0.5	+	−	−

[a]PHA-LyCM was prepared from pooled peripheral blood of normal individuals by Associated Biomedics Systems, Buffalo, New York, according to the procedures described by Prival et al. (44). CM (13 liters) was first heated at 56°C for 1 hr and the denatured protein and cell debris were removed by low-speed centrifugation (1000 × g, 10 min). All steps were performed at 4°C. The heated CM was then precipitated with ammonium sulfate between the concentrations of 25% to 65%. The ammonium sulfate precipitate was resuspended in 700 ml PBS and then dialyzed against PBS. The concentrated CM was then applied to a sephadex G-150 column with a bed volume of approximately 25 liters. The column was equilibrated and eluted with PBS. The eluate was collected with 25 ml per fraction. A total of 1300 fractions were collected. The void volume of the column was determined to be 7 liters with blue dextran. The molecular weight of the eluates was calibrated with aldolase, ovalbumin, and cytochrome C. Three types of activities were measured with every 20th fraction of the eluates. The activities are ability to stimulate (1) human colony formation in semisolid agar medium; (2) the growth of marrow cells in suspension culture; and (3) the ^{3}H-thymidine uptake of human marrow cells in suspension culture. (The procedures for agar colony formation are described in the footnote to Table 2.) The growth of marrow cells in suspension culture is measured by a viable cell count using the Trypan blue exclusion technique. For measurement of ^{3}H-thymidine uptake, cells were first exposed to thymidine-free medium; 10 μC/ml ^{3}H-thymidine was added that incubated at 37°C for 4 hours. After removal of medium, cells were disrupted with 1 N NaOH and vigorous vortexing. An aliquot of the disrupted cells was then used to measure trichloroacetic acid (TCA) precipitable radioactivity. The fractions were then pooled in five main fractions according to the content of the above-mentioned activities: (++) Potent activity; (+) moderate activity; (±) low activity; and (−) no activity.

TABLE 5 Fractionation of CS and GS by a Concanavalin A–Agarose Column[a]

FRACTION	COLONY-STIMULATING ACTIVITY (NO. OF COLONIES/MG/10^5 CELLS)		CLUSTER-STIMULATING ACTIVITY (NO. OF COLONIES/MG/10^5 CELLS)		STIMULATION OF CELL GROWTH IN SUSPENSION
	Agar	Methylcellulose	Agar	Methylcellulose	
Flowthrough	21υ	667	1020	598	Marginal
Borate buffer eluate	10	3900	4080	4100	Marginal
α-D-Methylmannoside eluate	0	1444	1200	1166	Moderate
Ethylene glycol and α-D-Methylmannoside eluate	0	nd[b]	0	nd	None

[a] An aliquot of Pool No. 4 from Sephadex column was applied to a concanavalin A–agarose column. Con A–agarose was prepared at pH 6 and equilibrated with a buffer containing 10 mM Na acetate pH 6.6, 1 mM CaCl$_2$, 1 mM MgCl$_2$, 0.5 mM MnCl$_2$, 0.2 M NaCl. After extensive washing with the sample buffer, the column was eluated sequentially with three solutions: (1) 0.3 M borate buffer pH 9; (2) 0.1 M α-D-methylmannoside; and (3) 50% ethylene glycol in 0.1 M α-D-methylmannoside. Extensive washing of the column was performed between each elution. The eluates were measured both for CS and GS activities. The procedures for CS activity in semisoft agar medium and GS in suspension culture were described in the footnotes to Tables 2 and 4. The CS activity in methylcellulose medium was performed according to a modified procedure of Ruscetti and Chervenick (48). Briefly, 10^5 nucleated cells were resuspended in 1 ml culture medium containing 1.4% methylcellulose, 5% fetal calf serum, 5% horse serum, and 10% conditioned medium containing GS. For each point, triplicate plates were prepared with a 35-mm plastic petri dish, and were incubated at 37°C in 10% CO$_2$ for 9–14 days.

[b] nd: Not done.

at 56°C for 30 min and at 4°C for approximately 4 months).

3. They are resistant to DNase, RNase, and neuraminidase, partially sensitive to pronase, completely sensitive to trypsin and periodate.

4. Some are absorbed to concanavalin A sepharose, which suggests that some are glycoproteins.

5. They are specific to primate bone marrow cells.

6. They are sensitive to high salt (0.5 M NaCl) treatment. This effect is reversible upon extensive dialysis, which suggests a possibility of subunit interactions.

7. They do not appear to require a cofactor for their activity.
It is not known yet which of these properties, if any, are shared by T-GS or other GS factors.

DERANGEMENT OF CELL-FACTOR INTERACTION

Derangements of cell-factor interactions could result in impaired blood cell function. The outcome may be a deficiency in functional mature cells. This could be accompanied by either an accumulation or a deprivation of earlier immature cells. In theory, the defectiveness of cell-factor interactions could be intrinsic or extrinsic to factor-responsive cells (FRC) (Table 6). For example, if FRC are infected by a tumor virus or are exposed to x-ray radiation or to a chemical car-

cinogen, altered FRC would not be able to respond to regulatory action of normal factors either *in vivo* or in culture. However, sometimes these modified cells may respond to other (specific) factors not present under normal conditions (see ref. 19 and 20 for examples). Alternatively, if FRC are normal but the extrinsic factor that regulates the differentiation of FRC is either deficient or abnormal, the proper cell-factor interaction will also be disturbed. If the factor-producing cells (FPC) are modified by

TABLE 6 Defect in Cell-Factor Interaction

CAUSE OF DERANGE-MENT OF CELL-FACTOR INTERACTION	RESPONSIVENESS TO REGULATOR(S)	
	In vivo[c]	*In culture*
Intrinsic to FRC[a]		
Normal factor	No	No
Specific factor[b]	Yes	Yes
Extrinsic to FRC		
Defective factor production	Yes	Yes
Inhibitor	No	Yes

[a] FRC: Factor response cell.

[b] Sometimes the altered factor-response cell can respond to some specific factor(s) that are not the normal factor(s).

[c] The *in vivo* response is more a prediction than an observation.

mechanisms such as those mentioned earlier for FRC, the production of the normal factors will be affected. Alternatively, inhibitors of the factors present either *in vivo* or in culture (these inhibitors may be antibodies, lipoproteins, or some chelating agents) could lead to the same consequences. FRC should be able to respond to factors in culture upon removal of inhibitors.

ARE LEUKEMIC CELLS AUTONOMOUS OR SUBJECT TO NORMAL REGULATING MECHANISMS?

Human leukemia includes each of the above-mentioned possibilities of cell-factor derangement. If the defect is intrinsic to the leukemic cells, the blockage of differentiation will not be corrected by normal factors (autonomous growth). Therapy of this type of leukemia therefore depends on destroying leukemic cell populations or replacing leukemic cell populations by normal competent cell populations. If the defect is caused by extrinsic factors, the leukemia should show a response to regulatory factors. The fact that most leukemias in the actue phase do not respond to factors in culture suggests that they are autonomous. This is particularly true of myelogenous leukemia. However, sometimes the defect is not completely autonomous, for in some cases, growth and differentiation of leukemic cells can be regulated in culture through the use of PHA and CS. For example, Aye et al. reported that a short-term suspension culture involves the interaction of a factor-producing population with the leukemic leukocytes to promote the proliferation of the leukemic cells when these cells are cultured in the presence of PHA (3, 4). Dicke et al. (16) reported that AML cells, which normally do not form colonies in the presence of a feeder cell layer in a semisoft agar medium, can be induced to do so with PHA. As a source for CS, the feeder layer does not appear to be essential for colony formation by PHA-stimulated leukemic cells. We have confirmed this observation, and have further observed responsiveness of leukemic cells to PHA to form colonies with *all* types of myelogenous leukemia in the acute phase, such as untreated AML, CML in blast crisis, and AMML. We have observed that only about 55% of AML cases respond to PHA stimulation. This difference between responding and nonresponding AML cells may be attributable to a difference in the stage of the leukemia sample studied. Among the responding AMLs, colony formation from about two-thirds of the cases was stimulated by CS in PHA-LyCM. The presence of PHA alone in the agar plates did not have the same effect. The main effect of CS on colony formation of PHA-stimulated leukemia leukocytes was the enhancement of the average size of the colonies and the degree of differentiation (Fig. 2). The effect of CS on the

FIGURE 2. Effect of lymphocyte conditioned medium on colony formation of leukocytes derived from blood of a patient with acute myelogenous leukemia presensitized with phytohemagglutinin. Leukocytes from an untreated patient with AML were treated overnight with 5 μg/ml PHA-P in a suspension culture. The PHA-sensitized cells were plated at 10^5 cells/plate in the upper layer of a two-layer semisoft agar medium. In addition to the presensitized cells, the upper layer contained 10% fetal calf serum and 10% horse serum and 0.3% agar in McCoy's 5A medium. The lower layer contained 10% each of fetal and horse serum, and 0.5% agar in McCoy's 5A medium. Conditioned medium at a concentration of 20%, when present, was placed at the lower layer of the culture system. (A) Representative colony derived from a plate that did not contain conditioned medium. (B) Representative colony derived from a plate that contained lymphocyte conditioned medium.

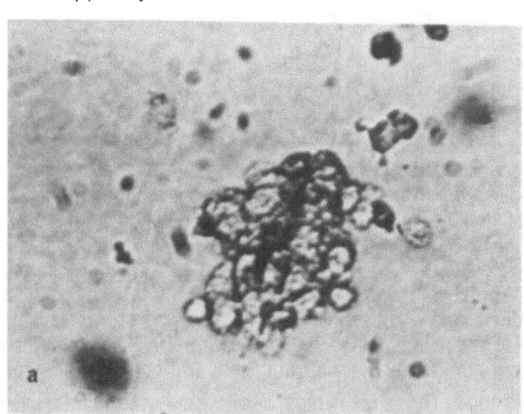
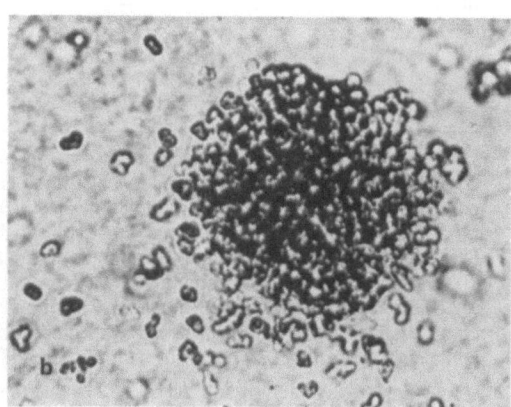

number of colony-forming units from PHA-stimulated leukemia cell populations was variable. Morphologic examination of the individual cells in the colonies showed the presence of mature macrophages and neutrophils.

In addition, some of the colonies were positive for nonspecific esterase, a marker for monocyte-macrophage cells (58) and some of the colonies was strongly positive for peroxidase, a marker for myeloid cells (58). No evidence could be found of T-lymphocyte colonies in semisoft agar medium, as reported by Rozenszajn et al. (47), who used human lymphocytes first sensitized with PHA and seeded on the upper layer of semisoft agar medium in the presence of PHA in the lower layer. At no time did we observe colony formation in our system when normal bone marrow cells were used.

In fact, based on the specific responsiveness of leukemic cells to PHA and then CS, myelogenous leukemic cells can be divided into four classes (shown in Table 7). Class I includes leukemic cells in which no response to any factor is seen; these cells therefore appear to be autonomous. Classes II–IV include leukemic cells in which proliferation and differentiation of varying degrees is observed. This classification implies that some of the population of AML patients contain semiautonomous leukemic cells. When such

cells are triggered by PHA for proliferation, these PHA-stimulated leukemic cells are then able to respond to normal regulators of differentiation. Under this new state, CS can further affect the proliferation and differentiation of the leukemic leukocytes. Figure 3 is a model of this two-step transition of AML leukocytes by PHA and CS. The PHA-sensitive step in myelogenous leukemic cells appears to precede the CS-sensitive step, because when we first treated AML cells with CS, removed CS, and plated the CS-treated AML cells in semisoft agar medium in the presence of PHA, no enhancement of colony formation was observed. The finding of the two-step response of leukemic cells to PHA and CS might have some clinical implication in developing a method in which to manage AML based on its response to regulatory factors *in vitro*.

An ideal situation whereby leukemic cells can be grown in culture is to find a specific "abnormal" factor that can stimulate only the growth of leukemic cells. In fact, Gallaher et al. discovered such a factor(s) from a specific conditioned medium derived from one human embryo that was specific for myelogenous human leukemia cells. Among 17 cases tested, all grew in suspension culture for a long period of time, some as long as 1 year (19) prior to the total loss of the conditioned medium.

TABLE 7 Effect of Mitogens and Colony Stimulators on the Colony Formation of Leukocytes from Leukemia Patients in Semisoft Agar Medium

CLASSIFICATION BASED ON TYPE OF RESPONSE	RESPONSE TO:		SOURCE OF COLONY-FORMING CELLS[b]				
	Mitogen[a]	Colony Stimulator (CS)[a]	AML	CML (BC)	AMML	ALL and CLL	NORMAL MARROW
I	−	−	16 (42.1%)	3 (10%)	0	12 (100%)	0
II	−	+	1 (2.6%)	10 (33%)	1 (20%)	0	15 (100%)
III	+	−	9 (23.7%)	6 (20%)	3 (60%)	0	0
IV	+	+	12 (31.6%)	11 (37%)	1 (20%)	0	n.a.
			38	30	5	12	15

[a]The procedures for CSA-dependent colony formation was described in the footnote of Table 2. Mitogen-stimulated colony formation was performed as follows: buffy coat leukocytes from each individual was at a concentration of 10^6 cell/ml was incubated overnight with either PHA-P (5 μ/ml DIFCO Lab, Detroit) in a suspension culture with RPMI 1640 medium with 10% fetal calf serum. The PHA-sensitized cells were plated at 10^5 cells/plate in the upper layer of a two-layer semisoft agar medium. Besides the presensitized cells, the upper layer contained 10% fetal calf serum and 10% horse serum and 0.3% agar in McCoy's 5A medium. The lower layer contained 10% each fetal and horse serum, and 0.5% agar in McCoy's 5A medium. Conditioned medium obtained from PHA-stimulated lymphocyte culture at a concentration of 20% when present, was placed at the lower layer of the culture system. Only those aggregates containing more than 50 cells were scored. The cell number per colony ranged from 50 to 1000 cells.

[b]All AML cases were untreated. All CML cases were in blast crisis. Those not in blast crisis were not used because most of them can respond to CSA.

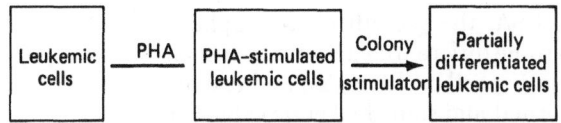

FIGURE 3. A model of proliferation of leukocytes present in the AML cell population. This model is based on the assumption made by Dicke and his colleagues that the colonies derived from AML stimulated by PHA are leukemic in origin (16; also *this volume*). Definite proof of this assumption is yet to be obtained.

DISCUSSION

Interactions between differentiating cells and the cellular environment appear to be essential processes for normal hemopoiesis. These interactions have long been recognized, but have been poorly measured. As a result of advances in tissue culture and biochemical techniques, some putative regulator factors were recently isolated; therefore, quantitative measurements of their function and more precise information on the differentiation process in culture will become feasible.

We have several methods with which to measure these functional proteins, i.e., CS activity in semisoft agar medium and in methylcellulose media and GS activity in suspension culture. Using these measurements, we fractionated conditioned medium obtained from short-term cultures of PHA-stimulated human lymphocytes by gel filtration with Sephadex G-150 and affinity chromatography with con A sepharose. Both CS and GS activities contain heterogeneous species of proteins with respect to molecular weight stimulating activities in various culture methods and to binding affinity to lecithin. Some of these proteins are similar and some appear quite distinct. Fractionation of each species of functional protein may lead to the identification of the correspondent FRC. This identification of a specific factor and its FRC is essential for studying specific cell-factor interactions at both the cellular and molecular levels.

Acute myelogenous leukemia has been proposed as the consequence of blockage of normal maturation. This may be visualized as a derangement in the interaction between regulatory factors and receptor sites in FRC. Some observations indicate that the defectiveness of cell-factor interactions could be caused by either alteration of FRC or defectiveness of the factors. Among those intrinsically defective myelogenous leukemic cells, about 50% appear to be autonomous, but they may be responsive to unknown growth conditions. Some can be stimulated to growth by PHA and to differentiate further by CS. Therefore, myelogenous leukemic cells might have at least two defects in the differentiation pathway. The first defect can be "unlocked" by PHA, which then sensitizes the cells to stimulation by CS. As the colony cells in this system obtained after CS stimulation are still not completely differentiated, it is likely that there are other defects still to be identified and corrected.

SUMMARY

Specific interactions between factors and progenitor cells are required for normal differentiation of hemopoietic cells. An impairment of cell-factor interaction may lead to various hemopoietic disorders. Leukemia could be a form of phenotypic alteration resulting from an impairment of factor-cell interaction. In culture, the granulocytic progenitor cells, also called granulocytic colony-forming units, are measured by their ability to form colonies in a semisoft medium system in the presence of colony-stimulating activity (CSA) or by their ability to proliferate and differentiate in suspension culture in the presence of growth-stimulating activity (GSA). Several studies on the heterogeneity of murine CSA have been reported. None of them have characterized individual CSA species for its specific function. In human, protein factors obtained from conditioned medium of short-term culture of lymphocytes (LyCM) contain both CSA and GSA. They are produced from cell populations enriched with T-lymphocytes and the production is stimulated 2 to 5-fold by phytohemagglutinin (PHA-LyCM). These factors are specific for primate hemopoietic cells. Both GSA and CSA contain heterogeneous species of proteins. At least three peak activities of CSA and GSA, respectively, are obtained when PHA-LyCM are fractionated in a Sephadex G150 column. Some GSA and CSA peaks are overlapped but some are not. Each CSA peak has a different ability to induce different types of colonies in semisoft medium. The peak with a molecular weight of 17,000 stimulates predominantly the growth of eosinophilic colonies, while the other peaks of CSA, the growth of macrophage-monocyte colonies. Similarly, each GSA peak has some differential activity in suspension culture with respect to their ability in stimulating cell proliferation and enhancing the number of colony-forming cells both of normal and some leukemic cells. One GSA is able to stimulate growth of T-type lymphoid cells in suspension culture. In addition, LyCM contains protein factors that enhance the growth of macrophage-type colonies in agar medium from actue myelocytic leukemia (AML) leukocytes pretreated with PHA. This is in contrast to the finding made by Dicke et al. that colony formation from PHA-stimulated AML leukocytes is not responsive to leukocyte feeder layer in semisoft agar medium (16). Further biochemical fractionation and functional characterization of these heterogeneous protein factors are obviously useful in understanding the biologic significance on the specificity of the interaction between factors and their responsive cells.

ACKNOWLEDGMENTS

The authors wish to thank Andrea Woods, Vicki Bear, Tom LeBon, and Ellen Hambleton for excellent technical assistance and useful discussions, and Dr. Morgan and Dr. Dean (Litton Bionetics, Inc., Bethesda, Maryland) for performance of E-rosette formation. This work was supported by Contract NO1-CP-33211 from the Virus Cancer Program, National Cancer Institute, National Institute of Health, Bethesda, Maryland.

REFERENCES

(1) Austin, P. E., McCulloch, E. A., and Till, J. E. Characterization of the factor in L-cell conditioned medium capable of stimulating colony formation by mouse marrow cells in culture. *J. Cell. Physiol.*, 77: 121, 1971.

(2) Axelrad, A. A., McLeod, D. L., Shreeve, M. M., and Heath, D. S. Properties of cells that produce erythrocytic colonies in plasma culture. In Robinson W. A., ed., *Proceedings of the Second International Workshop on Hemopoiesis in Culture, Airlie, Virginia*, New York: Grune & Stratton, 1973.

(3) Aye, M. T., Niho, Y., Till, J. E., and McCulloch, E. A. Studies of leukemic cell populations in culture. *Blood, 44*: 205, 1974.

(4) Aye, M. T., Till, J. E., and McCulloch, E. A. Growth of leukemic cells in culture. *Blood, 40*: 806, 1972.

(5) Barr, R. D., Whang-Peng, J., and Perry, S. Hemopoietic stem cells in human peripheral blood. *Science, 190*: 284, 1975.

(6) Boyum, A., Boecker, W., Carsten, A. L., and Cronkite, E. P. Proliferation of human bone marrow cells in diffusion chambers implanted into normal or irradiated mice. *Blood, 40*: 163, 1972.

(7) Bradley, T. R., and Metcalf, D. The growth of mouse bone marrow cells *in vitro. Aust. J. Exp. Biol. Med. Sci.*, 44: 287, 1966.

(8) Bradley, T. R., Stanley, E. R., and Sumner, M. A. Factors from mouse tissues stimulating colony growth of mouse bone marrow cells *in vitro. Aust. J. Exp. Biol. Med. Sci.*, 49: 595, 1971.

(9) Bull, J. M., Duttera, M. J., Stashick, E. D., Northup, J., Henderson, E., and Carbone, P. P. Serial *in vitro* marrow culture in acute myelocytic leukemia. *Blood, 42*: 679, 1973.

(10) Cerny, J. Stimulation of bone marrow haemopoietic stem cells by a factor from activated T-cells. *Nature (London), 249*: 63, 1974.

(11) Clarke, B. J., Axelrad, A. A., Shreeve, M. M., and McLeod, D. L. Erythroid colony induction without erythropoietin by Friend leukemia virus *in vitro. Proc. Natl. Acad. Sci. USA, 72*: 3556, 1975.

(12) Clarkson, B. Review of recent studies of cellular proliferation in acute leukemia. *Natl. Cancer Inst. Monog. 30*: 81, 1969.

(13) Cline, M. J., and Golde, D. W. Production of colony-stimulating activity by human lymphocytes. *Nature (London), 248*: 703, 1974.

(14) Cooper, M. C., Levy, J., Cantor, L. N., Marks, P. A., and Rifkind, R. A. The effect of erythropoietin on colonial growth of erythroid precursor cells *in vitro. Proc. Natl. Acad. Sci. USA, 71*: 1677, 1974.

(15) Dexter, T. M., and Testa, N. G. Differentiation and proliferation of haemopoietic cells in culture. In Prescott, D. M., ed., *Methods in Cell Biology*, Vol. 14, New York; Academic Press, 1976.

(16) Dicke, K. A., Spitzer, G., and Ahearn, M. J. Colony formation *in vitro* by leukaemic cells in acute myelogenous leukaemia with phytohaemagglutinin as stimulating factor. *Nature (London) 259*: 129, 1976.

(17) Fibach, E., Gerass, E., and Sachs, L. Induction of colony formation *in vitro* by human lymphocytes. *Nature (London)* 259: 127, 1976.

(18) Foster, R., Jr., and Metcalf, D. Bone marrow colony stimulating activity in human sera. *Br. J. Haematol.*, 15: 147, 1968.

(19) Gallagher, R. E., and Gallo, R. C. Continuous production of complete type-C virus by exponentially-growing cultured leukocytes from one to sixteen patients with myelogenous leukemia. *Proc. 2nd Int. Congr. Pathol. Physiol., Prague, Czechoslovakia.*

(20) Gallagher, R. E., Salahuddin, S. Z., Hall, W. T., McCredie, K. B., and Gallo, R. C. Growth and differentiation in culture of leukemic leukocytes from a patient with acute myelogenous leukemia and re-identification of type-C virus. *Proc. Natl. Acad. Sci. USA*, 72: 4137, 1975.

(21) Gallo, R. C. On the nature of the cellular defect in acute leukemia. *Med. Clin. North Am.*, 57: 343, 1973.

(22) Glass, J., Lavidor, L. M., and Robinson, S. H. Use of cell separation and short-term culture techniques to study erythroid cell development. *Blood*, 46: 705, 1975.

(23) Gordon, M. Y., Blackett, N. M., and Douglas, I. D. C. Colony formation by human haemopoietic precursor cells cultured in semi-solid agar in diffusion chambers. *Br. J. Haematol.*, 31: 103, 1975.

(24) Gregory, C. J., McCulloch, E. A., and Till, J. E. Transient erythropoietic spleen colonies: Effects of erythropoietin in normal and genetically anemic w/w^v mice. *J. Cell. Physiol.*, 86: 1, 1975.

(25) Ichikawa, Y. Differentiation of a cell line of myeloid leukemia. *J. Cell. Physiol.*, 74: 223, 1969.

(26) Iscove, N. N., Senn, J. S., Till, J. E., and McCulloch, E. A. Colony formation by normal and leukemic human marrow cells in culture: Effect of conditioned medium from human leukocytes. *Blood*, 37: 1, 1971.

(27) Iscove, N. N., Sieber, F., and Winterhalter, K. H. Erythroid colony formation in cultures of mouse and human bone marrow: Analysis of the requirement for erythropoietin by gel filtration and affinity chromatography on agarose-concanavalin A. *J. Cell. Physiol.*, 83: 309, 1974.

(28) Jacobson, L. D., Goldwasser, E., and Gurney, C. W. Transfusion induced polycythemia as a model for studying factors influencing erythropoiesis. In Wolstenholme, G. E. W. and O'Connor, M., eds., *Ciba Foundation Symposium on Haemopoiesis*, p. 423. London: Churchill, 1960.

(29) Lafleur, L., Underdown, B. J., Miller, R. G., and Phillips, R. A. Differentiation of lymphocytes: Characterization of early precursors of B lymphocytes. *Ser. Haematol.*, 5: 50, 1972.

(30) Landau, T., and Sachs, L. Characterization of the inducer required for the development of macrophage and granulocyte colonies. *Proc. Natl. Acad. Sci. USA*, 68: 2540, 1971.

(31) Levis, W. R., and Robbins, J. W. Methods for obtaining purified lymphocytes, glass-adherent mononuclear cells, and a population containing

both cell types from human peripheral blood. *Blood*, 40: 77, 1972.

(32) Lozzio, C. B., and Lozzio, B. B. Human chronic myelogenous leukemia cell-line with positive Philadelphia chromosome. *Blood*, 45: 321, 1975.

(33) McCulloch, E. A., Mak, T. W., Price, G. B., and Till, J. E. Organization and communication in populations of normal and leukemic hemopoietic cells. *Biochim. Biophys. Acta*, 335: 260, 1974.

(34) McCulloch, E. A., and Till, J. E. Regulatory mechanisms acting on hemopoietic stem cells. Some clinical implications. *Amer. J. Pathol.*, 65: 601, 1971.

(35) McLeod, D. L., Shreeve, M. M., and Axelrad, A. A. Induction of megakaryocyte colonies with platelet formation *in vitro*. *Nature (London)*, 261: 492, 1976.

(36) Messner, H. A., Till, J. E., and McCulloch, E. A. Specificity of interacting populations affecting granulopoiesis in culture. *Blood*, 44: 671, 1974.

(37) Metcalf, D., MacDonald, H. R., Odartchenko, N., and Sordat, B. Growth of mouse megakaryocyte colonies *in vitro*. *Proc. Natl. Acad. Sci. USA*, 72: 1744, 1975.

(38) Metcalf, D., Warner, N. L., Nossal, G. J. V., Miller, J. F. A. P., Shortman, K., and Rabellino, E. Growth of B lymphocyte colonies *in vitro* from mouse lymphoid organs. *Nature (London)*, 255: 630, 1975.

(39) Morgan, D. A., Ruscetti, F. W., and Gallo, R. C. Selective *in vitro* growth of T-lymphocytes from normal human bone marrow. *Science*, 193: 1007, 1976.

(40) Nooter, K., and Ghio, R. Hormone-independent *in vitro* erythroid colony formation by bone marrow from Rauscher virus-infected mice. *J. Natl. Cancer Inst.*, 55: 59, 1975.

(41) Perry, S., and Gallo, R. C. Physiology of human leukemic leucocytes kinetics and biochemical consideration. In Gordon, A., ed., *Regulation of Hemopoiesis*, p. 1221. New York: Appleton-Century-Crofts, 1970.

(42) Pluznik, D. H., and Sachs, L. The cloning of normal 'mast' cells in tissue culture. *J. Cell. Comp. Physiol.*, 66: 319, 1965.

(43) Pope, J. H., Scott, W., and Moss, D. J. Human lymphoid cell transformation by Epstein-Barr virus. *Nature [New Biol.]*, 246: 140, 1973.

(44) Prival, J. T., Paran, M., Gallo, R. C., and Wu, A. M. Colony-stimulating factors in cultures of human peripheral blood cells. *J. Natl. Cancer Inst.*, 53: 1583, 1974.

(45) Robinson, W. A., and Pike, B. L. Colony growth of human bone marrow cells *in vitro*. In Stohlman, F., ed., *Haemopoietic Cellular Proliferation*, pp. 249–259. New York: Grune & Stratton, 1970.

(46) Robinson, W. A., Stanley, E. R., and Metcalf, D. Stimulation of bone marrow colony growth *in vitro* by human urine. *Blood*, 33: 396, 1969.

(47) Rozenszajn, L. A., Shoham, D., and Kalechman, I. Colonal proliferation of PHA-stimulated human lymphocytes in soft agar culture. *Immunology*, 29: 1041, 1975.

(48) Ruscetti, F. W., and Chervenick, P. A. Release of colony-stimulating activity from thymus-derived lymphocytes. *J. Clin. Invest., 55*: 520, 1975.

(49) Sachs, L. Regulation of membrane changes, differentiation and malignancy in carcinogenesis. *Harvey Lect., 68*: 1, 1974.

(50) Sheridan, J. W., and Stanley, E. R. Tissue sources of bone marrow colony stimulating factor. *J. Cell. Physiol., 78*: 451, 1971.

(51) Stephenson, J. R., Axelrad, A. A., McLeod, D. L., and Shreeve, M. M. Induction of colonies of hemoglobin-synthesizing cells by erythropoietin *in vitro. Proc. Natl. Acad. Sci. USA, 68*: 1542, 1971.

(52) Till, J. E., and McCulloch, E. A. A direct measurement of the radiation sensitivity of normal mouse bone marrow cells. Radiat. Res., 14: 213, 1961.

(53) Watson, J., and Prichard, J. Characterization of a factor required for the differentiation of myeloid and lymphoid cells in vitro. *J. Immunol., 108*: 1209, 1972.

(54) Wood, W. G. Erythroid cell proliferation in human bone marrow suspension cultures. *Br. J. Haemat., 26*: 441, 1974.

(55) Wood, W. G. Haemoglobin synthesis in suspension cultures of human bone marrow. *Br. J. Haemat., 26*: 451, 1974.

(56) Wu, A. M., and Gallo, R. C. Biochemical basis of leukemia and lymphoma in man. In Hoffbrand, A. V., Brain, M. C., and Hirsch, J., eds., *Recent Advances in Haematology* Vol. 2. London: Churchill Livingstone, in press.

(57) Wu, A. M., and Gallo, R. C. The phenotypic abnormality in leukemia: A defective cell-factor interaction? In Neth, R., Gallo, R. C., Mannweiler, K., and Moloney, W. C., eds. *The Symposium on Modern Trends in Human Leukemia:* II; pp. 51–62. Veriag, Munich: J. F. Lehmanns, 1976.

(58) Yam, L. T., Li, C. Y., and Crosby, W. H. Cytochemical identification of monocytes and granulocytes. *Am. J. Clin. Pathol., 55*: 283, 1971.

PART V

Bone Marrow Transplantation Immunology

John J. Trentin Ph.D.

The history of bone marrow transplantation, from the days of Egon Lorenz, has been replete with surprising revelations and important contributions to the fields of immunology and hematology. I well remember the surprise and excitement caused by the discovery that the "secondary irradiation syndrome" or secondary disease of "homologous" irradiation chimeras was not a secondary rejection of the marrow graft, but rather a secondary reaction of the graft versus the host (17, 18, 19). This discovery additionally implied what was not yet then known, namely, that in addition to donor repopulation of the myeloid system, bone marrow transplantation resulted also in functional donor repopulation of the lymphoid system. In those days, although a mere 20 years ago, we did not yet know about the thymic-dependent and -independent portions of the lymphoid system. The role of the thymus in immunity was still to be discovered!

There have been so many important contributions of experimental bone marrow transplantation to the field of immunology, that it is impossible to do justice to them in the time allotted. Instead, I would like to skip now to one of the most exciting recent developments in the immunology and genetics of bone marrow transplantation: The long known exception to the laws of tissue transplantation, known originally as hybrid resistance to parental marrow transplantation, is emerging as a natural lymphoma-leukemia defense mechanism. For most of these past 20 years it has been known that the transplantation of bone marrow and lymphoid tissue presented a striking exception to the genetic laws of tissue transplantation that govern the take or rejection of the other tissues of the body (3). Although the F_1 hybrid mouse is ordinarily the universal recipient of grafts of all other tissues of the inbred parental strains, in some genetic combinations the lethally radiated F_1 hybrid would not accept inbred parental marrow or lymphoid grafts from one parent strain. Although some investigators believed this to represent a passive failure to "take," it is in fact an active and quick rejection, resembling specific presensitization but in the absence of prior exposure to donor tissue (2). The mechanism of resistance is most unusual and has no previously known counterpart. The effector cell is not a T-cell or B-cell, but has some of the properties of macrophages (13). It originates in the marrow but congregates in the spleen (12). Resistance "matures" abruptly at about 21 days of postnatal age in the mouse (1, 16). It is directed at unique antigens determined by noncodominant genes, and is

Bone Marrow Transplantation Immunology

John J. Trentin, Rolf Kiessling, Hans Wigzell, Michael T. Gallagher, Surjit K. Datta, and Sulabha S. Kulkarni

19

expressed only on hemopoietic and lymphoid tissues (1). It is extremely radioresistant but can be abrogated in a variety of ways. These include:

1. Extremely high single exposure to wholebody radiation (2200 to 5000 R), or lower exposures to fractionated radiation (16).

2. Very high cell dose of bone marrow alone, or a standard dose plus a high dose of donor type thymus cells. Resistance breaks in the marrow cavity at lower donor cell doses than in spleen, resulting in radiation survival in the absence of early spleen repopulation (15).

3. Pretreatment of recipient with cyclophosphamide, or *Corynebacterium parvum*, or silica particles, or carrageenan, or heterologous antisera against bone marrow or thymus (4, 13, 16). Thymectomy does not abrogate, whereas marrow ablation by ^{89}Sr does abrogate resistance, indicating that the abrogating activity of some antithymocyte sera is due to cross-reactivity with the marrow effector cells (12).

In recent years it has been shown that the same mechanism of resistance to parental marrow transplantation operates in some allogeneic combinations (2) and, most surprisingly, in some xenogeneic (rat to mouse) combinations (14). The many characteristics common to hybrid resistance, allogeneic resistance, and xenogeneic resistance are shown in Table 1. The designation of

genetic resistance (GR) to bone marrow transplantation (BMT) was therefore adopted to refer to hybrid resistance, allogeneic resistance, and xenogeneic resistance to BMT, either collectively or individually (20).

Now, what could possibly be the natural function, the survival value, of this unusual resistance mechanism? Abrogation of GR to BMT by ^{89}Sr ablation of the bone marrow also abrogates resistance to Friend virus-induced erythroleukemia (12). The C57 strain of mice is very resistant to spontaneous leukemia, even though the strain harbors murine leukemia virus. GR to BMT is strongly expressed in the C57 strain and its hybrids. Fractionated low dose radiation (four weekly exposures to 225 R) abrogates GR to BMT, and also abolishes resistance of C57 mice to their own endogenous leukemia virus, resulting in a high incidence of virus-mediated and virus-transmissible lymphoma-leukemia (7). These considerations led us to test whether GR to BMT represents a natural lymphoma-leukemia defense mechanism, as follows: Lethally radiated (C57 × AKR) F_1 hybrid mice have genetic resistance to C57 parental bone marrow cells, but not to AKR parental bone marrow cells (Fig. 1). Lethally radiated (C3H × AKR) F_1 hybrids do not have GR to bone marrow transplantation from either of their parental strains. However, intravenously transplanted AKR lymphoma cells produced lymphomatous spleen colonies and mortality in radiated (C3H × AKR) F_1 hybrids but *not* in radiated (C57 × AKR) F_1. Thus, "responder" (C57 × AKR) F_1 hybrids can recognize and reject AKR lymphoma cells, but *not* normal AKR bone marrow cells. A normal biologic role of lymphoma-leukemia surveillance was postulated for genetic resistance to marrow transplantation, directed at antigens which, like TL, are expressed on normal hemopoietic cells of some strains, but only on leukemic cells of other strains (5, 6).

TABLE 1 Common Characteristics of Hybrid Resistance, Allogeneic Resistance, and Zenogeneic Resistance to Bone Marrow Transplantation

1. All are contingent on genotype of donor and/or recipient.
2. All involve, but are not limited to, the C57 strain and its hybrids.
3. All mature abruptly at 3 weeks of age in the mouse.
4. All are extremely radioresistant (> 1100 R).
5. All can be abrogated by very high single-doze wholebody radiation:
   hybrid resistance; 2200 to 5000 R
   xenogeneic resistance; 6600 to 8000 R or lower doses of fractionated radiation.
6. None requries priming.
7. Resistance can be broken by very high donor cell doses (bone marrow alone or bone marrow + thymus).
8. Resistance breaks in marrow at lower cell doses than in spleen, resulting in radiation survival in absence of *early* spleen repopulation.
9. All can be abrogated by cyclophosphamide.
10. All can be abrogated by *Corynebacterium parvum*.
11. All can be abrogated by silica particles.
12. All can be abrogated by carrageenan.
13. All can be abrogated by heterologous antisera against bone marrow or thymus.

FIGURE 1. Schema of take (+) or rejection (−) of intravenously transfused inbred parental bone marrow (BM) or lymphoma cells in lethally irradiated (950R 137Cs, total body) (C_{57} × AKR) F_1 responder mice and (C_3H × AKR) F_1 nonresponder mice.

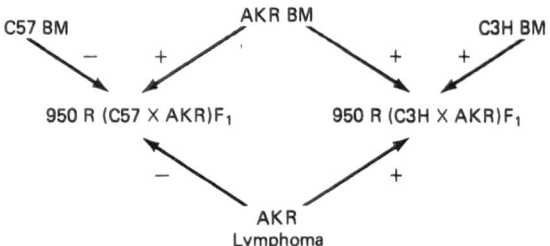

TABLE 2 Comparison of Genetic Resistance to C57 Bone Marrow Transplantation *In Vivo*, with Spleen NK[a] Cell-Mediated Lysis of YAC Lymphoma Cells *In Vitro*

AGE OF (C57 × A)F_1 MICE	TREATMENT OTHER THAN RADIATION	RESISTANCE TO B.M.T. IN VIVO		LYSIS OF YAC LYMPHOMA *IN VITRO*	
		Wholebody radiation (R)	Present, absent, or % of control resistance	Wholebody radiation (R)	Present, absent, or % of positive control lysis
12–18 days		1100	—[b]	0	—
22–30 days		1100	+[b]	0	+
Young adult		1100	+[b]	0	+
Young adult		1100	100[b]	1100	100
Young adult		2200	81[c]	2200	34
Young adult		4400	8[c]	4400	24
Young adult		6600	0[c]	6600	0
Young adult	Silica	1100	76[c]	0	74
Young adult	Cyclophosphamide	1100	41[c]	0	24
Young adult	Carrageenan	1100	23[c]	0	17

[a]NK = natural killer.

[b]Based on spleen colony counts of piror experiments.

[c]Based on spleen colony counts of concurrent experiments.

Attempts to develop an *in vitro* effector system for GR to BMT have been unfruitful. However, several *in vitro* natural killer cell systems for rodent tumor cells have been reported. While some of these have been shown to be T-cell-mediated or antibody-dependent, others, directed against mouse lymphoma cells, have been shown to be mediated by cells other than T- or B-lymphocytes, to be antibody-independent, and to have characteristics in common with GR to BMT (9, 11). One of these is directed against the Moloney virus-induced YAC lymphoma of A/Sn mice. Reactivity to this lymphoma is genetically determined, with the C57 strain and its hybrids being good responders, both by *in vitro* cytolysis (8) and by *in vivo* rejection of transplanted lymphoma cells (10). We compared the effect of a va-riety of factors on splenic natural killer (NK) cells which were directed against YAC lymphoma cells *in vitro*. These factors had a known effect on GR to BMT *in vivo*. Splenic NK cells were found to appear at about 3 weeks of age, and to be radioresistant to 1100 R wholebody radiation, but partially inhibited by 2200 R and totally inhibited by 6600 R. Treatment of mice with cyclophosphamide or carrageenan or silica suppressed both GR to BMT and the *in vitro* cytotoxic effect of splenic NK cells on YAC lymphoma cells (Table 2). Moreover, treatment of (C57 × A) F_1 mice with horse anti-mouse thymocyte serum or rabbit anti-mouse bone marrow serum, both of which have suppressive effects on genetic resistance to bone marrow transplantation, also suppressed splenic NK cell lysis of YAC lymphoma cells *in vitro* (Table 3).

TABLE 3 Effect of Treatment of (C57 × A)F_1 Mice with Horse Anti-Mouse Thymocyte Globulin (HAMTG) or Rabbit Anti-Mouse Bone Marrow Serum (RAMBMS) on Spleen NK Cell-Mediated Lysis of YAC Lymphoma Cells *In Vitro*[c]

TREATMENT	24-HR		48-HR	
	% Lysis[a]	% Inhibition of Lysis	% Lysis[a]	% Inhibition of Lysis
None (normal control)	28.5	—	21	—
15 mg HAMTG i.p.	1	96	4	80
25 mg HAMTG i.p.	0	100	1.5	92
0.3 ml RAMBMS i.v.	2.5	91	6	71
0.3 ml RAMBMS i.p.	nd[b]	—	8	61
0.5 ml RAMBMS i.p.	3	89	8	61

[a]Average of 2 animals.

[b]ND = Not done.

[c]Spleen cell to YAC cell ratio = 100:1.

Results of a similar nature have also been obtained by Drs. Wigzell and Kiessling in collaboration with Dr. G. Cudkowicz and Dr. G. M. Shearer.

It thus appears that GR to BMT may indeed be a manifestation of a natural resistance mechanism against lymphoma-leukemia.

SUMMARY

One of the most exciting recent developments in the immunology and genetics of bone marrow transplantation is that the long-known exception to the laws of tissue transplantation, known originally as hybrid resistance to parental bone marrow transplantation, is emerging as a natural leukemia-lymphoma defense mechanism.

Lethally radiated F_1 hybrid mice of certain genotypes only actively and specifically reject normal parental bone marrow or lymphoid cells from some parent strains, not others without prior immunization with donor cells. In recent years it has been shown that the same mechanism operates in some allogeneic and xenogeneic donor-recipient combinations, so the collective term of genetic resistance (GR) to bone marrow transplantation (BMT) has been adopted (20). The effector mechanism matures abruptly at about 21 days of age and is radioresistant and thymic independent. It is directed at unique antigens determined by non-codominant genes, and mediated by macrophage-like cells of the bone marrow and spleen, but not of peritoneal exudate.

Abrogation of GR to BMT by either ^{89}Sr (12) or fractionated radiation (16) also abrogates resistance to Friend virus-induced and to "spontaneous" leukemia, respectively. GR to BMT is strongly expressed in the C57 strain and its hybrids. The C57 strain is very resistant to spontaneous leukemia. These considerations led us to test whether GR to BMT represents a natural lymphoma-leukemia defense mechanism, as follows:

(C57 × AKR) F_1 hybrid mice show GR to C57 parental bone marrow cells, but not to AKR parental bone marrow cells (C3H × AKR) F_1 hybrids show no GR to bone marrow transplantation from either parental strain. However, transplantation of AKR lymphoma cells into lethally irradiated "resistant" (C57 × AKR) F_1 and "nonresistant" (C3H × AKR) F_1 hybrids produced lymphomatous spleen colonies in "nonresistant" hybrids but *not* in "resistant" hybrids. Thus "resistant" (C57 × AKR) F_1 hybrids can recognize and reject AKR lymphoma cells, but *not* normal AKR bone marrow cells. A normal biologic role of leukemia-lymphoma surveillance was postulated for genetic resistance to marrow transplantation, directed at antigens which, like TL, are expressed on normal hemopoietic cells of some strains, but only on leukemic cells of other strains (5).

More recently, Klein et al. (11a) have described a natural killer cell mechanism against YAC lymphoma cells *in vitro*, which appears to have the properties of GR to BMT (Hans Wigzell, personal communication).

ACKNOWLEDGMENTS

This investigation was supported by USPHS grants CA 12093, CA 03367, K6 CA 14219 and HL 17269.

REFERENCES

(1) Cudkowicz, G. Hybrid resistance to parental grafts of hematopoietic and lymphoma cells. In *The Proliferation and Spread of Neoplastic Cells*. Baltimore: The Williams and Wilkins Company, p. 661, 1968.

(2) Cudkowicz, G., and Bennett, M. Peculiar im-

munobiology of bone marrow allografts. I. Graft rejection by irradiated responder mice. *J. Exp. Med., 134*: 83, 1971.

(3) Cudkowicz, G., and Stimpfling, J. J. Deficient growth of C57 BL marrow cells transplanted in F_1 hybrid mice. Association with the histocompatibility-2 locus. *Immunology, 7*: 291, 1964.

(4) Gallaher, M. T., Rauchwerger, J. M., Lotzova, E., Monie, H. J., and Trentin, J. J. Genetic resistance: Lymphoid antisera and absorption studies. *Exp. Hematol. (Copenhagen) 2*: 303, 1974.

(5) Gallagher, M. T., Lotzova, E., and Trentin, J. J. Genetic resistance to marrow transplantation as a leukemia defense mechanism. *Biomedicine, 25*: 1, 1976.

(6) Gallagher, M. T., Lotzova, E., and Trentin, J. J. Genetic resistance to marrow transplantation as a leukemia defense mechanism. In Battisto, J. R., and Streilein, J. W., eds., *Proceedings of the Conference on Immuno-aspects of the Spleen.* Amsterdam: North-Holland Publishing Company, 1976.

(7) Kaplan, H. S. On the natural history of murine leukemias. *Cancer Res., 27*: 1325, 1967.

(8) Kiessling, R., Klein, E., and Wigzell, H. "Natural" killer cells in the mouse. I. Cytotoxic cells with specificity for mouse Moloney leukemia cells: Specificity and distribution according to genotype. *Eur. J. Immunol., 5*: 112, 1975.

(9) Kiessling, R., Klein, E., Pross, H., and Wigzell, H. "Natural" killer cells in the mouse. II. Cytotoxic cells with specificity for mouse Moloney leukemia cells. Characteristics of the killer cell. *Eur. J. Immunol., 5*: 117, 1975.

(10) Kiessling, R., Petranyi, G., Klein, G., and Wigzell, H. Genetic variation of *in vitro* cytolytic activity and in vivo rejection potential of non-immunized semi-syngeneic mice against a mouse lymphoma line. *Int. J. Cancer, 15*: 933, 1975.

(11) Kiessling, R., Petranyi, G., Karre, K., Jondal, M., Tracey, D., and Wigzell, H. Killer cells: A functional comparison between natural, immune T-cell and antibody-dependent in vitro systems. *J. Exp. Med., 143*: 772, 1976.

(11a) Klein, E., Becker, S., Svedmyr, E., Jondal, M., Vanky, F. Tumor infiltrating lymphocytes. *Ann. N.Y. Acad. Sci. 276*:207, 1976.

(12) Kumar, V., Bennett, M., and Eckner, R. J. Mechanism of genetic resistance to Friend virus leukemia in mice. I. Role of ^{89}Sr-sensitive effector cells responsible for rejection of bone marrow allografts. *J. Exp. Med., 139:* 1093, 1974.

(13) Lotzova, E., Gallagher, M. T., and Trentin, J. J. Macrophage involvement in genetic resistance to bone marrow transplantation. *Transpl. Proc.* (in press).

(14) Rauchwerger, J. M., Gallagher, M. T., and Trentin, J. J. "Xenogeneic resistance" to rat bone marrow transplantation. I. The basic phenomenon. *Proc. Soc. Exp. Biol. Med., 143*: 145, 1973.

(15) Rauchwerger, J. M., Gallagher, M. T., and Trentin, J. J. "Xenogeneic resistance" to rat bone marrow transplantation. II. Relationship of hemopoietic regeneration and survival. *Biomedicine, 18*: 109, 1973.

(16) Rauchwerger, J. M., Gallagher, M. T., Monie, H. J., and Trentin, J. J. "Xenogeneic resistance" to rat bone marrow transplantation. III. Maturation age, and abrogation with cyclophosphamide, *Corynebacterium parvum* and fractionated irradiation. *Biomedicine, 24*: 20, 1976.

(17) Trentin, J. J. Mortality and skin transplantability in X-irradiated mice receiving isologous, homologous, or heterologous bone marrow. *Proc. Soc. Exp. Biol. Med., 92*: 688, 1956.

(18) Trentin, J. J. Induced tolerance and "homologous disease" in X-irradiated mice protected with homologous bone marrow. *Proc. Soc. Exp. Biol. Med., 96*: 139, 1957.

(19) Trentin, J. J. Grafted-marrow-rejection mortality contrasted to homologous disease in irradiated mice receiving homologous bone marrow. *J. Nat. Cancer Inst., 22*: 219, 1959.

(20) Trentin, J. J., Rauchwerger, J. M., and Gallagher, M. T. Genetic resistance to marrow transplantation. *Biomedicine, 18*: 86, 1973.

INTRODUCTION

The graft-versus-host (GvH) reaction is a complex biologic phenomenon, the underlying cellular events of which are not fully defined (for reviews, see refs. 8 and 14). Studies in congenic mice indicate an important relationship between the Ir region of the H-2 gene complex and GvH reactivity (9, 18, 20). Because Ir gene products are known to influence lymphocyte recognition and proliferation in the mixed lymphocyte reaction (MLR) (1, 2, 19), it has been suggested that GvH and MLR genes are identical (18, 25). There are considerable data to support this concept as well as data to suggest an important role for non-MLR and non-Ir region genes in initiating GvH disease. Isolated K-region and possibly D-region differences can produce a weak GvH reaction (18) as can non-H-2 loci (5, 10). Demonstration of GvH reactivity of non-Ir region loci depends on the assay employed and frequently occurs in the absence of a detectable MLR or the generation of cytotoxic lymphocytes. GvH disease frequently develops in dogs and primates grafted with MLR nonreactive allogeneic bone marrow (7, 27, 28) without acquisition of MLR reactivity (22). Finally, there are preliminary data, which suggest that GvH and MLR reactive cells represent separate T-lymphocyte subpopulations expressing quantitative differences in Thy-1 antigen and in Ly alleles (15, 17).

To investigate the relationship between GvH and MLR reactive cells, we studied the GvH reactivity of lymphocyte populations specifically depleted of histocompatibility alloantigen reactive cells as defined in the MLR. We also investigated the effects of these *in vitro* manipulations of committed granulocyte-monocyte precursors (CFU-c).

20

Modulation of Graft-versus-Host (GvH) Disease in the Rat; Effect of Hydroxyurea on the Mixed Lymphocyte Reaction and Graft-versus-Host Reactivity

Robert Peter Gale, Caprice Rutkosky, and David W. Golde

MATERIALS AND METHODS

ANIMALS

Inbred rats, Lewis (Lew, Ag-B[1]); Buffalo (BU, Ag-B[6]); Brown Norway (BN, Ag-B[3]); (Lew × BU) F1; and (Lew × BN) F1, were obtained from Microbiological Associates (Bethesda). Animals were kept in "clean" rooms in autoclaved cages and received tetracycline in their drinking water. Nonsterile food was used.

MIXED LYMPHOCYTE REACTION

Responder cell populations were prepared by harvesting spleens from 12 to 16-week-old rats. Spleens were teased apart with sterile forceps and a single cell suspension prepared by aspiration through a syringe and passage through a loose-weave nylon cloth. Red cell contamination was removed by treatment with NH_4Cl–Tris buffer. Stimulator cells were inactivated by incubation with mitomycin C (Sigma, St. Louis; 50 mg/ml × 30 min) and are designated by the subscript m. Cell viability determined by 0.2% Trypan Blue dye exclusion was ≥95%. Responder cells (1.5 × 10^6) and equal numbers of mitomycin C-treated stimulator cells were cultured in 1 ml RPMI-1640 (Gibco, Grand Island, N.Y.) supplemented with 5% glutamine, 5% penicillin, and streptomycin stock solution (Gibco), and 20% heat-inactivated fetal calf serum previously screened for its ability to support rat MLR. Cultures were set up in 16 × 125 mm plastic screwtop tubes (Falcon, Oxnard, Ca.) in a humidified environment of 6% CO_2 in air at 37°C. After 120 hr of culture, 5 μCi tritiated thymidine (^{3}H-TdR; specific activity 6.7 Ci/mM; New England Nuclear, Boston, Mass.) was added for 16 hr. Cultures were terminated by adding cold media and TCA precipitable counts determined as described (6). One tube from each experimental group was assayed for viability and cell recovery. ^{3}H-TdR incorporation was expressed as counts per minute (cpm) ± SEM from quadruplicate cultures. Stimulation indices (SI) were determined by dividing counts per minute of stimulated cultures by counts per minute of controls containing mitomycin-treated autologous stimulator cells.

HYDROXYUREA

Hydroxyurea, 5 × 10^{-4} M (Sigma) was added to selected MLRs for 16 hr at predetermined intervals (24–96 hr). Culture tubes were centrifuged, cells washed, and fresh media added. MLRs were harvested and processed as described, between 120 and 140 hr.

TWO-STAGE MLR

MLRs were initiated with primary mitomycin treated stimulator cells. At 48 hr, selected cultures were pulsed with 5 × 10^{-4} M hydroxyurea for 16 hr and were then pooled and washed after which cell count and viability were determined. Second-stage MLRs were initiated with 1.5 × 10^6 viable hydroxyurea-treated responder cells and an equal number of fresh secondary mitomycin-treated stimulator cells. Second-stage MLRs were cultured for 96 to 120 hr and processed as described (vide supra).

POPLITEAL LYMPH NODE ASSAY

The popliteal lymph node assay was used to evaluate GvH reactivity (11). Cultured, or hydroxyurea-treated spleen cells (1.5–3.0 × 10^6) were injected into the footpads of 6 to 8-week-old (Lew × BU)F1 or (Lew × BN)F1 rats. Animals were sacrificed 7–8 days later and the popliteal lymph nodes were carefully dissected and weighed. Groups of four to eight recipients were used, and data were expressed either as $\overline{X}$ ± SEM or in terms of relative GvH potency. The latter was determined by extrapolating the dose of inoculated cells required to produce a 10-mg popliteal lymph node as compared to controls.

COLONY-FORMING CELLS (CFU-c)

Granulocyte-monocyte precursors capable of colony formation in vitro (CFU-c) were assayed using a modified methylcellulose culture technique (16). Spleen cells, freshly prepared or recovered from MLR cultures, were suspended in 1 ml 0.8% methylcellulose with alpha medium and 15% fetal calf serum and plated over an underlayer of 1 ml 0.5% agar containing 40 μl post-endotoxin rat serum (12, 13). The cultures were incubated at 37°C in a humidified environment of 7.5% CO_2 in air for 7–10 days. Colonies containing 50 or more cells were enumerated with a dissecting microscope.

RESULTS
EFFECTS OF HYDROXYUREA

In preliminary experiments optimal conditions for MLR reactivity were determined. Proliferative response measured by ^{3}H-TdR incorporation was first detected at 48 hr, peaked at 96–120 hr, and declined to baseline by 140 hr. Subsequent cultures were assayed at 120 hr. The addition of 16 hr pulses of hydroxyurea to ongoing MLRs significantly inhibited ^{3}H-TdR incorporation. Inhibition was most marked at 24–48 hr (Table 1). By 96 hr, only modest inhibition was observed. To determine the specificity of the hydroxyurea-induced MLR inhibition, treated cells were rechallenged with secondary stimulator cells related or unrelated to the primary stimulator cell. Results are indicated in Table 2. In three or four experiments, inhibition of MLR reactivity was specific for histocompatibility antigens of the primary stimulator cell.

TABLE 1 Time Kinetics of Hydroxyurea Effects on MLC[a]

| TIME (HR) | ^{3}H-TdR INCORPORATION (CPM + SEM) | | | |
	BU × BU$_m$	BU × (Lew × BU)F1$_m$	SI[b]	% INHIBITION[c]
None	3210 ± 676	7214 ± 792	2.25	—
48	2951 ± 113	1383 ± 113	0.47	79.2
72	1797 ± 182	1795 ± 177	0.99	56.0
96	1748 ± 189	3322 ± 351	1.90	15.6

[a] MLCs were cocultivated for 120-hr as described in Materials and Methods. HU, 5×10^{-4} was added for 16-hr pulses at the time indicated.

[b] Stimulation index, ratio of CPM, stimulated/unstimulated (BU × BU$_m$) cultures.

[c] Calculated as a function of maximal SI of cultures BU × (Lew × BU)F1$_m$, without HU pulse.

GvH REACTIVITY

The GvH potency of hydroxyurea-treated spleen cells recovered from MLRs was assayed in (Lew × BU) F1 recipients. *In vitro* sensitization increased GvH potency by a factor of 1.7–3.0 (Figs. 1 and 2). At low doses ($0.5, 1.0 \times 10^6$), increased reactivity was consistently observed. At higher doses ($\geq 3.0 \times 10^6$), results were variable. In three experiments, hydroxyurea treatment resulted in a significant decrease in GvH reactivity below the baseline reactivity of unstimulated controls (Fig. 1). GvH potency decreased by a factor of 12.5 compared to unstimulated cells and 21.4 compared to sensitized cells. Similar results were obtained when GvH reactivity was corrected for cell recovery. In one of four experiments (Fig. 2), hydroxyurea-treated cells were 3.3 times less potent than sensitized cells but exhibited 3.8 times greater GvH reactivity than unsensitized cells. The specificity of the increased GvH reactivity was demonstrated by testing in unrelated (BU × BN) F1 recipients (not shown).

EFFECTS OF HYDROXYUREA ON CFU-c

Hydroxyurea-treated cells recovered from MLR's were assayed for CFU-c activity. Results from three of five experiments are indicated in Table 3. Stimulated cultures showed a significant increase over autologous control cultures. Hydroxyurea produced a modest net decrease in CFU-c in three experiments and an increase in two compared with unstimulated cultures in terms of both cloning efficiency and CFU-c recovery/culture.

DISCUSSION

Precise identification of the cell(s) responsible for GvH reactivity has been difficult. Although it is clear that T-lymphocytes are centrally involved, interrelationships between the subpopulations responsible for MLR reactivity, cytotoxicity, and GvH reactivity are poorly understood. Several observations suggest identity between the MLR and

TABLE 2 Specificity of Hydroxyurea Treatment on MLC Reactivity[a]

RESPONDER CELL	PRIMARY STIMULATOR CELL	SECONDARY STIMULATOR CELL	^{3}H-TdR INCORPORAITON (CPM ± SEM)	SI[b]
BU	BU$_m$	BU$_m$	9,377 ± 153	—
BU	(Lew × BU)F1$_m$	BU$_m$	8,674 ± 184	—
BU	(Lew × BU)F1$_m$	(Lew × BU)F1$_m$	5,102 ± 49	0.59
BU	(Lew × BU)F1$_m$	BN	23,423 ± 614	2.70

[a] MLCs were initiated as described in Materials and Methods. At 48-hr, HU, 5×10^{-4} M was added for 16 hr. Cultures were pooled, washed and fresh stimulator cells added. They were pulsed with ^{3}H-TdR at 120 hr and CPM determined.

[b] Stimulation index.

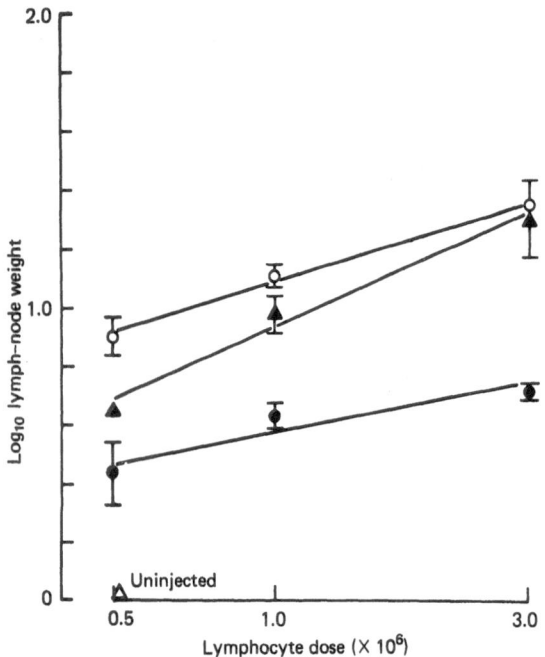

FIGURE 1. Effects of hydroxyurea (HU) on GvH reactivity. BU cells were cultured with BU_m (▲) or (Lew × BU) $F1_m$ cells (○) for 48 hr and pulsed with $5 × 10^{-4}$ M HU (●) for 16 hr. Recovered cells were injected into (Lew × BU) F1 recipients. Popliteal lymph nodes were removed and weighed on day 8. Datum points represent $\bar{X} ± SEM$; $N = 6-8$.

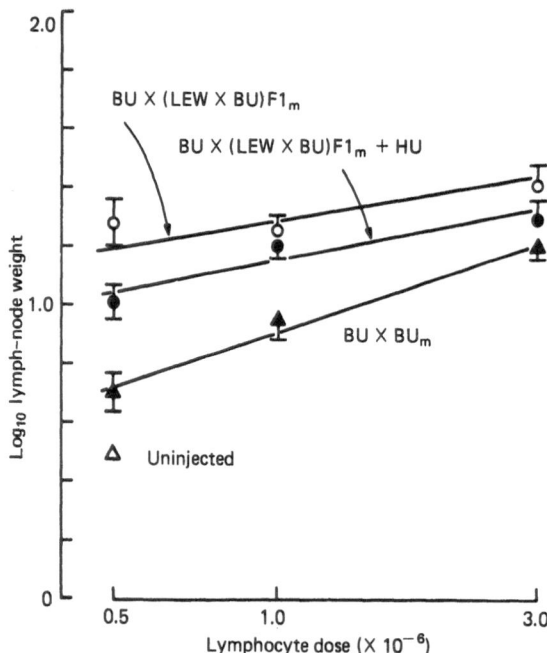

FIGURE 2. Effects of hydroxyurea treatment of MLRs cultures on GvHR. MLRs were pulsed with HU $5 × 10^{-4}$ for 16 hr at 48 hr of culture, cells were harvested, and injected into (Lew × BU) F1 recipients. Popliteal lymph nodes were removed and weighed on day 8. Datum points represent $\bar{X} ± SEM$; $N = 4-6$. (○) BU × (Lew × BU) $F1_m$; (●) BU × (Lew × BU) $F1_m$ + HU; (▲) BU × BU_m; (△) Uninjected.

GvH reactive cell (18, 25). It is also clear, however, that MLR nonreactive cells with only non-*Ir* or non-*H-2* region differences can initiate a GvH reaction (5, 10). MLR reactivity is not acquired under conditions of *in vivo* sensitization (22).

To investigate the relationship between MLR and GvH reactive cells, we evaluated the GvH reactivity of lymphocytes deleted of MLR reactive cells by hydroxyurea-induced suicide. Our results indicate that hydroxyurea-treated

TABLE 3 Effects of Hydroxyurea on CFU-c[a]

EXPT.	RESPONDER CELL	STIMULATOR CELL	HU	CFU-c[b]	
				$2 × 10^5$	Culture
1	Lew	Lew $_m$	−	13	25
	Lew	(Lew × BU)$F1_m$	−	12	43
	Lew	Lew $_m$	+	8	18
	Lew	(Lew × BU)$F1_m$	+	7	41
2	BU	BU_m	−	12	41
	BU	(Lew × BU)$F1_m$	−	24	86
	BU	BU_m	+	40	80
	BU	(Lew × BU)$F1_m$	+	57	80
3	BU	BU_m	−	7	12
	BU	(Lew × BU)$F1_m$	−	15	27
	BU	BU_m	+	2	7
	BU	(Lew × BU)$F1_m$	+	3	7

[a] MLRs were performed as in Materials and Methods. At 48 hr selected cultures were pulsed with $5 × 10^{-4}$ M HU. Cultures were either immediately harvested (experiments 1, 2) or cultured for an additional 48 hr (expt. 3) and CFU-c determined.
[b] CFU-c expressed either as a function of $2 × 10^5$ cells plated or corrected for the cell recovery per MLR culture.

lymphocytes display decreased but persistent GvH reactivity. Although it is possible that this residual GvH reactivity reflects survival of MLR reactive cells at a level below the sensitivity of the assay, a more likely explanation is the interaction of two or more GvH reactive cells in either an additive or synergistic fashion. Cantor and Asofsky (3, 4) and Tigelaar and Asofsky (29, 30) have demonstrated synergistic cooperation between T-lymphocyte subpopulations in inducing GvH disease.

The hydroxyurea-induced suicide of MLR reactive cells confirms the clonal restriction of histocompatibility antigen-reactive cells reported by others using bromodeoxyuridine (BUdR; ref. 31) and ^{3}H-TdR (26). Furthermore it suggests that the persisting GvH reactive cells are not progeny of MLR reactive clones. Decreased GvH reactivity of MLR reactive cells treated with BUdR has been reported by Rich et al. (24).

Hydroxyurea treatment appeared to have little significant effect on committed CFU-c. This is consistent with data using ^{3}H-TdR suicide (21) and suggests that splenic CFU-c, in contrast to bone marrow CFU-c, have either a longer cycle time or alternatively that a greater proportion of cells are in G_0. The increased CFU-c levels of MLR-stimulated cultures have been previously reported (23).

The present study suggests a diversity of lymphocytes capable of inducing a GvH reaction and suggests the feasibility of modulation of GvH reactivity without significant loss of CFU-c. The clinical implications of these findings are clear.

SUMMARY

Precise definition and identification of the cell(s) responsible for the graft-versus-host reaction has been difficult. The mixed lymphocyte reaction may be an *in vitro* analogue of the GvH reaction. We addressed the problem of whether MLR and GvH reactive cells are identical by specific deletion of MLR reactive cells with hydroxyurea. Hydroxyurea treated cells were assayed for GvH reactivity and committed granulocyte-monocyte precursors. Results suggest that both MLR reactive and nonreactive cells are capable of inducing a GvH reaction and that the former can be specifically decreased or prevented by *in vitro* incubation with hydroxyurea. Hydroxyurea treatment had no adverse effect on splenic CFU-c under these conditions. It appears possible to modulate GvH reactivity without adversely affecting hematopoietic restorative potential as measured by CFU-c.

ACKNOWLEDGMENTS

Robert Peter Gale is a Scholar of the Leukemia Society of America. This study was supported by Grants CA-12800 and CA 15866 from the National Cancer Institute, National Institutes of Health, Bethesda, Maryland. We thank Grace McMurray for excellent secretarial assistance.

REFERENCES

(1) Bach, F. H., Bach, M. L., and Klein, J. Genetic and immunological complexity of major histocompatibility regions. *Sciences*, *176*:1024, 1972.

(2) Bach, F. H., Widmer, M. B., Bach, M. L., and Klein, J. Serologically defined and lymphocyte-defined components of the major histocompatibility complex in the mouse. *J. Exp. Med.*, *136*:1430, 1972.

(3) Cantor, H., and Asofsky, R. Synergy among lymphoid cells mediating the graft-versus-host response. *J. Exp. Med.*, *131*:235, 1970.

(4) Cantor, H., and Asofsky, R. Synergy among lymphoid cells mediating the graft-versus-host response. *J. Exp. Med.*, *135*:764, 1972.

(5) Cantrell, J. L., and Hildemann, W. H. Characteristics of disparate histocompatibility barriers in congenic strains of mice. *Transplantation*, *14*:761, 1972.

(6) Canty, T. G., and Wunderlich, J. R. Quantitative *in vitro* assay of cytotoxic cellular immunity. *J. Natl. Cancer Inst.*, *45*:761, 1970.

(7) Cline, M. J., Gale, R. P., and Stiehm, E. R. Bone

marrow transplantation in man. *Ann. Int. Med.,* *83*:691, 1975.

(8) Elkins, W. L. Cellular immunology and the pathogenesis of graft versus host reactions. *Prog. Allergy, 15*:78, 1971.

(9) Elkins, W. L., Kavathas, P., and Bach, F. H. Activation of T cells by H-2 factors in the graft-vs-host reaction. *Transplant. Proc., 5*:1759, 1973.

(10) Festenstein, H., Abbasi, K., Sachs, J. A., and Oliver, R. T. D. Serologically undetectable immune responses in transplantation. *Transplant. Proc., 4*:219, 1972.

(11) Ford, W. L., Burr, W., and Simonsen, M. A lymph node weight assay for the graft-versus-host activity of rat lymphoid cells. *Transplant., 10*:258, 1970.

(12) Golde, D. W., Bersch, M., and Cline, M. J. Potentiation of erythropoiesis *in vitro* by dexamethason. *J. Clin. Invest., 57*:57, 1976.

(13) Golde, D. W., and Cline, M. J. Identification of the colony-stimulating cell in human peripheral blood. *J. Clin. Invest., 51*:2981, 1972.

(14) Grebe, S. C., and Streilein, J. W. Graft-versus-host reactions: A review. *Adv. Immunol., 22*:119, 1976.

(15) Howe, M. L., and Cohen, L. Participation of lymphoid cell subpopulations in the mixed lymphocyte interaction, Lindahl-Kiessling, L. and Osoba, D., eds. In *Proc. 8th Leukocyte Culture Conf.,* p. 611. New York: Academic Press, 1974.

(16) Iscove, N. N., Senn, J. S., Till, J. E., and McCulloch, E. A. Colony formation by normal and leukemic human marrow cells in culture: Effect of conditioned medium from human leukocytes. *Blood, 37*:1, 1971.

(17) Kisielow, P., Hirst, J. A., Shiku, H., Beverley, P. C. L., Hoffmann, M. K., Boyse, E. A., and Oettgen, H. F. Ly antigens as markers for functionally distinct subpopulation of thymus-derived lymphocytes of the mouse. *Nature (London), 253*:219, 1975.

(18) Klein, J., and Park, J. M. Graft-versus-host reaction across different regions of the H-2 complex of the mouse. *J. Exp. Med., 13*:1213, 1973.

(19) Klein, J., Widmer, M. B., Segall, M., and Bach, F. H. Mixed lymphocyte culture reactivity and H-2 histocompatibility loci differences. *Cell Immunol., 4*:442, 1972.

(20) Livnat, S., Klein, J., and Bach, F. H. Graft-versus-host reaction in strains of mice identical for H-2K and H-2D antigens. *Nature [New Biol.], 243*:42, 1973.

(21) Metcalf, D. Effect of thymidine suiciding on colony formation *in vitro* by mouse hematopoietic cells. *Proc. Soc. Exp. Biol. Med., 139*:511, 1972.

(22) Opelz, G., Gale, R. P., and the UCLA Bone Marrow Transplant Team. Absence of specific mixed leukocyte culture reactivity during graft-versus-host disease ad following bone marrow transplant rejection. *Transplantation, 22*:474, 1976.

(23) Parker, J. W., and Metcalf, D. Production of colony-stimulating factor in mixed leukocyte cultures. *Immunology, 26*:1039, 1974.

(24) Rich, R. R., Kirkpatrick, C. H., and Smith, T. K. Simultaneous suppression of responses to allogeneic tissue *in vitro* and *in vivo. Cell Immunol., 5*:190, 1972.

(25) Rodey, G. E., Bortin, M. M., Bach, F. H., and Rimm, A. A. Mixed leukocyte culture reactivity and chronic graft-versus-host reaction (secondary disease) between allogenic H-2^k mouse strains. *Transplantation, 17*:84, 1974.

(26) Salmon, S. E., Krakaner, R. S., and Whitmore, W. F. Lymphocyte stimulation: selective destruction of cells during blastogenic response to transplantation antigens. *Science, 172*:490, 1971.

(27) Storb, R., Welden, P. L., Schroeder, M. L., Graham, T. C., Lerner, K. G., and Thomas, E. D. Marrow grafts between canine littermates homozygous for lymphocyte defined histocompatibility antigens. *Transplantation, 21*:299, 1976.

(28) Thomas, E. D., Storb, R., Clift, R. A., Feber, A., Johnson, F. L., Neiman, P. E., Lerner, K. G., Glucksberg, H., and Buckner, C. D. Bone marrow transplantation. *New Engl. J. Med., 292*: Part I, 832; and Part II, 895; 1975.

(29) Tigelaar, R. E., and Asofsky, R. Synergy among lymphoid cells mediating the graft-versus-host response. *J. Exp. Med., 135*:1059, 1972.

(30) Tigelaar, R. E., and Asofsky, R. Synergy among lymphoid cells mediating the graft-versus-host response. *J. Exp. Med., 137*:239, 1973.

(31) Zosche, D. C., and Bach, F. H. Specifity of antigen recognition of human lymphocytes *in vitro. Science, 170*:1404, 1970.

Mechanism of Donor to Host Tolerance in Rat Bone Marrow Chimeras

P. Tutschka, R. Schwerdtfeger,
R. Slavin, and G. Santos

INTRODUCTION

It is a well-established fact that individuals (animals and humans) may ultimately recover from acute graft-versus-host disease (GvHD), which complicates otherwise successful allogeneic bone marrow transplantation. Such individuals may show complete lymphohematopoietic chimerism and yet no clinical or histological evidence of continuous GvHD (2, 25, 30, 31).

The mechanisms responsible for the recovery from GvHD are, however, incompletely understood; various reports in the literature on this subject remain controversial (5). The purpose of this study was to define the mechanism or mechanisms responsible for recovery from GvHD (donor-to-host "tolerance") in conventionally reared inbred rats given Ag-B incompatible or Ag-B compatible bone marrow cell infusions following high-dose cyclophosphamide (CY) conditioning.

MATERIALS AND METHODS

ANIMALS

21

Female Lewis (Ag-B1), F344 (Ag-B1), ACI (AgB4), and BN (AgB3) rats 10–12 weeks of age were obtained from Microbiological Associates (Bethesda, Md). Animals were housed in polycarbonate cages, four to a cage. They were provided with tap water and Purina Chow *ad libitum*.

DRUGS AND TBI

Busulfan (BU), generously supplied by Dr. George Hitchings (Burroughs Wellcome and Company, Research Triangle Park, N.C.), was prepared in 2.5% carboxymethylcellulose in water. CY, generously supplied by Dr. Walter A. Zygmunt (Mead Johnson, Evansville, Ind.), was prepared in saline. All drug injections were given intraperitoneally (i.p.) a few minutes after preparation of the drug in a volume of 10 ml/kg of body weight. A dual-source ^{137}Ce small animal irradiator delivering 136R/min was used for total body radiation (TBI).

PREPARATION OF CELL INOCULA

Under light ether anesthesia donor animals were killed by cervical dislocation. Marrow was collected in cold Tyrode's solution from femurs, tibias, and humeri, cell counts were made, and viability was estimated with Trypan Blue as described previously (26). Marrow was given intravenously (i.v.) in a constant 1-ml volume at the desired cell concentration. Marrow preparations showed 90% to 95% viability, as estimated by Trypan Blue exclusion.

Whole blood was collected by cardiac puncture from rats under ether anesthesia employing heparin as anticoagulant. The white blood cell and differential counts were determined by routine hematologic methods.

HISTOLOGIC EXAMINATION

Skin, gut, spleen, lymph nodes, liver, tongue, and marrow were processed with 10% Formalin, sectioned, and stained with hematoxylin and eosin for microscopic examination and documentation of lesions characteristic of GvHR reactions (28).

ASSESSMENT OF CHIMERISM

Antisera specific for ACI and F344 immunoglobulin alloytypes (7) were used in a gel-diffusion test (18) to assess lymphoid chimerism indirectly. In addition, animals were implanted with 1-cm circular skin grafts from the donor, recipient, and a third-party (BN) strain by techniques described previously (24).

The clinical course of animals in various experiments was followed by daily observation and periodic weight determinations. Diarrhea is not a feature of GvHD in this model system, but dermatitis is. The presence or absence of clinical GvHD was judged on the basis of dermatitis, weight loss, and generally unkempt appearance. These clinical observations were all confirmed by histologic examination.

A more direct documentation of chimerism was employed in ACI-Lewis (donor-host) chimeras. Peripheral blood nucleated cells of animals were typed with strain-specific cytotoxic isoantisera, using a Trypan Blue dye exclusion test modified after AMOS et al. (1). Animals were typed with host specific (ACI-anti-Lewis) and donor specific (Lewis-anti-ACI) sera.

PREPARATION OF LONG-TERM CHIMERAS

Lewis rats were given 200 mg/kg CY i.p. followed in 24 hr by the i.v. infusion of 64×10^6 nucleated ACI or F344 bone marrow cells. No subsequent drug was given to prevent GvHD. GvHD was severe and mortality high (70–80%) in recipients of ACI marrow. GvHD was far less severe in recipients of F344 marrow with lower mortality (20–30%). Forty surviving animals of each type were examined at 250 days postgrafting. Clinically they appeared healthy and showed normal hematologic parameters without evidence of GvHD. All animals showed the presence of donor immunoglobulin allotype in their serum, they accepted donor type skin grafts but rejected BN skin allografts in a normal fashion (11.0 ± 1.6 days). In addition, the peripheral lymphocytes of ACI-Lewis chimeras were shown to be of ACI origin with the use of specific cytotoxic typing sera.

Four chimeras in each group were sacrificed for histological studies. There was no evidence of florid or quiescent GvHD. The bone marrows were cellular and lymphoid organs showed well developed T- and B-cell areas. These long-term chimeras (ACI-Lewis and F344-Lewis) were the subject of our studies.

RESULTS
TRANSFER OF CHIMERIC MARROW INTO NEW HOSTS

Ten long-term ACI-Lewis chimeras were sacrificed, their marrow pooled, and 64×10^6 nucleated marrow cells injected into normal ACI, Lewis or BN rats (10 animals per group) 24 hr after conditioning with 950R TBI. Control animals were prepared identically but injected with normal ACI marrow cells. Five animals of each group were sacrificed 17 days after transplantation for histologic examination. The five remaining animals in each group were observed for 100 days and were then sacrificed for histologic examination.

Lewis and ACI recipients of chimeric bone marrow showed no evidence of GvHD on histologic examination at 17 days. BN recipients of chimeric bone marrow showed evidence of florid GvHR (Table 1).

No evidence of clinical GvHD was seen in ACI or Lewis recipients observed for 100 days. Moderate to severe clinical GvHD was seen in BN recipients. These clinical observations were confirmed by histologic examination (Table 1).

In experiments of similar design F344-Lewis chimeras served as bone marrow donors for F344, Lewis, and BN recipients prepared with 950R TBI. GvHD was demonstrated clinically and histologically in BN animals but not in Lewis or F344 recipients (Table 2).

TABLE 1 Transfer of Chimeric Cells Into New Hosts: AgB Incompatible System[a]

GROUP	RECIPIENT	DONOR	Histology Day 17	CUMULATIVE MORTALITY (FRACTION WITH CLINICAL GvHD)				
				Day 14	Day 21	Day 35	Day 56	Day 100
1	ACI	(ACI-L)	No Gvhr	0/5(0/5)	0/5(0/5)	0/5(0/5)	1/5(0/4)	1/5(0/4)
2	Lewis	(ACI-L)	No Gvhr	0/5(0/5)	0/5(0/5)	0/5(0/5)	0/5(0/5)	1/5(0/4)
3	BN	(ACI-L)	Florid Gvhr	1/5(0/4)	2/5(1/3)	2/5(3/3)	3/5(2/2)	4/5(1/1)
4	ACI	ACI	No Gvhr	0/5(0/5)	0/5(0/5)	1/5(0/4)	2/5(0/3)	2/5(0/3)
5	Lewis	ACI	Florid Gvhr	2/5(2/3)	3/5(2/2)	4/5(1/1)	5/5	—
6	BN	ACI	Florid Gvhr	2/5(3/3)	4/5(1/1)	5/5	—	—
7	Lewis	—	—	4/5(0/1)	5/5	—	—	—

[a]ACI, Lewis, or BN recipients, 10 per group, conditioned with 950R TBI and injected with 64×10^6 ACI or (ACI-L) chimeric marrow cells.

TRANSFER OF DONOR TYPE CELLS INTO CHIMERAS

ACI-Lewis chimeras were given no drug, BU 30 mg/kg, CY 50 mg/kg, CY 150 mg/kg, or 950R TBI prior to i.v. infusion of 64×10^6 nucleated marrow cells from normal ACI donors. All animals were examined for histologic evidence of GvHD 17 days later. In no case was there evidence for GvHD in animals that received no drug, BU, or CY 50 mg/kg (Table 3, groups 3 and 4). The addition of 10^7 peripheral blood lymphocytes to the marrow inoculum did not alter these results (Table 3, group 2). Histologic evidence of florid GvHD was seen in recipients pretreated with CY (150 mg/kg) and TBI (950R) (Table 3, groups 5 and 6).

SUPPRESSION OF GvHD WITH CHIMERIC CELLS

Normal Lewis rats were exposed to 950R TBI and given 64×10^6 normal ACI bone marrow cells 24 hr later. These animals were divided into five groups of 10 animals each and received either no additional cells, 30×10^6 peripheral blood lymphocytes pooled from five normal ACI donors, 30×10^6 peripheral blood lymphocytes pooled from five ACI-Lewis chimeras or 30×10^6 peripheral blood lymphocytes pooled from another five ACI-Lewis chimeras but irradiated (in vitro) with 1500R just prior to transfer. The last group of Lewis rats received 3 ml serum from ACI-Lewis chimeras. Five animals were sacrificed for histologic evidence of GvHD at 17 days and five were observed for 100 days for clinical evidence of GvHD. The results of this experiment are shown in Table 4.

Transplantation of incompatible ACI marrow into TBI-conditioned Lewis recipients resulted in severe GvHD (Table 4, group 1). The addition of peripheral blood lymphocytes increased the severity of GvHD (Table 4, group 2).

The addition of unirradiated peripheral blood lymphocytes from (ACI-L) chimeras to the inoculum of normal donor cells suppressed GvHD clinically and histologically (Table 4, group 3). Of

TABLE 2 Transfer of Chimeric Cells Into New Hosts: AgB Compatible System[a]

GROUP	RECIPIENT	DONOR	Histology Day 17	CUMULATIVE MORTALITY (FRACTION WITH CLINICAL GvHD)				
				Day 14	Day 21	Day 35	Day 56	Day 100
1	F344	(F344-L)	No Gvhr	0/5(0/5)	0/5(0/5)	1/5(0/4)	1/5(0/4)	1/5(0/4)
2	Lewis	(F344-L)	No Gvhr	0/5(0/5)	0/5(0/5)	1/5(0/4)	1/5(0/4)	1/5(0/4)
3	BN	(F344-L)	Florid Gvhr	0/5(0/5)	1/5(2/4)	2/5(2/3)	3/5(2/2)	3/5(2/2)
4	F344	F344	No Gvhr	0/5(0/5)	0/5(0/5)	0/5(0/5)	1/5(0/4)	1/5(0/4)
5	Lewis	F344	Florid Gvhr	0/5(0/5)	0/5(1/4)	1/5(2/4)	2/5(1/3)	2/5(0/3)
6	BN	F344	Florid Gvhr	2/5(3/3)	3/5(2/2)	5/5	—	—
7	Lewis	—	—	—	—	—	—	—

[a]F344, Lewis, or BN recipients, 10 per group, conditioned with 950 R TBI and injected with 64×10^6 F344 or (F344-L) chimeric marrow cells.

TABLE 3 Injection of Normal ACI Donor Cells into Untreated or Drug-Treated (ACI-L) Chimeras

GROUP	RECIPIENT	DONOR	PREPARATIVE REGIMEN	CELL INOCULUM	HISTOLOGY DAY 17
1	(ACI-L)[a]	ACI	no	64×10^6 marrow	No Gvhr 5/5
2	(ACI-L)	ACI	no	64×10^6 marrow 10×10^6 blood lymphocytes	No Gvhr 5/5
3	(ACI-L)	ACI	BU 30 mg/kg	64×10^6 marrow	No Gvhr 5/5
4	(ACI-L)	ACI	CY 50 mg/kg	64×10^6 marrow	No Gvhr 3/5 Lymphoid hyperplasia 2/5
5	(ACI-L)	ACI	CY 150 mg/kg	64×10^6 marrow	Florid Gvhr 5/5
6	(ACI-L)	ACI	950R TBI	64×10^6 marrow	Florid Gvhr 5/5
7	Lewis	ACI	950R TBI	64×10^6 marrow	Florid Gvhr 5/5

[a](ACI-L) chimeras were given normal ACI cells i.v. without or with conditioning with BU, CY, or TBI.

the five animals examined at 17 days from this group four of five showed no histologic evidence of GvHD and one showed only lymphoid hyperplasia. Three animals survived until day 100. It was impossible to determine the cause of death of the two animals dying during the observation period because marked autolysis prevented adequate histologic examination. The three survivors sacrificed at 100 days showed only minimal lesions of GvHR in the liver. Addition of irradiated peripheral blood lymphocytes from (ACI-L) chimeras (Table 4, group 4) or serum from (ACI-L) chimeras (Table 4, group 5) to the inoculum of normal donor cells, however, did not suppress GvHD.

DISCUSSION

The mechanisms of recovery from GvHD and maintenance of the "tolerant" state in a long-term chimera are not clearly known. The theoretical possibilities include specific or nonspecific central depletion of donor cells capable of reacting against the host or peripheral "tolerance"

maintenance, whereby potentially reactive donor cells are blocked by humoral factors or suppressor cell populations. A third possibility might be the failure of reactive donor cells to respond because of depletion of appropriate host target cells.

A nonspecific central depletion would result in a more generalized immunologic deficiency of these long-term chimeras, operationally resulting in "tolerance" of the grafted cells toward the host. It has been shown that chimeras up to a certain time post transplantation are indeed immunologically deficient, but ultimately recover fully (11, 12).

The chimeras used in the studies reported here showed intact T- and B-cell areas of the lymphoid organs and were able to reject third party skin grafts in a normal fashion, suggesting the presence of functioning thymus dependent cells (8, 19). In the present studies, as well as previous studies from this laboratory (26), cells from "tolerant" rat chimeras were unable to cause GvHD when transferred to host-type recipients prepared with CY (26) or TBI (present studies), but they were able to cause GvHD when transferred to a suitably prepared unrelated rat strain. This is

TABLE 4 Attempts to Supress GvH

	INOCULUM		HISTOLOGY		CUMULATIVE MORTALITY (FRACTION WITH CLINICAL GvHD)	
Group[a]	Marrow ($\times 10^6$)	Additional blood products	Day 17	Day 100	Day 21	Day 100
1	ACI 64	—	Gvhr	Gvhr	2/5(3/3)	4/5(1/1)
2	ACI 64	ACI, 30×10^6 blood lymphocytes	Gvhr	—	3/5(2/2)	5/5
3	ACI 64	(ACI-L) 30×10^6 blood lymphocytes	No Gvhr	Minimal Gvhr	1/5(0/4)	2/5(0/3)
4	ACI 64	(ACI-L) 30×10^6 blood lymphocytes irradiated with 1500R	Gvhr	—	3/5(2/2)	5/5
5	ACI 64	(ACI-L) 3 ml serum	Gvhr	—	2/5(3/3)	5/5

[a]Normal Lewis rats, 10 per group, conditioned with 950R TBI and injected with normal ACI cells or a mixture of ACI cells and chimeric (ACI-L) cells or (ACI-L) serum.

most compatible with a specific "tolerance" of the chimeras rather than unspecific immune deficiency. These studies also suggest that the depletion or absence of host target cells is not an important mechanism in the recovery from GvHD.

Induction and maintenance of the "tolerant" state by specific deletion or specific and non-reversible inactivation of reactive cells has been the postulate of the classical clonal selection theory (4, 5), a concept derived from the inability of cells from tolerant animals to react in local graft-versus-host reactions and mixed leukocyte reactions to donor or host strain antigens (3, 14, 17, 35–37). The studies reported here, which involve the infusion of donor-type lymphohematopoietic cells into established chimeras, provide evidence against a central depletion hypothesis as well as confirm recent studies in canines (35), in which infusion of large amounts of donor lymphocytes did not reactivate GvHD in long-term chimeras.

A number of workers, however, have reported that tolerant individuals do possess reactive cells against host-type antigens. The majority of these data were obtained in neonatally tolerant animals, using the colony inhibition test (10, 37) as well as MLR and local GvHR assays, but have been extended to allograft-tolerant chimeric animals (16, 29) and tetraparental animals (34).

Field et al. (13) and Voisin et al. (33) provided evidence that humoral factors might transfer refractoriness to GvHD. Subsequently, it was demonstrated that serum from operationally tolerant animals, including radiation chimeras (16, 21, 34), blocked in vitro responses, which led to the concept of tolerance maintenance by serum blocking factors (15).

A number of workers have challenged this concept by failing to demonstrate this mechanism in rodent chimeras (5,17), agammaglobulinemic chickens (23) and most recently in canine radiation chimeras (32). In the study presented here, chimeric serum was unable to "block" GvHD, which suggests that serum factors were not a prerequisite for maintenance of stable donor-to-host "tolerance."

Philips and Wegman (22) noted that lymphoid cells from tetraparental chimeras do not respond to parental cells in vitro. These cells, however, were able to prevent an interaction between the two corresponding parental cell lines in the MLR which suggests the existence of a suppressor cell population. Recently transplantation "tolerance" could be actively transferred with thoracic duct lymphocytes providing evidence that suppressor T-cells were operating (9), a concept supported by Gengozian et al. (14) with mixed transfer systems in murine radiation chimeras.

In the present report we were successful in preventing serious GvHD when peripheral blood lymphocytes were transferred together with normal donor-type bone marrow cells to suitably prepared recipients. Although we did not provide evidence that the cells responsible were T-cells, we were able to demonstrate their sensitivity to TBI and CY (immunosuppressive agents) but not to BU (nonimmunosuppressive) (27).

The results presented here, together with the evidence summarized earlier, in conventionally reared (nongermfree) animals strongly suggest the existence of suppressor cell populations being of prime importance in the initiation and maintenance of the "tolerant" state.

SUMMARY

Lewis rats were conditioned with cyclophosphamide and grafted with AgB incompatible bone marrow. They were examined 250 days after transplantation and demonstrated to be healthy complete chimeras. Marrow cells from these chimeras were infused into lethally irradiated ACI, Lewis and BN recipients. Graft-versus-host disease occurred only in the BN rats.

Other chimeric rats were given no treatment, busulfan, CY, or total body irradiation prior to the infusion of normal ACI BM. GvHD occurred only in animals given CY or TBI.

Normal Lewis rats were conditioned with TBI and given ACI BM. In addition, they received whole blood, irradiated blood, or serum from chimeric rats. GvHD developed in all animals except those given unirradiated chimeric blood.

These studies suggest that suppressor cell populations, sensitive to immunosuppression, are likely the fundamental mechanism of recovery from GvHD.

ACKNOWLEDGMENT

This work was supported by Public Health Service research grants CA06973 and CA15396 from the National Cancer Institute, National Institutes of Health, Bethesda, Maryland.

REFERENCES

(1) Amos, D. B., Bashir, H., Boyle, W., Mac Queen, M., and Trilikainen, A. A simple micro cytotoxicity test. *Transplantation, 7*:220, 1969.

(2) Barnes, D. W. H., Ford, C. E., and Ilberg, P. L. T. Tolerance in the radiation chimera. *Transplant. Bull., 6*:101, 1958.

(3) Beverly, P. C., Brent, L., Brooks, C., Medawar, P. B., and Simpson, E. *In vitro* reactivity of lymphoid cells from tolerant mice. *Transplant. Proc., 5*:679, 1973.

(4) Billingham, R. E., Brent, L., and Medawar, P. B. Quantitative studies on tissue transplantation immunity. II. Actively acquired tolerance. *Philos. Trans. R. Soc. London Ser. B, 239*:357, 1956.

(5) Brent, L., Brooks, C. F., Lubling, N., and Thomas, A. V. Attempts to demonstrate an *in vivo* role for serum blocking factors in tolerance mice. *Transplantation, 14*:382, 1972.

(6) Brent, L., Brooks, C. G., Medawar, P. B., and Simpson, E. Transplantation tolerance. *Br. Med. Bull., 32*:101, 1976.

(7) Campbell, D. H., Garvey, Y. S., and Cremer, N. E. eds., *Methods in Immunology*, pp. 143–149. New York: W. A. Benjamin, 1963.

(8) Cooper, M. D., Peterson, R. D. A., South, M. A., and Good, R. A. The function of the thymus system and the bursa system in the chicken. *J. Exp. Med., 123*:75, 1966.

(9) Dorsch, S., and Roser, B. T cells mediate transplantation tolerance. *Nature (London), 258*:233, 1975.

(10) Droege, W., and Major, K. Graft versus host reactivity and inhibitory serum factors in allograft-tolerant chickens. *Transplantation, 19*:517, 1975.

(11) Elfenbein, G. J., Anderson, P. N., Humphrey, R. L., Mullins, G. M., Sensenbrenner, L. L., Wands, J. R., and Santos, G. W. Immune system reconstitution following allogeneic bone marrow transplantation in man: A multi-parameter analysis. *Transplant. Proc., 18*:641, 1976.

(12) Elkins, W. L. The cellular basis of transplantation tolerance. *Transplant. Proc., 5*:685, 1973.

(13) Field, E. O., Cauchi, M. N., and Gibbs, J. E. The transfer of refractoriness to graft-versus-host disease in F_1 hybrid rats. *Transplantation, 5*:241, 1967.

(14) Gengozian, N., and Urso, P. Functional activity of T and B lymphocytes in radiation chimeras. *Transplant. Proc., 8*:631, 1976.

(15) Hellström, I., Hellström, K. E., Storb, R., and Thomas, E. D. Colony inhibition of fibroblasts from chimeric dogs mediated by the dogs' own lymphocytes and specifically abrogated by their serum. *Proc. Natl. Acad. Sci. USA, 66*:65, 1970.

(16) Hellström, I., Hellström, K. E., and Trentin, J. J. Cellular immunity and blocking serum activity in chimeric mice. *Cell. Immunol., 7*:73, 1973.

(17) Heron, J. Transplantation tolerance in the rat serum—mediated? *Transplantation, 15*:534, 1973.

(18) Humphrey, R. L., and Santos, G. W. Serum protein allotype markers in certain inbred rat strains. *Fed. Proc. Fed. Am. Soc. Exp. Biol., 30*:314, 1971.

(19) Miller, J. F. A., and Osaba, D. Current concepts of the immunological function of the thymus. *Physiol. Rev., 47*:437, 1967.

(20) Ochs, H. D., Storb, R., Thomas, E. D., Kolb, H. J., Graham, T. C., Mickelson, E., Parr, M., and Rudolph, R. H. Immunologic reactivity in canine marrow graft recipients. *J. Immunol., 113*:1039, 1974.

(21) Philips, S. M., Martin, W. J., Shaw, A. R., and Wegmann, T. G. Serum-mediated immunological non-reactivity between histoincompatible cells in tetraparental mice. *Nature (London), 234*:146, 1971.

(22) Phillips, S. M., and Wegmann, T. G. Active suppression as a possible mechanism of tolerance in tetraparental mice. *J. Exp. Med., 137*:291, 1973.

(23) Rouse, B. T., and Warner, N. L. Induction of T cell tolerance in agammaglobulinemic chickens. *Eur. J. Immunol., 2*:102, 1972.

(24) Santos, G. W., and Owens, A. H., Jr. A comparison of the effects of selected cytotoxic agents on allogeneic skin graft survival in rats. *Bull. Johns Hopkins Hosp., 116*:327, 1965.

(25) Santos, G. W. Marrow transplantation in cyclophosphamide treated rats: Early donor-to-host tolerance and long lived chimeras. *Exp. Hematol., 13*:36, 1967.

(26) Santos, G. W., and Owens, A. H., Jr. Syngeneic and allogeneic marrow transplants in the cyclophosphamide pretreated rat. In Dausset, J., Hamburger, J., and Mathé, G., eds., *Advances in Transplantation*, pp. 432–436. Copenhagen: Munksgaard, 1968.

(27) Santos, G. W., and Tutschka, P. J. Marrow transplantation in the busulfan-treated rat: Preclinical model of aplastic anemia. *J. Natl. Cancer Inst., 53*:1781, 1974.

(28) Slavin, R. E., and Santos, G. W. The graft-verus-host reaction in man following bone marrow transplantation: Pathology, pathogenesis, clinical features and implications. *Clin. Immunol. Immunopathol., 1*:472, 1973.

(29) Sprent, J., vonBoehmer, H., and Nabholz, M. Association of immunity and tolerance to host H-2 determinants in irradiated F_1 hybrid mice reconstituted with bone marrow cells from one parental strain. *J. Exp. Med., 142*:321, 1975.

(30) Storb, R., Rudolph, R. H., Kolb, H. J., Graham, T. C., Mickelson, E., Erickson, V., Lerner, K. G., Kolb, H., and Thomas, E. D. Marrow grafts between DL-A matched canine littermates. *Transplantation, 15*:92, 1973.

(31) Thomas, E. D., Storb, R., Clift, R. A., Fefer, A., Johnson, F. L., Neiman, P. E., Lerner, K. E., Glucksberg, H., and Buckner, C. D. Bone marrow transplantation. *N. Engl. J. Med., 292*:832, 1975.

(32) Tsoi, M. S., Storb, R., Weiden, P. L., Graham, T. C., Schroeder, M. L., and Thomas, E. D. Canine marrow transplantation: Are serum blocking factors necessary to maintain the stable chimeric state? *J. Immunol., 114*:531, 1975.

(33) Voisin, G. A., Kinsky, R., and Maillard, Y. Protection against homologous disease in hybrid mice by passive and active immunological enhancement facilitation. *Transplantation, 6*:187, 1968.

(34) Wegmann, T. G., Hellström, I., and Hellström, K. E. Immunological tolerance: "Forbidden Clones" allowed in tetraparental mice. *Proc. Natl. Acad. Sci. USA, 68*:1644, 1971.

(35) Weiden, P. L., Storb, R., Tsoi, M. S., Graham, T. C., Lerner, K. G., and Thomas, E. D. Infusion of donor lymphocytes into stable canine radiation chimeras. Implications for mechanisms of transplantation tolerance. *J. Immunol., 116*:1212, 1976.

(36) Wilson, D. B., and Nowell, D. C. Quantitative studies on the mixed lymphocyte interaction in rats. IV. Immunologic potentiality of the responding cells. *J. Exp. Med., 131*:391, 1970.

(37) Wright, P. W., Bernstein, I. D., Hamilton, D., Gluckman, J. C., and Hellström, K. E. Cell-mediated reactivity and serum blocking activity in tolerant rats. *Transplantation, 18*:46, 1974.

The differentiation of uncommitted lymphocytes into specifically immune effector T-cells is a multifaceted and complex process. The mechanisms by which the immature lymphocytes are programmed into defined specific pathways, and their subsequent commitment to restricted expression of antigen receptors, are two intriguing questions confronting biologists today. Significant progress has been made, though, in the definition of a number of the steps involved in the development of the precommited lymphocytes into effector T-cells. While specific activation of precommited effector cell precursors depends upon interaction with a particular antigenic determinant, the biochemical events that follow antigen-stimulated or mitogen-stimulated activation probably are quite similar for cells within a given subpopulation. The differentiation processes within lymphocytes confined to each known subpopulation of cells, however, are quite distinct. This is most readily exemplified by a comparison between the effector mechanisms of mature cells from the B-cell limb and the T-cell limb of immunity. Whether the helper cells that mediate T-cell dependent B-cell responses function in a manner identical to those auxiliary cells required for killer T-cell differentiation is not known. However, preliminary reports suggest similar mechanisms. It is quite apparent, though, that differentiation of such helper cells and killer T-cells precursors involved different pathways.

Effects of a Cell-Free Helper Factor(s) on the Kinetics of T-Cell Responses to Histocompatibility Antigens

Janet M. D. Plate

22

We have utilized an antiserum directed against an H-2-associated antigen(s) that is expressed on the surfaces of some C57BL/10 cell subpopulations to remove auxillary cells that are required for both the replication and differentiation of the killer cell precursors (4). The antiserum is probably directed against an H-2I region-determined antigen that is expressed on helper T-cells. Until the appropriate genetic studies are completed, we have utilized caution and termed the antigen Ha, for helper cell-associated antigen. The functions of the anti-Ha-depleted cell suspensions cannot be replaced by nonT-cells (cells remaining after lysis with anti-θ serum) (5). It is clear from these studies that the functionally important Ha-bearing cell subpopulation must reside in the θ-positive or T-cell population.

The Ha-bearing cell subpopulation functions in the regulation of the response of killer T-cell precursors to histocompatibility antigens. In fact, both the replication and differentiation of the Ha-negative precursor cells are dependent upon some function of the Ha-positive cell subpopulation(s). Our recent data suggest that Ha-positive cells, upon activation with antigen, produce and release a protein substance(s) that mediates at least the initiation of T-killer cell differentiation (4).

The activation of the Ha-negative killer T-cell precursors requires two signals. One signal is presented by the foreign antigen, but this in itself is not sufficient to trigger differentiation. The second signal is provided by the protein factor that is released, probably by the antigen-activated Ha-positive cells. We have demonstrated that, in the presence of foreign antigen, Ha-negative cells fail to respond. The Ha-negative precursor T-cells are fully capable of differentiating into effector cells, however, upon addition of the soluble protein factor harvested from activated lymphocyte cultures. Thus, this cell-free helper substance(s) can functionally substitute for the Ha-positive cell subpopulation and effect the generation of specific killer T-cells. The helper factor alone, in the absence of foreign antigen, does not induce killer T-cell development (6). Hence, both signals, the antigen and the helper factor, are required in order to activate the proliferative and differentiation processes within the precursor T-cells.

We have also examined the effects of the helper factor on the kinetics of killer T-cell differentiation. Our data suggest that: (1) antigen binding by specific receptors on precursor T-cells is substantially greater than binding by receptors on helper cells; (2) under normal circumstances, activation of precursor T-cells must await activation of helper cells and their production of a collaborative, helper factor; (3) memory cells may be helper cells that contain reservoirs of substantial quantities of helper factor(s) which can be released virtually immediately upon contact with antigen, thus resulting in a shift in the kinetics of effector cell generation; and (4) the levels of response attained in a primary response are limited by the helper cells and the quantity of helper substance(s) they can produce rather than by the numbers of specific killer cell precursors available to respond to the specific antigen.

MATERIALS AND METHODS
ANIMALS

Inbred mice of the C57BL/10, B10.D2, and DBA/2J strains and the (C57BL/6 × A/J) F1 hybrids were purchased from the Jackson Laboratories, Bar Harbor, Maine.

SENSITIZATION IN VITRO (SIV)

Mouse lymphoid cell suspensions were cultured in RPMI 1640 medium supplemented to 0.01 M Hepes buffer and 6 percent human serum (heated at 56℃ for 30 min), and with 100 units penicillin G, 100 μg L-arginine, 146 μg L-glutamine, and 75 μg kanamycin sulfate/ml. One-way cell proliferation was achieved by submitting the "stimulating" strain cells to 1000 rad of α-radiation from a ^{137}Cs source at a rate of approximately 950 rad/min. The lymphoid cell suspensions, 2×10^6 of each, were incubated in 12 × 75 mm tubes for 5 days at 37°C in a humidified atmosphere of 5% CO_2.

CELL MEDIATED CYTOTOXICITY (CMC)

DBA/2 mastocytoma P815-X2 cells (maintained in DBA/2 female mice) were labeled with ^{51}Cr and mixed with varying proportions of viable cells harvested from the 5-day SIV cultures. The cell mixtures were centrifuged at 225 g for 5 min and then incubated at 37°C for 4 hr. The cells were resuspended and centrifuged at 900 g for 10 min before the supernatants were decanted into scintillation tubes and counted. Specific ^{51}Cr release was determined by dividing the experimental values minus spontaneously released ^{51}Cr by the total amount of radioactivity detected in both the supernatant and pellets of ^{51}Cr-labeled P815-X2 cell cultured alone (Na$_2$ ^{51}CrO$_4$, sp. act. 1 mCi/ml; New England Nuclear, Boston, Mass.).

PRODUCTION OF
HELPER FACTOR SUPERNATANTS

Supernatants containing active helper substances were produced in a procedure involving two steps of cell culture. After 4 or 5 days in culture the viable cells from a number of replicate tubes of allogeneic lymphoid cell mixtures were pooled and washed. Fresh lymph node cell (LNC) suspensions of the stimulating strain were prepared. Equal numbers of activated and stimulating strain cells were mixed and the suspensions diluted to a concentration of 8×10^6 cells/ml with RPMI 1640 medium containing 0.01 M Hepes buffer. Neither antibiotics nor serum was added. Aliquots of 1.0 ml each were distributed into 12 × 75 mm tubes and the cells pelleted by centrifugation at 200 g for 5 min and incubated at 37°C for 6 hr. The cell pellets were then resuspended, pooled, and centrifuged at 900 g for 10 min. To ensure complete removal of all cells, the supernatant fluid was filtered through a 0.45 μm Millipore filter.

RESULTS

EFFECT OF HELPER FACTOR ON THE ANTIGEN DOSE REQUIRED TO STIMULATE THE GENERATION OF EFFECTOR T-CELLS

Various numbers of radiated B10.D2 LNC (stimulating cells) were mixed with a constant number of B6AF1 responding cells. Twelve replicated cultures for each stimulating cell dose were prepared. The final volume was adjusted to 2.0 ml in one-half of the cultures with fresh medium while the other half received a dilution of helper factor. The helper factor was produced from mixtures of B6AF1 and B10.D2 lymphoid cells. The final concentration of helper factor supernatant in these cultures was 22.5%. After 5 days of incubation, the viable cells from each set of cultures were pooled, washed, and adjusted to 4.0 ml. The cells were counted and four serial dilutions, starting with 1.0 ml of SIV cells, were set up in duplicate in 12 × 75 mm tubes. A 1.0 ml volume of ^{51}Cr -P815-X2 cells, 5 × 10⁴/ml, was added to each tube, and the generation of specific cytotoxic killer cells assessed. The killing activities of the 1.0 ml dilution of 25% concentration of the total yield of viable SIV cells are plotted in Fig. 1.

Substantially fewer antigens (stimulating cells) were required to stimulate the generation of significant numbers of killer T-cells when the cell cultures were supplemented with an exogeneous source of helper factor (Fig. 1). The generation of killer T-cells in the helper factor-supplemented cultures was greatly enhanced. No killing activity was detectable in cultures without stimulating cells. These data demonstrate that the specific precursor T-cells readily interacted with or bound to the foreign antigenic determinant on the stimulating cells. The intrinsic level of helper factor in the normal control cultures, however, was limiting. Under these circumstances low levels of killer cells were generated. Clearly, the concentration of helper factor in the unsupplemented cultures had a limiting effect on the differentiation of killer cell precursors that could readily recognize and respond to quite low doses of antigen.

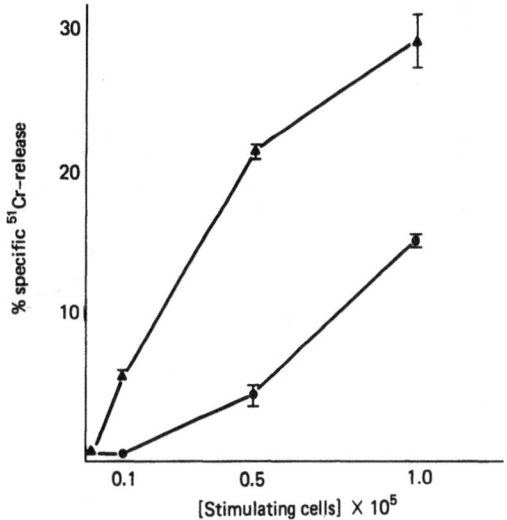

FIGURE 1. Effect of helper factor supernatants on the generation of effector T-cells stimulated by limiting numbers of radiated cells. B6AF1 LNC, 2 × 10⁶/ml/culture, were mixed with either fresh medium (●——●), or a 25% dilution of helper factor supernatants (▲——▲). The average and ranges of the percent specific ^{51}Cr -release caused by the B6AF1 cells cultured for 5 days with no stimulating cells, 0.1 × 10⁵, 0.5 × 10⁵, and 1.0 × 10⁵; radiated B10.D2 cells are plotted. Percent spontaneous ^{51}Cr -release in this experiment was 13.8 ± 0.3.

EFFECT OF HELPER FACTOR ON THE RATE OF KILLER T-CELL DEVELOPMENT *IN VITRO*

Equal numbers of C57BL/10 lymph node cells and radiated B10.D2 spleen cells, 2 × 10⁶ of each per culture, were mixed and cultured either in the presence of fresh medium or in a final concentration of 25% supernatant produced from mixtures of C57BL/10 + B10.D2 cells. Six replicate tubes of each set were harvested and assessed for the development of killer cell activity in two separate experiments designed to cover a 7-day period.

The data presented in Fig. 2 demonstrate that the helper factor can effect a shift in the kinetics of the generation of killer T-cells. Both the timing of response and the level of killer T-cell activity attained were affected. The precursor T-cells in cultures supplemented with helper factor supernatants differentiated into killer T-cells at a faster rate than did the cells in unsupplemented cultures. Also, the helper factor effected an increase in the numbers of T-cells generated.

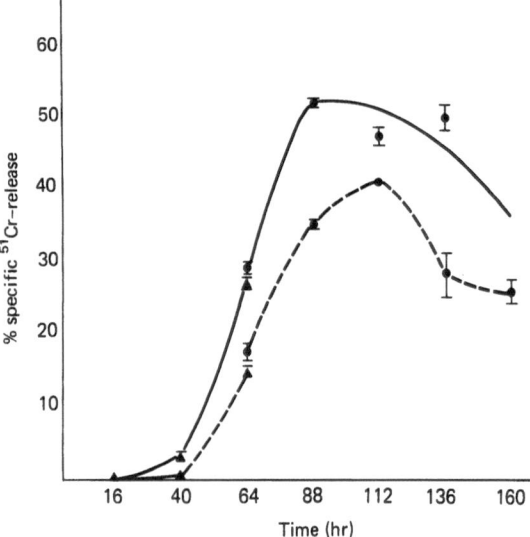

FIGURE 2. Effect of helper factor supernatants on the kinetics of killer T-cell development. Equal numbers of C57BL/10 LNC and B10.D2 radiated spleen cells were mixed and cultured with either fresh medium (---), or with a 25% concentration of helper factor supernatant (——). Six replicate tubes of each set were harvested and assessed for the generation of killer cell activity at various times after intial culturing. Two separate experiments were performed so that the results could be observed over a 7-day period. Experiment 1 (●) covers the 64–160-hr period, while experiment 2 (▲) covers the 16–64-hr period. The killing activities of the 1.0 dilution or 25% concentration of the total yield of viable SIV cells are plotted.

DISCUSSION

It appears, then, that the function of auxiliary "helper" cells in the generation of killer T-cells is to control the activation of the killer cell precursors. Specificity of the response that is generated is determined by the intitial interaction of precursor T-cells with the foreign antigen. The selection of the appropriate precursor T-cells depends, of course, upon the nature of the binding receptors on their cells surfaces (7). Our data demonstrate (Fig. 1) that the killer cell precursors readily bind to foreign antigen. In fact, the "dose-limiting" study suggests that stimulating cell antigens are bound preferentially or more strongly by the precursor T-cells than by the auxiliary cells. If the precursor T-cells bind as effectively to antigen in the presence or absence of soluble helper substances, then the low levels of killer cells generated in the unsupplemented cultures suggest either that the auxiliary cells, which presumably produce helper factors *in situ*, were not readily activated under these dose-limiting conditions, or

that only relatively few cells were activated. The activation of the auxiliary cells must also be a result of antigen recognition or binding. The results from this experiment indicate then that the binding affinities of receptors on the auxiliary cells are considerably weaker than that of receptors on the precursor T-cells. Cell absorption experiments performed on cell monolayers also indicate that precursor T-cells and their differentiated descendents, the effector or killer T-cells, readily bind to specific antigens (1,2,9).

The kinetic studies performed in the presence or absence of exogenous helper factor contribute to our knowledge of the events that follow antigen recognition. The shift in the kinetics of precursor T-cell differentiation into killer T-cells that is effected in the helper factor-supplemented cultures indicates that the elaboration of helper factrs n the unsupplemented cultures is rate-limiting. Since the differentiation of the killer cell precursors is dependent upon this second signal, their development, at least in a primary immune response, must await the activation and subsequent production of helper substances by the auxiliary cells.

A primary cell-mediated immune response, then, involves a sequence of steps beginning with: (1) specific antigen recognition and binding by the precursor T-cells; (2) antigen recognition and activation of auxiliary (helper) cells; (3) the elaboration of soluble helper substances by the auxiliary cells; (4) the interaction of helper substances with precursor T-cells; (5) the activation of the specific precursor T-cells that have now received both signals, antigen and helper substance; and ending with (6) differentiation of these precursor T-cells into effector T-cells.

It is of interest, also, to compare the kinetic data obtained from the generation of killer T-cells in the presence of helper factor (Fig. 2) with that published previously for the induction of secondary versus primary responses *in vitro* (3,8). Both the timing and levels of responses generated in the cultures supplemented with helper factor are very similar to the results reported for secondary immune responses. Evidently, immune cells readily provide the second signal required for the activation of killer T-cell precursors. These results have contributed to our suggestions that memory cells may be auxiliary or helper cells that contain reservoirs of helper substances; that preformed helper substances may be released by these cells virtually immediately upon interaction with antigen; and that the quantity of helper substances available to interact with the killer cell precursors determines the levels of killing activity obtained.

SUMMARY

We have examined the effects of helper factor(s) on the kinetics of killer T-cell differentiation. Our data suggest that: (1) antigen binding by specific receptors on precursor T-cells is substantially greater than binding by receptors on helper cells; (2) under normal circumstances, activation of precursor T-cells must await activation of helper cells and their subsequent production of a collaborative, helper factor; (3) memory cells may be helper cells that contain reservoirs of substantial quantities of helper factor(s) which can be released virtually immediately upon contact with antigen, thus resulting in a shift in the kinetics of effector cell generation; and (4) the levels of response attained in a primary response are limited by the helper cells and the quantity of helper substance(s) they can produce rather than by the numbers of specific killer cell precursors available to respond to the specific antigen.

ACKNOWLEDGMENT

This work was supported by USPHS grant CA-17754-01.

REFERENCES

(1) Bach, F. H., Segall, M., Zier, K. S., Sondel, P. M., Alter, B. J., and Bach, M. L. Cell mediated immunity: Separation of cells involved in recognition and destructive phases. *Science, 180*: 403, 1973.

(2) Berke, G., and Levey, R. Cellular immunoabsorbents in transplantation immunity: Specific in vitro deletion and recovery of mouse lymphoid cells sensitized against allogeneic tumors. *J. Exp. Med., 135*: 972, 1972.

(3) Cerottini, J. C., Engers, H. D., MacDonald, H., and Brunner, K. T. Generation of cytotoxic T-lymphocytes in vitro. I. Response of normal and immune mouse spleen cells in mixed leukocyte cultures. *J. Exp. Med., 140*: 703, 1974.

(4) Plate, J. M. D. Soluble factors substitute for T-T cell collaboration in the generation of T-killer lymphocytes. *Nature (London), 260*: 329, 1976.

(5) Plate, J. M. D. Cellular responses to murine alloanti-gens of the major histocompatibility complex. The role of cell subpopulations that express different quantities of H-2 associated antigenic markers. *Eur. J. Immunol., 6*: 180, 1976.

(6) Plate, J. M. D. Two signals required for the initiation of killer T-cell differentiation. Submitted for publication.

(7) Ramseier, H., and Lindenmann, J. Cellular receptors: Effect of antialloantiserum on the recognition of transplantation antigens. *J. Exp. Med., 134*: 1083, 1971.

(8) Rollinghoff, M., and Wagner, H. Secondary cytotoxic allograft response *in vitro*. I. Antigenic requirements. *Eur. J. Immunol., 5*: 875, 1975.

(9) Wekerle, H., Lonai, P., and Feldman, M. Fractionation of antigen reactive cells on a cellular immunoadsorbent: Factors determining recognition of antigens by T-lymphocytes. *Proc. Nat. Acad. Sci. U.S.A., 69*: 1620, 1972.

PART VI

Experimental Models of Clinical Conditions in Hematology

Georges Mathé, M.D.

INTRODUCTION

Effectiveness of active immunotherapy with systemic administration of BCG has been confirmed in many animal tumor systems as well as in the treatment of acute lymphocytic leukemias and melanomas (reviewed by Mathé, ref. 5). Nevertheless, it must be admitted that little is known about the BCG-induced mechanisms that lead to tumor-growth control. It is furthermore evident that the clinical aim to combine administration of immune adjuvants with conventional cancer chemotherapy is an effort to optimize an eventual synergistic effect of these two antineoplastic weapons.

Much experimental research is therefore needed, before rational protocols can be designed; it is equally evident that this experimental research will be valid only if relevant animal models are employed.

The results obtained with murine 1210-Leukemia—probably the most widely employed experimental leukemia—have very restricted value, as this leukemia differs considerably from human leukemias with respect to its immunology and growth characteristics.

We have therefore searched for another murine model of acute lymphoid leukemia. The AkR-leukemia is a good model, but it is very time-(and space-) consuming, often difficult to handle, and not useful in experiments, when a known tumor load is desirable from the start of the experiment. A more appropriate model was found in a leukemia induced in the C57Bl/6-mouse by several weekly injections of AkR-leukemia cells (1) and which is isogeneic to the C57B1/6-mouse.

This leukemia, known as E(AkR)-leukemia grows when inoculated subcutaneously, intravenously (splenic), and intraperitoneally (ascites). When inoculated by the latter route, only 10 leukemic cells kill 95–100% of the mice in 25–50 days. The leukemic cells disseminate to various organs, including the central nervous system, with a tumor load of only 10^6–10^7 cells. The animals die anemic, with pancytopenia, and often with infections and intrinsic hemorrhagias.

The cytology of the tumor cell was shown to be typical for a lymphoid leukemia (7,8). Furthermore, it was found that the cytology is rather heterogeneous with a dominance of blastic cells in the exponential growth phase and a dominance of "prolymphocytic" cells in the saturation phase, a phenomenon known in human leukemias (3).

The cytokinetics of the E(AkR)-leukemia also have characteristics similar to human

The E(AkR)-Leukemia as Murine Model of Human Acute Lymphoid Leukemia for Immunotherapy Trials

L. Olsson, I. Florentin, N. Kiger, and G. Mathé

23

TABLE 1 E(AkR) Tumor Model—Immunotherapy by BCG, BCG Plus Irradiated Tumor Cells, or Irradiated Tumor Cells Alone[a]

GROUP I	GROUP II	GROUP III	GROUP IV (CONTROLS)
1 mg BCG i.v.[b] per week	1 mg BCG i.v.[b] 10^7 radiated E(AkR) tumor cells s.c. per week	10^7 radiated E(AkR)[b] tumor cells s.c. per week	No treatment

[a] 10^5E(AkR) tumor cells were grafted i.p. on day 0 into 3-month-old C57B1/6 mice.
[b] Treatment started at day +1.

leukemias, as it was demonstrated that this leukemia always contains at least about 40% so-called G_0-cells (8). This is extremely important in trials of cytostatic protocols as these G_0-cells are not susceptible to conventional chemotherapy, and therefore must be eradicated by other mechanisms.

These growth characteristics thus not only increase the value of the E(AkR)-leukemia as a model for cytostatic thereapy, but also as a model in immunotherapy. It was therefore used to study the effect of BCG immunotherapy, associated or not with injections of radiated tumor cells, on various immunologic parameters of the host antitumor response. An attempt was made to correlate host immunologic response with tumor-growth characteristics. The modalities of immunotherapy are presented in Table 1 and results are summarized in Table 2.

A significant prolongation of the survival time was observed in mice treated with BCG alone or with BCG plus radiated tumor cells.

These treatments cured 22% and 10% of the animals, respectively. The growth rate of the tumor in untreated mice and in mice treated with radiated tumor cells alone followed the same Gompertz curve. Mice given BCG alone or BCG plus radiated tumor cells could be separated into subgroups according to the tumor evolution—subgroup A, in which the tumor grew as rapidly as in the controls, and subgroup B, in which the tumor began to regress on day 8 after grafting. Concomitant decreases in the number of tumor cells with increases in the number of lymphocytes (which constituted almost the entire peritoneal cell population at day 22), as well as an increase of the number of immunoblast-like cells in the ascitic fluid, were observed.

Cytotoxic antibodies were evidenced on day 8 in the sera of all mice given immunotherapy. The titer reached its maximum level on day 13 in mice treated with BCG or BCG plus irradiated cells, whereas it increased during the entire observation period in mice given

TABLE 2 Tumor Characteristics in Relation to the Tumor Cell Cytotoxic Potential of Antibodies and Lymphoid Cells[a]

PARAMETER	I A	I B	II A	II B	III	IV
Tumor size	↓	→	↓	→	→	Controls
% TC in peritoneal fluid	↓	→	↓	→	→	Controls
% Lymphocytes at days 13 and 18	↑	→	↑	→	→	Controls
% Immunoblast-like cells at days 13 and 18	↑	→	↑	→	→	Controls
Complement-dependent antibody-mediated cytotoxicity (CDAC)	++	++	++	++	+(+)	0
Inhibitor of CDAC	++	++	++	++	++	−
Spleen	++	++	++	++	++	0
Direct lymphocyte cytotoxicity (DLC)						
Lymph nodes	0	0	0	0	0	0
Spleen	++	±	++	±	±	(+)
Antibody-dependent cell-mediated cytotoxicity (ADCM) (lymph nodes)	++	0	++	0	0	0

[a] Group I: BCG alone. Group II: BCG plus irradiated tumor cells. Group III: irradiated tumor cells. Group IV: tumor-bearing controls. Subgroup A: Tumor size below controls. Subgroup B: Tumor size equal to controls. ↓, lower than control values. ↑, higher than control values. →, equal to control values.

radiated tumor cells alone. Moreover, all sera exerted an inhibitory effect on complement-dependent antibody-mediated cytotoxicity (CDAC) when used at low dilutions. Similar blockage of CDAC with slightly diluted sera was recently reported in three other tumor models (9). No cytotoxic activity was evidenced in the sera of tumor-bearing controls.

Spleen cells from mice treated with BCG or BCG plus irradiated cells were directly cytotoxic to tumor cells *in vitro* 4 and 8 days after tumor grafting. Spleen cells from irradiated cell-treated mice were only cytotoxic at day 4. No lymphocytotoxicity was observed with lymph-node cells from these animals as well as with spleen- or lymph-node cells from tumor-bearing controls.

FIGURE 1. Percentage distribution of ascitic tumor cells as a function of single-cell DNA content at various times after i.p. inoculation of 10^5 tumor-cells. 2N indicates the mean DNA-content of G_1-phase cells, and 4N the mean DNA-content of mitoses (and G_2-phase cells). Each value is the mean of three to five mice. BCG I: BCG-treated mice with tumor load and tumor-cell mitotic activity no different from controls. BCG II: BCG-treated mice with a lower tumor-cell number and a higher tumor-cell mitotic activity than in controls.

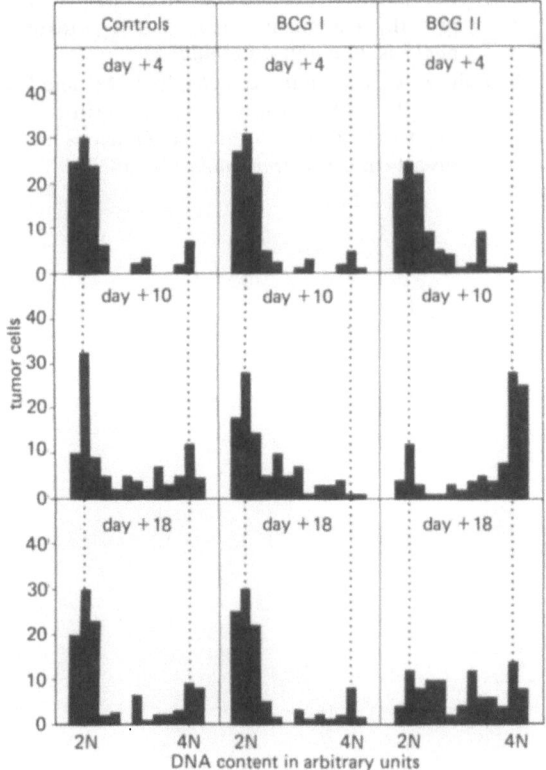

Antibody-dependent cell-mediated cytotoxicity (ADCMC) was evaluated by incubating tumor cells with the lymphoid cells and the serum of the same animal and by subtracting the cytotoxic activity of the lymphoid cells alone. A weak ADCMC was observed with spleen cells of mice given immunotherapy and of tumor-bearing controls at days 4 and 8. It increased considerably with time in mice treated with BCG or BCG plus radiated tumor cells for cases in which the tumor was regressing. Similar observations were made with lymph-node cells (subgroup IIB and IIIB).

A cytokinetic analysis of the E(AkR) ascitic tumor was performed during normal growth and during treatment by weekly injections of BCG alone. Tumor-cell mitotic activity, as expressed by both mitotic rate and mitotic index, was seen to be much higher in some BCG-treated mice (designated BCGII mice), compared to tumor-bearing controls and other BCG-treated mice (BCGI mice). This supranormal mitotic activity was confined to BCG-treated mice with a low tumor mass as compared to controls, whereas BCG I mice had tumor load similar to controls.

Cytophotometric studies (Fig. 1) have shown that tumor-bearing controls and BCGI mice are characterized by a high number of G_1 cells and few S cells and G_2 cells at day 4, a decrease in G_1 cells and an increase in S-cells and G_2 cells at day 10, and an increase in G_1 cells and to a small extent in G_2 cells and a decrease in S cells at day 18. BCGII histogram profiles are completely different with an increase in S cells seen as early as day 4. On day 10, a large number of cells were in late S phase or G_2 phase, and on day 18 nearly equal distribution of cells throughout the cell cycle could be observed.

In conclusion, the E(AkR) leukemic cells elicit no immune response in untreated recipients. Immunotherapy with BCG or BCG plus irradiated tumor cells cures a percentage of the animals and induces production of cytotoxic antibodies and cytotoxic lymphocytes. Nevertheless, only ADCMC activity is closely related to tumor size, which suggests that this immune mechanism plays a major role in BCG-induced tumor regression. A similar importance of ADCMC in tumor-growth control has been reported in other tumor systems both in mice (2,4,6) and cancer patients. From cytokinetic observations in mice in which immunotherapy led to tumor regression, it appears that BCG induces a selective loss of tumor cells in G_0/G_1 and in the first part of S phase. Thus, successful BCG immunotherapy eliminates cells that escape conventional immunotherapy.

We therefore consider E(AkR)-leukemia to be superior to any other transplantable murine leukemia; its cytokinetic and immunologic properties make it suitable for experimental studies on protocols that combine chemotherapy and immunotherapy.

SUMMARY

EAkR leukemia, transplanted in isogeneic C57Bl/6 mice, is a useful model for immunotherapy assays for the following reasons: (a) cellular kinetics of the tumor seems closer to the kinetics of human leukemias than the cytodynamics of L1210 leukemia. Particularly, the presence of a considerable number of so-called G_0 cells in the tumor, irrespective of tumor age, offers the opportunity to attack cells that escape conventional cancer chemotherapy; (b) the tumor may be transplanted in an ascitic form, which makes it possible to obtain a suspension of more than 90% living tumor cells for *in vitro* testing of tumor immunity.

REFERENCES

(1) Amiel, J. L., and Berardet, M. Induction d'une leucémie isogénique virale de Gross chez des C$_{57}$Bl/6 adultes par des injections répétées de cellules leucémiques AkR. *Rev. Fr. Etud. Clin. Biol., 14*: 587, 1969.

(2) Cadler, E. A., Irvine, W. J., and Ghaffer, A. K-cell cytotoxic activity in the spleen and lymph nodes of tumour-bearing mice. *Clin. Exp. Immunol., 19*: 393, 1975.

(3) Clarkson, B. D. Review of recent studies of cellular proliferation in actue leukemia. *Natl. Cancer Inst. Monogr., 30*: 89, 1969.

(4) Lamon, E. W., Andersson, B., Wigzell, H., Fenyö, E. M., and Klein E. The immune response to primary Moloney Sarcoma Virus tumors in BalB/c mice: Cellular and humoral activity of long-term regressors. *Int. J. Cancer 13*: 91, 1974.

(5) Mathé, G. *Introduction to Active Immunotherapy of Cancer*, Vol. 1. New York/Heidelberg: Springer-Verlag, 1976.

(6) Pollack, S., Heppner, G., Brawn, R. J., and Nelson K. Specific killing of tumor cells in vitro in the presence of normal lymphoid cells and sera from hosts immune to tumor antigens. *Int. J. Cancer, 9*: 316, 1972.

(7) Olsson, L., Florentin, I., Kiger, N., and Mathé G. The effects of BCG on the growth of a mouse leukemia: Various cellular and humoral parameters of the effect of antibodies and lymphoid cells to tumor cells and their correlation to tumor growth *in vivo*. *(Submitted for publication.)*

(8) Olsson, L., and Mathé, G. The effect of BCG on the growth of a murine leukemia: A cytokinetic analysis. *(Submitted for publication.)*

(9) Stolfi, R. L., Fugmann, R. A., Stolfi, C. M., and Martin D. S. Development and inhibition of cytotoxic antibody against spontaneous murine breast cancer. *J. Immunol., 114*: 1824, 1975.

INTRODUCTION

Significant advances have been made in the management of human actue myelocytic leukemia (AML) over the last decade (5,6). However, the overall remission induction rate and survival time are still less than obtained in other types of leukemia, e.g., acute lymphocytic leukemia (ALL) (3,8). An animal leukemia model analogous to human AML would be helpful in the design of experimental therapy regimens. Two prerequisites for the relevance of such a model are (1) a slow overall growth rate of the leukemia, and (2) a suppression of normal hemopoiesis that runs the same course as that in the human disease (10,11).

Various authors have reported studies on transplantable myelocytic leukemias in both mice (12,14) and rats (Shay) (7,9). These models are characterized by a very rapid growth rate, i.e., 10^7 leukemic cells kill animals within 6–10 days, and normal hemopoiesis is often rather well preserved. In this chapter, we describe a transplantable promyelocytic leukemia in rats that was developed in our laboratory. In addition to a slow growth rate, the pattern by which normal hemopoiesis decreases makes this particular model very comparable to human AML.

24

After a description of the major growth characteristics of this model, its relevance to human AML is further substantiated by evaluating the therapeutic response to clinically used AML and ALL treatment regimens. Finally, we report on our search for optimal remission-induction schedules in combination with hematologic supportive care, including bone-marrow transplantation.

MATERIALS AND METHODS
RATS

All experiments were carried out in female inbred Brown Norway (BN) rats. The recipient rats for the inoculation of the leukemia were used for rats 12–15 weeks of age, at which time they weighed from 140–170 g. The animals were routinely housed five to a cage and given food and water *ad libitum*.

LEUKEMIA

The leukemia was chemically induced by dimethylbenzanthracene. It can be transplanted by means of cellular transfer. Cytology and cytochemistry proved the myeloid nature of this

Experimental Chemotherapy: A Rat Model for Human Acute Myeloid Leukemia

Loes P. Colly and
Ton Hagenbeek

leukemia, which is abbreviated BNML (Brown Norway rat myelogenous leukemia). In order to follow the development of the leukemia, peripheral blood samples were taken from the tail for routine hematologic follow-up at various intervals after intravenous inoculation with 10^7 BNML cells. Follow-up consisted of total and differential leukocyte counts, platelet, erythrocyte and reticulocyte counts, and determination of hematocrit values.

A group of 42 rats was sacrificed at various intervals after inoculation to determine the growth rate of leukemia in the liver, spleen, and thymus by means of measuring fresh organ weights.

The spleen-colony assay (13) was used to determine the fate of the normal hemopoietic stem cells. With this technique, the number of colony forming units (CFU-s) in spleen is scored as a measure of the number of total bone-marrow stem cells. The assay was modified for a xenogeneic system. Cell suspensions from rat bone marrow, spleen, or blood were injected into lethally radiated (1025 rady) F_1 hybrid female mice from matings between the C57BL/Rij and the CBA/Rij strains. Spleen colonies were scored at day 9 after injection. These colonies were derived from normal rat CFU-s. The evidence for this is (1) endogenous colony formation is absent at this high total-body radiation dose, (2) there is a linear relationship between the number of injected rat cells and the number of colonies observed, and (3) the results to be presented are quite similar to those reported by van Bekkum et al. (1) with the spleen-colony assay in the rat-to-rat system. In addition, leukemic cells, possibly because of their slower growth rate, only give rise to colonies (LCFUs) 20 days after inoculation into nonirradiated normal rats. During leukomia development, CFU-s were monitored in the bone marrow, the buffy coat of the peripheral blood, and in the spleen. In order to determine if there was a linear dose–effect relationship, three concentrations of cell suspensions were injected. After the mice were sacrificed 9 days later, the spleen was excised and fixed in Tellyesniczky's solution and the macroscopic colonies were counted.

PREPARATION AND INJECTION OF CELL SUSPENSIONS

Bone marrow cell suspensions and spleen suspensions were prepared according to standard procedure (1). Nucleated cells in these suspensions were counted in a hemocytometer after red cell lysis using Türck's solution. Determination of eosin-resistant cells (0.2%) was also routinely

done. Only cell suspensions containing less than 10% dead cells were used for injection. Chemotherapeutic treatment was started in the period from days 11 to 18 after inoculation of the leukemia. At these times, the rat tumor load is comparable to that seen in the clinical situation. Cytosine arabinoside, abbreviated ara-c (Cytosar) was purchased from Upjohn-Netherland (Ede). Adriamycin was kindly supplied by Farmitalia Milan (Italy). Vincristine sulfate (Oncovin) was purchased from Lilly (Indiana). Prednisole (Di-Adreson-F aquosum) was purchased from Organon-Netherland (Oss).

HEMATOLOGIC SUPPORTIVE CARE

Platelet-enriched plasma was infused when the thrombocyte count fell below 20,000/μl blood. Packed erythrocytes were infused when the hematocrit fell below 15%. Isogeneic bone marrow cells were injected in a dosage of 1×10^8 cells/kg body weight.

RESULTS

GROWTH CHARACTERISTICS OF THE BNML

Inoculation with 10^7 leukemic spleen cells results in the death of a rat in 26–30 days. In Figure 1, the number of total white blood cells

FIGURE 1. Changes in the total and differential white blood cell counts after inoculation with 10^7 BNML spleen cells. Each point represents means ± SE of 5–7 rats. Day 0: Nonleukemic controls.

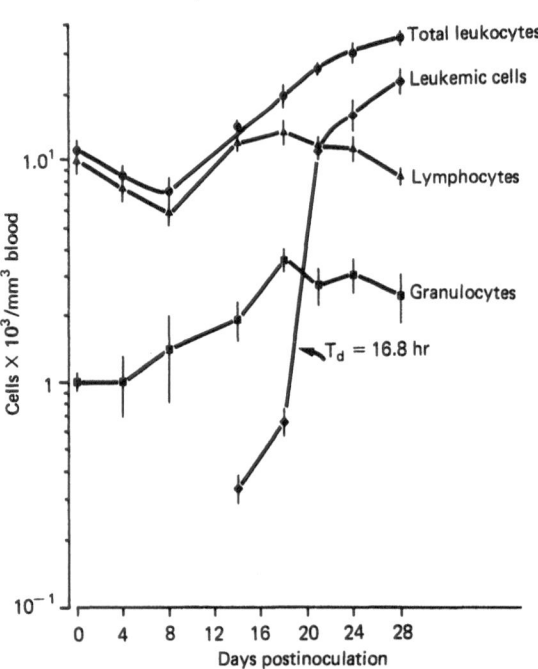

is plotted against the time after inoculation. The rise seen at day 12 is primarily attributable to the appearance of leukemic cells in the blood. Most of these early leukemic cells are leukemic promyelocytes. Their doubling time during the exponential phase of growth is 16.8 hr. A rise in the number of granulocytes can be seen on day 4; this may reflect a myeloid reaction against the leukemia. The changes in organ weight during leukemia development are plotted in Fig. 2. The liver weight shows an increase after day 15 with a doubling time of 9.8 days; the terminal weight is about 2.5 times the normal weight. The spleen weight shows an early increase starting at day 8, with a doubling time of 6.2 days and a final weight of about 8 times normal. The thymus weight tends to decrease during the terminal stage of the disease.

FATE OF THE NORMAL HEMOPOIETIC STEM CELL DURING THE PROGRESSION OF THE LEUKEMIA

The changes in the distribution of CFUs in bone marrow and spleen are plotted in Fig. 3 against the time after inoculation. The bone marrow of normal rats contain 750 CFUs per 10^7 nucleated cells. No remarkable changes are observed in the first 2 weeks after inoculation. Dur-

FIGURE 2. Changes in liver, spleen, and thymus weight after inoculation with 10^7 BNML spleen cells. Each point represents means ± SE of 6 rats. Day 0: Organ weights of nonleukemic controls.

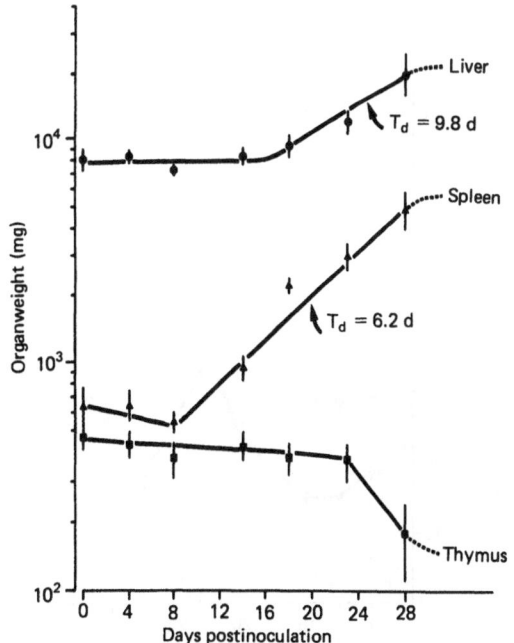

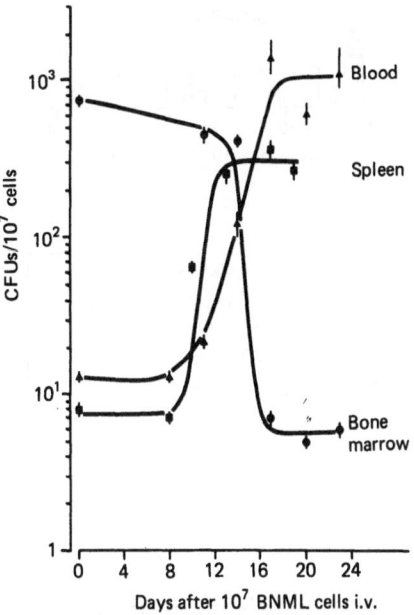

FIGURE 3. Changes in the number of bone marrow, peripheral blood, and spleen CFUs during progression of the leukemia. Day 0: Nonleukemic controls. Each point represents means of 2–3 experiments with 8–10 recipient mice per spleen colony assay. Vertical bars indicate SE values.

ing the subsequent 3 days, a steep fall ending in a plateau phase where finally less than 1% of the CFUs originally present in the bone marrow can be recovered is noted. This is an agreement with results of van Bekkum et al. (2) using the colony-forming unit culture. (CFU-c) and the rat-to-rat CFU-s technique. In both the blood and the spleen, the changes in CFUs number follow an inverse pattern to that seen in the bone marrow.

THERAPEUTIC RESPONSE TO CHEMOTHERAPY PROTOCOLS USED IN CLINICAL AML AND ALL

In an attempt to stimulate an AML schedule, the Ad-OAP regimen (Fig. 4) (4) was compared with a proved remission induction treatment for acute lymphocytic leukemia (ALL) (3,8). The ALL schedule was vincristine at a dosage of 0.18 mg/kg i.v. given at days 15 and 21. The dosage of prednisone was 5 mg/rat i.m. daily. Both schedules were adapted to the rat leukemia, with respect to time and drug dosages. Figure 5 shows the effect of the Ad-OAP regimen on the number of peripheral white blood cells. Shortly after treatment begins, a great decrease in total white blood cells is observed. During the terminal phase, there is an increase in total white blood

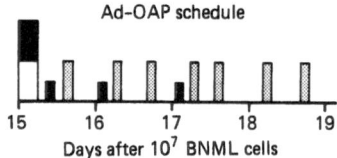

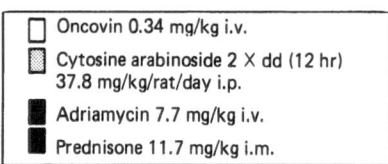

FIGURE 4. The clinically used Ad-OAP chemotherapy regimen for human AML adapted to the BN myelocytic leukemia (4).

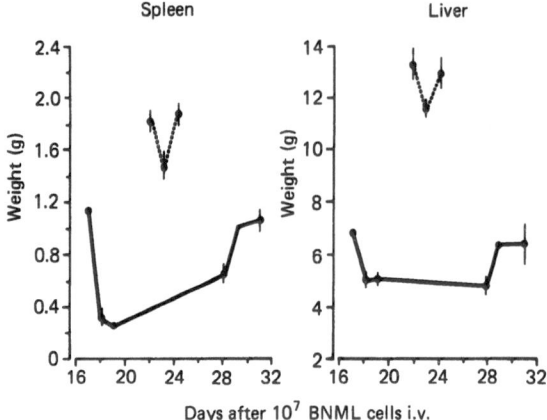

FIGURE 6. Changes in spleen and liver weights after the Ad-OAP regimen. The fresh weights were determined after spontaneous death. Each point represents means ± SE of 1–5 rats. (——) Treated rats; (---) controls.

cells, and the animals die in relapse. The efficiency of the Ad-OAP regimen is demonstrated by the decrease in spleen and liver weight (Fig. 6). The reduction in spleen weight was fourfold on the average; for the liver it was twofold.

Figure 7 shows the effect on the white blood cells during treatment with the ALL regimen. Although some depression of cell counts is observed, a rapid increase is noted soon after the first injection with vincristine. The reduction in spleen and liver weight is much less impressive than that seen in the Ad-OAP-treated animals (Fig. 8). Both spleen and liver weight reduction is about 1.2-fold. With respect to survival time, no significant differences were noted between the treated groups and their respective controls.

FIGURE 7. Changes in the number of leukocytes after application of a clinical ALL chemotherapy regimen adapted to the BN myelocytic leukemia vincristine, 0.18 mg/kg i.v. days 15 and 21; prednisone, 5 mg/rat i.m. daily. Vertical bars indicate SE values. ——, Treated rats: N = 7; ---, controls: N = 6.

FIGURE 5. Changes in the number of leukocytes during the Ad-OAP regimen. Vertical bars indicate SE values. (——) Treated cells: N = 7; (---) controls: N = 9.

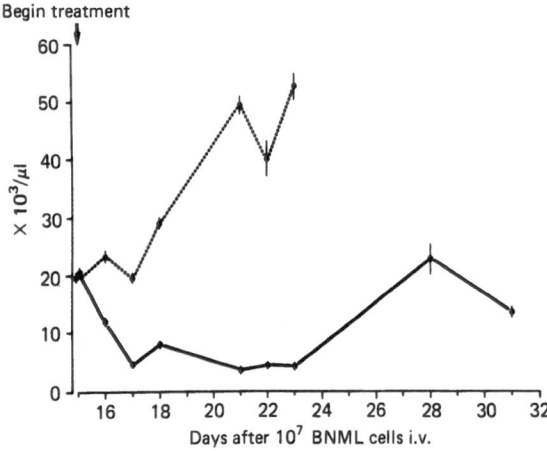

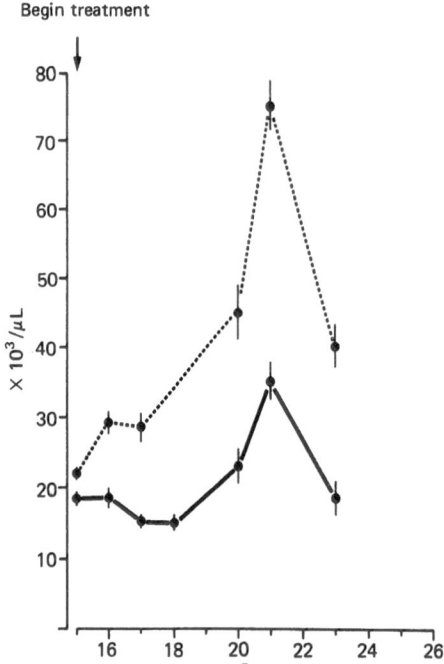

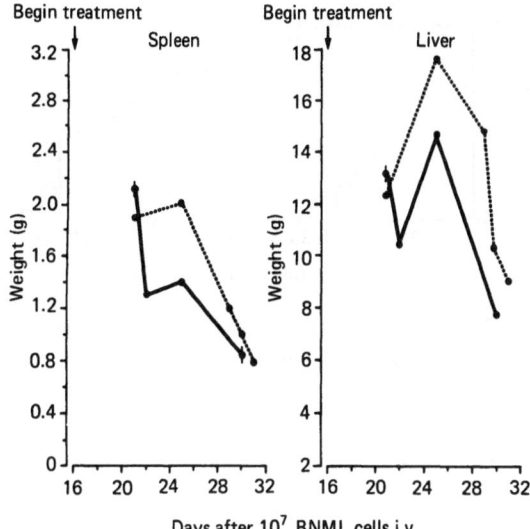

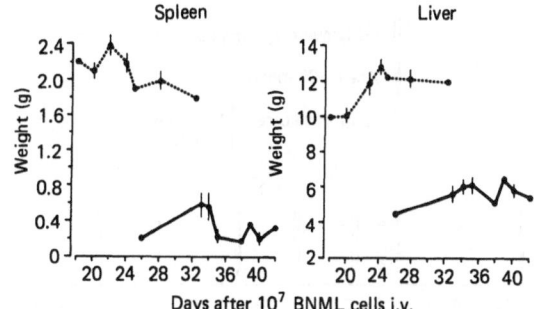

FIGURE 8. Changes in spleen and liver weights after the ALL regimen. Fresh weights were determined after spontaneous death. Each point represents means ± SE of 1–3 rats. ——, Treated rats; ---, controls.

FIGURE 10. Changes in spleen and liver weights after repeated schedules of ara-C and adriamycin. Fresh weights were determined after spontaneous death. Each point represents means ± SE of 1–4 rats. ——, Treated rats, ---, controls.

REMISSION INDUCTION CHEMOTHERAPY WITH MODIFIED AML SCHEDULES

Because of the complexity of the Ad-OAP schedule, a simplified regimen was studied. It combined two of the most effective drugs of the former regimen, ara-C (100 mg/kg i.v.) and adriamycin (7.7 mg/kg i.v.) (regimen I). Based on cell-cycle parameters, the treatment starts with ara-c, which is followed 16 hr later by adriamycin. This regimen was repeated as soon as an increase in the peripheral white blood cells was observed.

Figure 9 shows the change in the white blood cell counts during repeated therapy. His-

tologically, the bone marrow shows a hypoplastic to aplastic picture, but some foci of leukemic cell nests remain. Histological preparations of the liver show that, after two courses of treatment, the leukemic infiltration present on day 15 after inoculation disappears and the normal microarchitecture of the liver is restored. Increased macrophage activity is observed in the greatly enlarged leukemic spleen after treatment is started and, although the tumor load is greatly diminished, the normal microarchitecture of the spleen is not reestablished. Figure 10 shows the reduction in leukemic spleen weight to be 5-fold and that for the liver to be about twofold. Figure 11 shows the effect of repeated schedules of ara-C and adriamycin on the survival of the treated

FIGURE 9. Changes in the number of leukocytes after repeated schedules of ara-C and adriamycin. Vertical bars indicate SE values. ——, Treated rats; $N = 15$; ---, controls: $N = 8$.

FIGURE 11. Survival curves after repeated courses of Ara-C and Adriamycin in the BN myelocytic leukemia. $\overline{X}$ survival treated rats 36.8 days. X survival controls 30.2 days ($p < 0.001$.) ——, Treated rats, ---, controls.

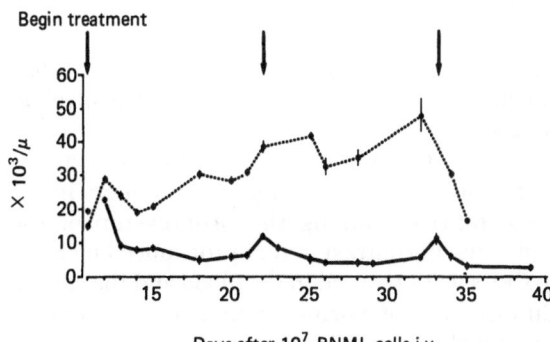

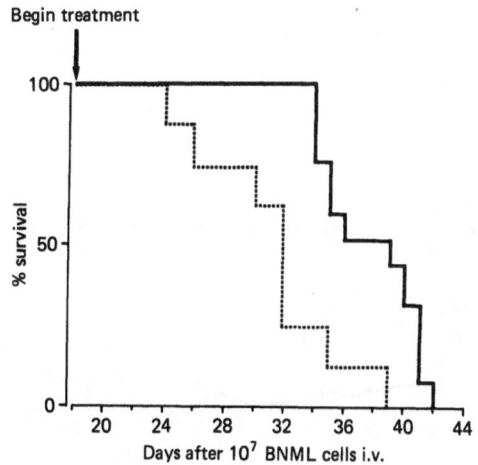

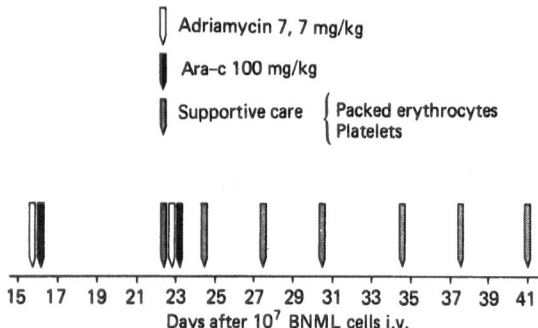

FIGURE 12. Treatment regimen with repeated courses of adriamycin and ara-C in the BN myelocytic leukemia in combination with hematologica supportive care.

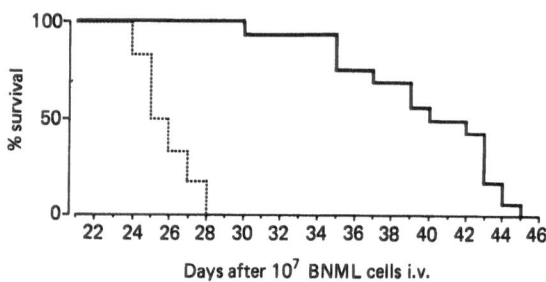

FIGURE 14. Survival curves after repeated courses with adriamycin and ara-C in combination with hematologic supportive care. $\bar{x}$ survival treated rats 30.2 days; $\bar{x}$ survival rats 24.8 days, $p < 0.00001$. ——, Treated rats: $N = 16$; ---, controls: $N = 5$.

group as compared to the controls. The treated animals die as a result of bone marrow aplasia, often in combination with cardiomyopathy caused by the toxic side effects of adriamycin. The differences in survival are statistically significant ($p < 0.001$). Injection of 10^7 cells of this remission bone marrow (taken at day 27 after two cycles of therapy) into normal young BN rats did not give rise to leukemia during 130 days of follow-up. Therefore, these rats may be regarded as being free of tumors, i.e., the inoculum apparently did not contain clonogeneic leukemic cells. As TD_{50} studies on BNML bone marrow cells have indicated that 1 out of 33 cells is capable of inducing leukemia, it is clear that the surviving fraction of clonogeneic leukemic cells per 10^7 total bone marrow cells is less than 3×10^{-5}.

A modified remission induction schedule (regimen II) differed from the ara-c/adriamycin combination only in that the two compounds were given in the reverse order at the same dosages—adriamycin followed 16 hr later by

ara-c. In this experiment, repeated hematologic supportive care (transfusions of platelets and packed erythrocytes) was provided during the remission phase. Figure 12 shows the days on which the various treatments were applied. The drug combination adriamycin and ara-c was given two times only. Figure 13 shows the change in white blood cell counts during this experiment. Most of the rats died in relapse. The effect of this regimen on the survival is presented in Fig. 14. There is a significant difference between the control group and the treated group ($p < 0.00001$), but all rats eventually die from leukemia.

DISCUSSION

The available data suggest that the BNML is one of the most realistic models for human leukemia. This statement is based on the following results.

1. There is cytologic and cytochemical evidence for the myelocytic nature of the disease.

2. In contrast with any other animal models, the BNML has a rather slow growth rate pattern and shows central and peripheral spread which is comparable with the human disease.

3. Normal hemopoiesis gradually disappears during the development of leukemia, similar to the situation in human AML.

4. Comparison of the results of the clinical AML versus the ALL schedule, provides additional evidence that the BNML is of a myeloid nature.

As shown in Fig. 3, normal CFUs were detected only in markedly reduced numbers in the bone marrow during the progression of the leukemia. In contrast, a very pronounced increase in CFUs was observed in the blood and the spleen. Suppression of normal hemopoiesis might be explained by several mechanisms.

FIGURE 13. Changes in the number of peripheral leukocytes after repeated courses of adriamycin and ara-C in combination with hematologic supportive care. Vertical bars indicate SE values. ——, Treated rats; $N = 16$); ---, controls: $N = 5$; ☆, supportive care.

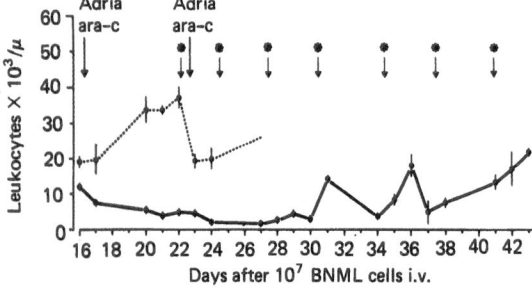

1. Direct cell–cell interactions between leukemic and normal cells might inhibit normal hemopoiesis—this possibility was ruled out by performing mixed culture studies on leukemic and normal bone marrow cells (2).

2. Leukemic cells might produce humoral factors that inhibit proliferation of normal CFU-s; however, diffusion chamber experiments did not support this hypothesis.

3. Normal precursor cells might be transformed into leukemic cells, but there is no evidence for this at present.

4. The normal CFU-s might migrate from the bone marrow when this tissue is being populated by leukemic cells. This study has provided strong evidence for this statement (Fig. 3).

Therefore, our present hypothesis is that CFU-s are forced to leave the bone marrow when the number of leukemic cells exceeds a certain critical level. After their circulation in the blood, they lodge at sites that offer more favorable conditions for further proliferation, i.e., the spleen and possibly also the liver. There is histologic evidence for the development of extramedullary hemopoiesis in these organs. From preliminary calculations, it appears that the total number of CFU-s in the body does not change markedly, but that they are simply redistributed throughout the body during leukemia development. Whether these redistributed stem cells in the blood, spleen, and liver are less capable of producing functional progeny remains to be established.

During the chemotherapy experiments a better therapeutic response was observed after the AML Ad-OAP, schedule (Figs. 5 and 6) than with the ALL regimen (Figs. 7 and 8); this again provides evidence for the myeloid character of the BNML.

With ara-c and adriamycin treatment, complete remission was said to have been achieved when no leukemic cells were seen in the peripheral blood and when there was less than 5% leukemic cells in a bone marrow smear. However, complete remission is a rather relative diagnosis, because it depends on the site of the bone marrow puncture and, in our experience, some sites have been found to be completely empty or to contain only a few clusters of hemopoietic elements; in the same rats, other sites showed residual foci of damaged leukemic cells.

In those studies in which treatment with ara-C and adriamycin was repeated as soon as there was an increase in the peripheral white blood cells (regimen I), the leukemia seemed to be cured, as 10^7 remission bone marrow cells did not give rise to leukemia in normal recipients. However, the rats in which remission was achieved died as a result of aplasia. This aplasia was caused by the chemotherapeutic regimen as well as by the leukemia itself.

In contrast, after one schedule, the duration of the remission is short and the animals die of leukemia relapse. Hematological supportive care provided by regular infusion of platelets and packed erythrocytes relieved the aplastic state for a rather long period of time (Figs. 13 and 14). In this experiment (regimen II), however, the treatment schedule of adriamycin followed by ara-c seems to be less effective than the reverse schedule used in regimen I, because the animals die of a relapse of the leukemia.

When adriamycin was given more than twice in the dosages used, an accumulated toxic effect was observed. It consisted of focal pyknosis of cardiac muscle cells and resulted in signs of cardiomyopathy (15).

Thus far, in this experimental leukemia model, the most effective regimen to effect a complete remission appears to be ara-C followed by adriamycin. However, this curative treatment regimen results in a lethal aplasia. By combining this regimen with highly effective supportive care (see regimen II), one would expect complete remission for a prolonged period of time, or even cures.

Another means of supportive care which would allow even more aggressive chemotherapy would be the infusion of isogeneic bone marrow. We are presently investigating this possibility. Preliminary data indicate highly beneficial effects from this kind of support. Autologous hemopoietic cell transplantation with stem cells separated from remission bone marrow or from peripheral blood will be also be applied in future experiments.

In conclusion one rat leukemia model for human AML seems to be ideally suited for the development of new therapeutic modalities and for the comparison of clinically used regimens.

SUMMARY

A transplantable rat leukemia has been developed in our laboratory which is being studied for its feasibility as an experimental model for human acute myeloid leukemia (AML). The leukemia was induced by dimethylbenzanthracene in the Brown

Norway (BN) rat. Cytologically and cytochemically, it has proved to be a leukemia of the myeloid type. The leukemia has a slow proliferation rate—an i.v. inoculum of 10^7 cells results in death after 30 days. Suppression of normal hemopoiesis in the bone marrow occurs as the leukemia develops—bone marrow colony-forming unit-spleen (CFU-s) decrease to less than 1% of normal during the terminal phase of the leukemia. However, a reverse pattern is observed for the course of CFU-s in blood and spleen. A redistribution of normal hemopoietic stem cells throughout the body is suggested.

In the present study, the response of the leukemia to various chemotherapy schedules has been evaluated using peripheral nucleated cell counts, survival time, liver and spleen weights and histology as criteria. A typical effective regimen for human acute lymphocytic leukemia (ALL) (weekly doses of vincristine and daily doses of prednisone) proved less effective than schedules employed successfully in human AML [a.o. the Ad-OAP regimen: adriamycin, oncovin, cytosine arabinoside (ara-C) and prednisone]. The two most effective drugs adriamycin and ara-C have been studied in more detail. With different time-spaced regimens, the rats died either from aplasia or from relapse of the leukemia. With repeated schedules of ara-c first, followed by adriamycin 16 hr later, it was possible to induce a complete remission. The results are discussed in conjunction with the feasibility of hematologic supportive care, either by regular infusions of packed erythrocytes and platelets or by bone-marrow transplantation with isologous or autologous cells.

ACKNOWLEDGMENTS

This work was supported by the Koningin Wilhelmina Funds of the Dutch National Cancer League, under Grant.:nr. WLV III-1 a. Adriamycin was generously supplied by Farmitalia, Italy. The authors wish to thank Miss A. Töns, A. C. M. Martens, Mrs. E. S. Dingjam-Hirschi, and Miss. J. N. Soekarman for expert technical assistance. We are also indebted to Professors D. W. van Bekkum, M. B. Edelstein, and G. Wagemaker for their scientific support.

REFERENCES

(1) van Bekkum, D. W., and de Vries, M. J. in *Radiation Chimeras.* New York: Academic Press, 1967.

(2) van Bekkum, D. W., van Oosterom, P., and Dicke, K. A. In vitro colony formation of transplantable rat leukemias in comparison with human acute myeloid leukemia. *Cancer Res., 36*:941, 1976.

(3) Berry, D. H., Pullen, J., George, St., Vietti, T. J., Sullivan, M. P., and Frenbach, D. Comparison of prednisone, vincristine, methotrexate, and 6-mercaptopurine vs. vincristine and prednisone induction therapy in childhood acute leukemia. *Cancer, 36*:98, 1975.

(4) McCredie, K. B., Bodey, G. P., Gutterman, J. U., Rodriquez, V., and Freireich, E. J. Sequential Adriamycin-Ara-C (A-OAP) for remission induction of adult acute leukemia (AAL). *Proc. Am. Assoc. Cancer Res., 15*:62, 1974.

(5) European Organization for research on Treatment of Cancer (EORTC). (Leukemia and hematosarcoma cooperative group.) A second comparative trial of remission induction (by Cytosine Arabinoside given every 12 hours, or CAR and thioguanine, or CAR and daunorubine) and

maintenance therapy (by CAR or methylgag) in Acute Myeloid leukemia. *Eur. J. Cancer, 10*:413, 1974.

(6) Gutterman, J. U., Rodriquez, V., Marligit, J. Burgess M. A., Geham, E., Hersch, M., McCredie, K. B., Reed, R., Smith, T., Bodey, Sr. S. P., and Freireich, E. J. Chemoimmunotherapy of adult acute leukemia. Prolongation of remission in myeloblastic leukemia with BCG. *Lancet, 1*:1405, 1974.

(7) Hoelzer, D. Growth characteristics of a transferable acute leukemia in rats. *J. Natl. Cancer Inst., 50*:1321, 1973.

(8) Mauer, A. M., and Simone, J. V. The current status of the treatment of childhood acute lymphoblastic leukemia. *Cancer Treatment. Rev., 3*:17, 1976.

(9) Moloney, W. C., Dorr, A. D., Dowd, G., and Boschetti, A. E. Myelogenous leukemia in the rat. *Blood, 19*:45, 1962.

(10) Moore, M. A. S., Spitzer, G., Williams, N., Metcalf, D., and Buckley, J. Agar culture Studies in 127 cases of untreated acute leukemia: The prognostic value of reclassification of leukemia accord-

ing to in vitro growth characteristics. *Blood,* *44*:2, 1974.

(11) Moore, M. A. S. In vitro studies in the myeloid leukemias. In Cleton, F. C. Crowther, D. C., and Malpas, J. S., eds., *Advances in Acute leukemia,* Chapter 7. Amsterdam: North-Holland—American Elsevier, 1974.

(12) Skipper, H. E., Schnabel, F. M., Trader, M. W., Laster, W. R., Simpson-Herren, L., and Lloyd, H. H. Basic and therapeutic trial results obtained in the spontaneous AK leukemia (lymphoma) model-End of 1971. *Cancer Chemother. Rep.,* *56*:273, 1972.

(13) Till, J. E., and McCulloch, E. A. A direct measurement of the radiation sensitivity of normal mouse bone marrow cells. *Radiat. Res.,* *14*:213, 1961.

(14) Wodinsky, J., Suiniorski, J., and Kensler, C. J. Spleen colony studies of leukemia L1210. 1. Growth kinetics of lymphocytic L1210 cells in vivo as determined by spleen colony assay. *Cancer Chemother. Rep.,* *51*:415, 1976.

(15) Zbinden, G., and Bränle, E. Toxicologie screening of daunorubicin (NSC-82151), adriamycin (NSC-1213127) and their derivatives in rats. *Cancer Chemother. Rep.,* *59*:707, 1975.

INTRODUCTION

Clinical graft-versus-host disease (GvHD) occurs in approximately 70% of patients with successful allogeneic marrow grafts (24). The diagnosis of GvHD is made on the basis of clinical impression with pathologic confirmation of involvement in any of the three target organs—the skin, the liver, or the gut. Clinical and histologic criteria have been proposed for staging GvHD on a scale ranging from + to ++++ (24). The pathologic assessment of mild, early GvHD is often difficult. Histologic changes in skin are minimal, and their interpretation is complicated by alterations produced by cytotoxic drugs and radiation. The patient's clinical condition often precludes a liver or bowel biopsy. The present study was initiated as an attempt to find an easily measurable marker for GvHD not dependent on target-organ damage.

The existence of colony-stimulating factor (CSF) was demonstrated by studies on the growth of mouse bone marrow and spleen cells in agar culture (2). It was found that cells in mouse bone marrow could proliferate *in vitro* and form colonies of granulocytic cells when a feeder layer of cells was placed beneath the hemopoietic cells (21). It was later shown that the feeder layer could be replaced with cell-free material, such as supernatant from cultures of feeder layer cells (3,22), small amounts of rodent serum (19), human serum (18), or human urine (18, 23). The stimulus to colony growth in these materials was termed CSF. Many cellular sources of CSF were found; they included embryonic cells, fibroblasts, and macrophages (6). Studies by Bradley and Metcalf showed that most organs contained cells that can secrete or release CSF on culture in agar (21). Recently, several groups showed that cultures of lymphocytes stimulated with either phytohemagglutinin (PHA), concanavalin A, or allogeneic cells release CSF into the culture medium (7,13,14,17,20). This appears to be a thymus-derived (T)-lymphocyte-produced mediator dependent for release upon protein synthesis but not on DNA replication; thus it possesses requirements similar to those necessary for the release of other lymphocyte factors, such as macrophage migration inhibition factor (MIF) and lymphotoxin (15,20). Parker and Metcalf demonstrated that removal of macrophages from stimulated lymphocyte cultures does not decrease the amount of CSF released into the culture medium (14). Present evidence suggests that T- and perhaps B-lymphocytes are capable of producing CSF. In this context, two groups of workers have reported that mice under-

Serum Colony Stimulating Factor: A Marker for Graft-versus-Host Disease in Humans

J. W. Singer, M. C. James, and E. D. Thomas

25

going GvHD have spontaneous CSF production from cultures of spleen cells (13) and have high serum levels of CSF irrespective of granulocyte counts (10).

We have studied serial CSF serum levels of patients undergoing allogeneic bone marrow transplantation with and without clinical GvHD. Patients undergoing syngeneic transplantation were studied as controls. The results of these studies indicate that high CSF levels are present in serum during GvHD, and a rise in CSF may precede the clinical diagnosis of GvHD by several days. Levels of CSF were also elevated during the granulocytopenic period before the graft became functional. Levels generally fell to near zero as the peripheral granulocyte count rose. A marked secondary rise in CSF was noted only in those patients in whom GvHD developed. Neither the patients undergoing syngeneic transplantation nor the allogeneic patients without GvHD demonstrated this marked secondary rise.

METHODS
PATIENTS

Serial 1-ml serum specimens were obtained from patients undergoing bone marrow transplantation at the Fred Hutchinson Cancer Research Center. Details of transplantation methodology, criteria of engraftment, and histologic criteria for the diagnosis of GvHD have been published elsewhere (24).

SERUM PREPARATION

Serum specimens were separated and frozen in glass tubes at $-20°C$ until assayed. All samples from a single patient were run on the same day. The serum samples were thawed and extracted with chloroform to remove nonspecific inhibitors of granulopoiesis (9). The serum was removed, placed in sterile tubes, and diluted 1 : 3 with Eagle's minimum essential medium (MEM). The interface containing lipoprotein inhibitors was removed from the residual chloroform and the chloroform was allowed to evaporate to dryness. The chloroform extractable residue (CER) was reconstituted with 0.5 ml MEM and stored at $-20°C$ until assayed.

ASSAY SYSTEM

Mouse Femurs were removed from ether-anesthestized LAF-1 male mice (Jackson Laboratories, Bar Harbor, Maine), and the cells were harvested by flushing with MEM. Cell dispersion was achieved by single passage through a 25-gauge needle. The number of Trypan Blue dye-excluding cells was counted in a hemocytometer, and the cell concentration was adjusted. The plating mixture consisted of 0.8% carboxymethylcellulose (Dow Chemical) in alpha medium (Flow Laboratories, Rockville, Md.), 10% fetal calf serum (Rehatuin, Reheis Chemical Company, Kankakee, Ill.), and 10% of the diluted, extracted serum to be tested. Mouse bone marrow cells (50,000/ml) were plated in plastic tissue-culture plates (Falcon Plastics) and placed in a $37°C–5\%$ CO_2, high-humidity incubator for 7 days. All samples were plated in duplicate. Colonies containing more than 50 cells were enumerated at $×25$ on an inverted microscope. Controls containing extracted normal serum and controls without test serum were included with each run. Any colonies formed by the medium control were subtracted from the results of all test sera.

In several experiments the ability of extracted serum or CER to enhance colony growth was examined. For these experiments, pooled serum from mice given 20 μg *Escherichia coli* lipopolysaccharide (026:B6; Difco) was diluted to a concentration so that fewer than 15 colonies per 50,000 cells were formed. The ability of CER from marrow transplant recipients to enhance colony growth was examined.

Human Normal marrow specimens were obtained from transplant donors. The buffy coat was removed after centrifugation (400 g for 8 min) and was washed twice in MEM. To eliminate autostimulation, adherent cells were removed by the method of Aye (1). A 5-ml syringe containing a wad of polyester fibers was incubated in MEM at 37°C for 1 hr. Cells were layered above the fibers, and the syringe was placed in the incubator at 37°C for 1 hr. The fibers were then washed with 30 ml MEM; the effluent cells were then concentrated by centrifugation and counted. With this treatment, marrow formed fewer than two colonies/10^5 cells unless an additional source of CSF was included. The final plating mixture consisted of 0.8% of carboxymethylcellulose, 20% fetal calf serum, $2 × 10^5$ cells/ml, and 10% diluted chloroform-extracted test serum. The ability of CER to stimulate growth was also tested in this system.

RESULTS

Chloroform extracted serum from hematologically normal volunteers produced $2 ± 2$ colonies per 50,000 nucleated mouse marrow cells plated. Figures 1–7 show CSF values and absolute

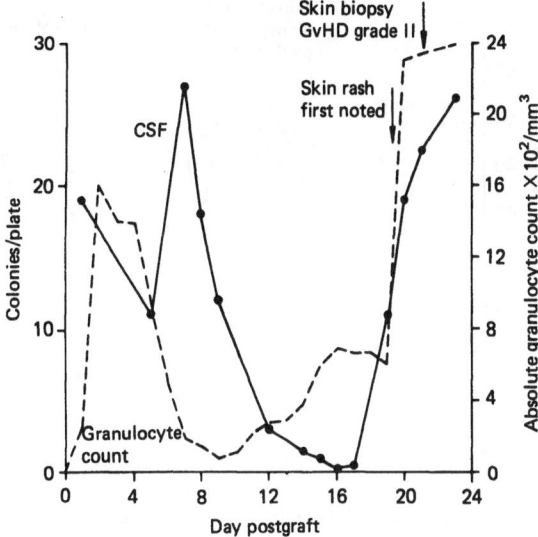

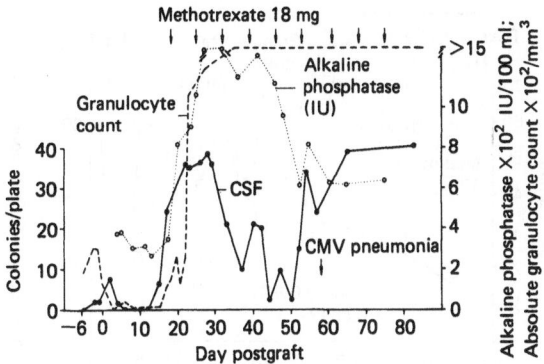

FIGURE 1. Postgrafting granulocyte counts and CSF levels of patient 571. CSF activity is plotted as the number of mouse bone marrow colonies formed per 50,000 cells plated. Early engraftment developed in this patient with an absolute granulocyte count of > 2000 mm³ by day 20. A skin rash was first noted on day 19, later biopsied and found to be grade II GvHD. The fall in CSF level preceded the rise in granulocyte count by several days. The secondary rise in CSF levels occurred on the same day on which the skin rash was first noted.

FIGURE 2. CSF levels and absolute granulocyte counts for patient 544. The fall in CSF level preceded the rising granulocyte count by four days. CSF levels fell to zero and remained there until day 23 when a narrow peak was noted. Following this, CSF fell to zero for 3 days, but was significantly elevated on day 29. At this time, early gut and liver GvHD had developed. He was initially treated with goat ATG, as indicated by the arrows. His condition deteriorated and treatment was changed to horse ATG. CSF levels fell following the treatment with horse ATG. The horse ATG was clinically effective.

FIGURE 3. Granulocyte counts, alkaline phosphatase, and CSF levels for patient 532. Striking abnormalities in alkaline phosphatase levels developed in this patient simultaneously with the rise in his granulocyte count. CSF levels rose on day 15 and reached a peak on day 28. At the same time, a peak in alkaline phosphatase activity was noted. Levels fell progressively until day 50 when a third peak of CSF was noted. At that time, the patient developed a CMV pneumonia. A liver biopsy performed on day 119 documented grade III GvHD.

FIGURE 4. CSF levels and absolute granulocyte counts following second marrow infusion in patient 549. A transitory marrow take and GvHD following a second marrow infusion on day 23 developed in this patient. Granulocyte counts exceeded 1500 mm³ by day 12 postsecond infusion and day 35 after initial marrow graft. At this time, a rash appeared which became more pronounced. Skin biopsy showed Grade I–II GvHD. A CSF rise occurred concurrently. There was a marked dip in granulocyte counts between days 48 and 54. The patient was given two doses of ATG on days 65 and 67 and subsequently rejected his graft.

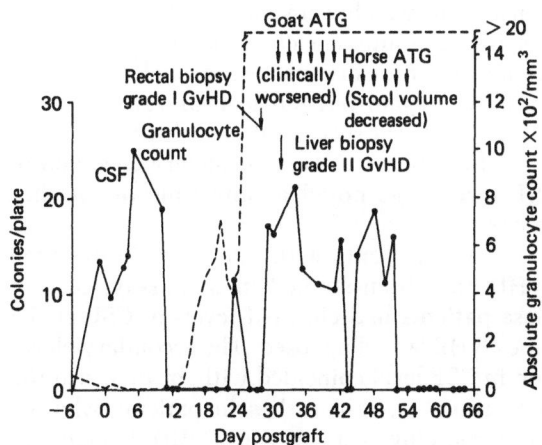

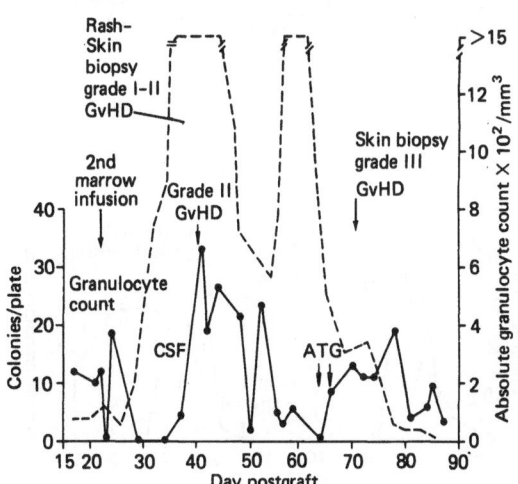

223

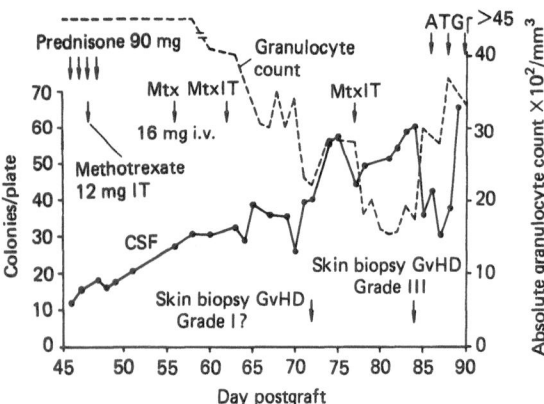

FIGURE 5. CSF levels and absolute granulocyte counts of patient 567. This figure shows part of the course of a patient in whom exceptionally late GvHD developed. The clinical diagnosis was made on day 84. CSF levels rose from day 44 through 90 and reached a peak of 66 colonies/plate. On day 58, granulocyte counts began to drop from a peak of > 10,000 and reached 1500 by day 81.

granulocyte counts of seven patients undergoing allogeneic marrow transplantation, in whom GvHD developed, as documented by histologic examination of either skin, liver, or gut. Each of these patients had two peaks of CSF activity. The first peak occurred during the period of aplasia following marrow grafting, and the second peak in CSF occurred during the period after the

FIGURE 6. CSF levels and absolute granulocyte counts for patient 550. An early rise in CSF levels occurred at the time of radiation and transplantation. Early engraftment ensued with a granulocyte count > 2000 by day 15. A skin rash was noticed on day 17, which was biopsy negative. On day 20 a skin biopsy was repeated and read as Grade II GvHD. Liver biopsy on day 21 also showed GvHD grade II. A secondary CSF rise accompanied the GvHD. Treatment with ATG appeared to suppress CSF activity.

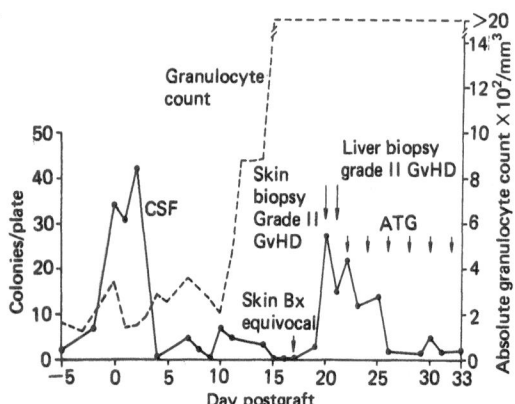

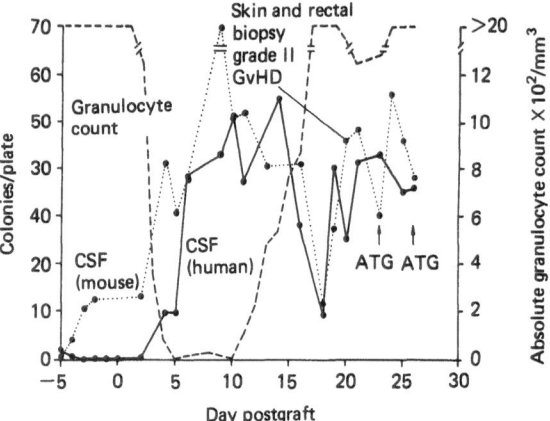

FIGURE 7. CSF levels as determined against mouse and human bone marrow with absolute granulocyte counts for patient 576. CSF levels began to rise as the granulocyte count fell postgrafting. When the granulocyte count reached 2000 by day 18, CSF levels measured on both human and mouse marrow declined to their nadir. Grade II GvHD was diagnosed by day 20, at which time secondary rises were noted in both mouse and human marrow-determined CSF.

granulocyte counts had risen toward normal and CSF levels had fallen to normal for several days. Capsule summaries of each of these patients' courses are included in the appendix to this chapter.

Figure 12 shows CSF values and absolute granulocyte counts from a patient undergoing syngeneic marrow transplantation. In this patient, unlike the allogeneic patients studied, the fall in CSF level occurs several days after the rise in granulocytes. Neither of two other syngeneic transplant patients had this delayed fall in CSF. Figures 8–11 show the courses of patients undergoing allogeneic transplantation in whom GvHD did not develop. In Fig. 10, initial high levels of CSF appeared to be suppressed by granulocyte transfusions which were successful in raising the absolute granulocyte count above 2000. After granulocyte transfusions became ineffective, CSF levels rose again but fell to near zero after the patient achieved successful engraftment. No evidence for either GvHD or secondary elevation in CSF levels was noted at any time during her course.

Ten patients with clinically documented GvHD have been studied in this assay; each of these patients had elevated levels of CSF at the time GvHD was diagnosed. The secondary elevation in CSF level coincided with, or preceded the appearance of, skin rashes in patients who acquired skin involvement with GvHD. In eight pa-

224

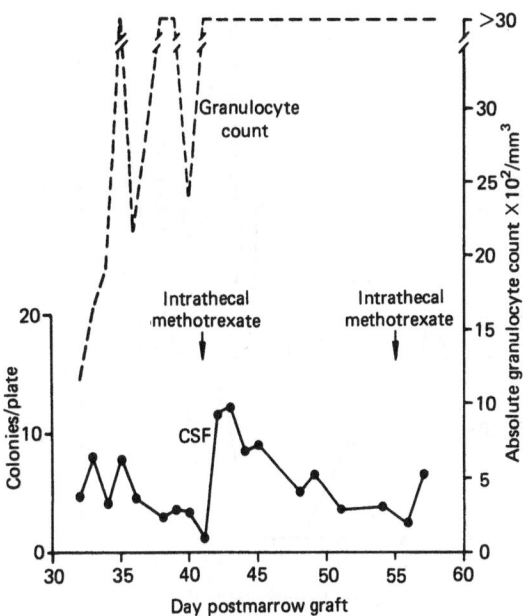

FIGURE 8. CSF values and absolute granulocyte counts for patient 574 between days 32 and 57 postgrafting. This patient never showed any evidence for GvHD. CSF values were low with the exception of one small rise occurring on day 42 and 43, probably in response to intrathecal methotrexate.

FIGURE 9. Absolute granulocyte counts and CSF levels for patient 568 from day 15 to 95 postmarrow grafting. Granulocyte counts surpassed 2000 on day 35. CSF levels declined below 10, as the granulocyte count rose and remained below 10 for the duration of the course. A decline in granulocyte count occurred during prophylactic therapy for *Pneumocystis carinii*. No evidence for GvHD was noted at any time.

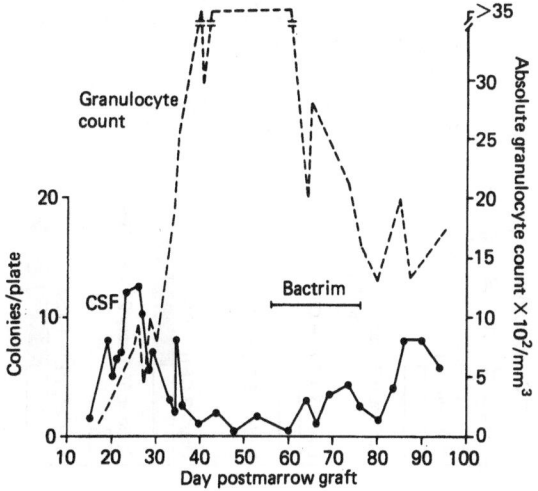

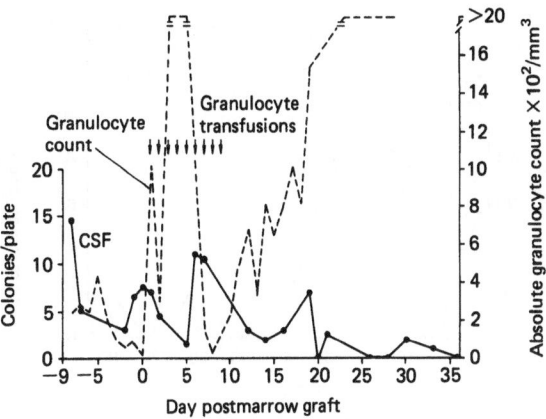

FIGURE 10. Absolute granulocyte counts and CSF values for patient 558. This patient received prophylactic granulocyte transfusions which resulted in exceptionally good granulocyte increments with the peripheral granulocyte count supported at > 2000/mm³ for several days. CSF starting at a value of 14 dropped to a low of 2 during the period of granulocyte support. Early engraftment with a granulocyte count in excess of 1500 developed by day 19. CSF levels remained low for the remainder of her posttransplant course. She never developed any evidence of GvHD.

tients, the mean CSF levels on the day of pathologic documentation of GvHD was 34 ± 4 colonies/5×10^4 marrow cells. This rise in CSF was not dependent on infection. Patients with similar granulocyte counts without GvHD had mean levels of 4 ± 1 colonies ($N = 6$).

FIGURE 11. Absolute granulocyte counts and CSF levels for patient 564. This patient had a single modest CSF peak during the first postgrafting week, which fell to zero shortly before the granulocyte count rose on day 17. CSF values through day 38 stayed below 10. No evidence for GvHD was noted during the posttransplant course.

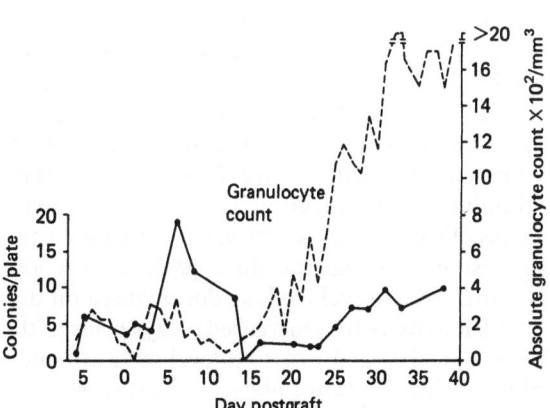

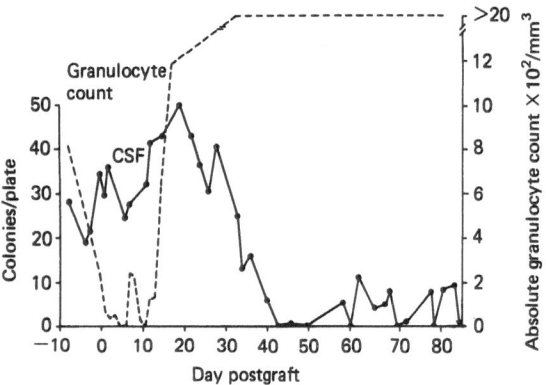

FIGURE 12. Granulocyte counts and CSF values for patient 545. This figure follows the course of a syngeneic transplant patient in whom early engraftment with a granulocyte count > 1000 by day 14 developed. CSF values remained elevated until day 40. The delayed fall in CSF, which was not seen in any other patients, may have resulted from lymphocyte infusions in a host with a disseminated cytomegalovirus infection.

When human marrow cells were plated with extracted serum from patients with GvHD, marked stimulation occurred. Neither normal human serum nor serum from bone marrow transplant recipients with absolute granulocyte counts in excess of 1500 mm³ without clinical GvHD exhibited this effect. CSF activity was present in serum during the leukopenic phase of marrow transplantation. Figure 7 shows the results of serum CSF from patient 576 assayed against both human and mouse marrows. Markedly similar patterns of CSF elevation can be seen.

CER from normal serum did not stimulate either mouse or human marrow granulocytic colony growth. By itself, CER from patients with GvHD did not stimulate murine colony growth. Figure 14, shows, however, that marked enhancement of murine CFU-c growth occurred when CER was added to cultures containing minimal stimulatory amounts of endotoxin-treated mouse serum. CER from normal serum did not stimulate human CFU-c growth. However, CER from patients with active GvHD had marked stimulatory effect on granulocytic colony growth. Figure 13 illustrates serial values of CER tested against human marrow in a patient in whom hepatic GvHD developed, as documented on day 63. CER activity first increased 4 days prior to the onset of GvHD and rose to a level of 42 colonies/plate during the active phase of the disease.

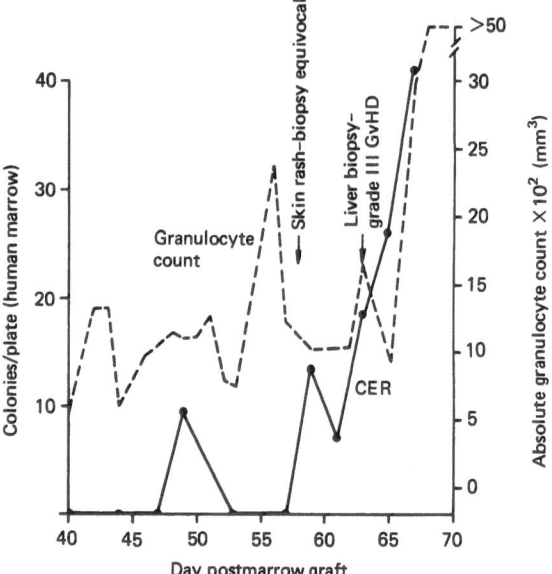

FIGURE 13. Values for absolute granulocyte counts and colony-stimulating activity of CER from patient 572 measured on human marrow. GvHD documented by skin biopsy and liver biopsy developed in this patient on day 63. Engraftment remained tenuous until after day 65, when the granulocyte count rose. The stimulatory effect of CER rose when grade III GvHD developed.

DISCUSSION

GvHD may be defined as an assault by immunologically competent allogeneic cells infused into a host who is unable to reject them. The effec-

FIGURE 14. The effect of CER from patients with and without GvHD on mouse marrow granulocytic colony growth in the presence of suboptimal levels of mouse CSF. CER samples from patient 576 prior to the development GvHD showed neither stimulation without, nor enhancing effect with, the mouse CSF. CER, however, markedly enhanced the response to CSF after GvHD developed on day 22. CER from patient 550 who had active GvHD on day 24 had a small amount of activity by itself and markedly enhanced the response to mouse CSF.

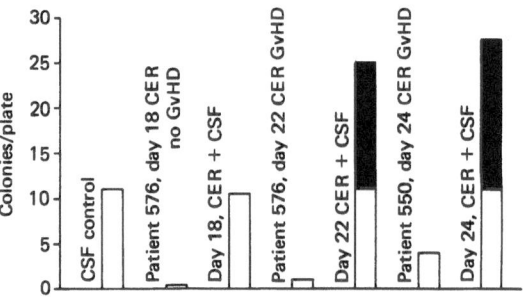

tor cell-mediating GvHD is probably a T-cell. However, involvement of other cell populations is likely (4,25,27). There is ample evidence that T-cells are capable of producing CSF *in vitro* in response to proliferative stimuli, such as allogeneic cells; recall antigens, such as PPD; or plant lectins, such as PHA or concanavalin A. Hara et al. (10) reported that sublethally irradiated F_1 hybrid mice receiving parental lymph node cells exhibited enhanced CSF activity *in vitro*. The sera of mice receiving parental lymph node and bone marrow cells also showed enhanced activity in spite of granulocytosis. Parker and Metcalf (13) also demonstrated CSF activity was released from spleen cells cultured from mice undergoing GvHD. These observations suggested that human serum CSF might reflect GvHD activity.

The measurement of CSF activity in human sera is complicated by the presence of inhibitors. These may be removed by dialysis (5), by ether or chloroform extraction (9), or by placing serum in a feeder layer which is then overlaid by marrow (13). Studies in humans have not yielded consistent patterns of CSF level elevation in specific disease states. For example, in a study of neutropenic patients, CSF levels varied independently of the granulocyte count (26). Futhermore, multiple studies of CSF levels in acute leukemia have yielded conflicting results (8,11,12,13,18). The lack of agreement may be partially accounted for by differences in methodology and the analysis of only single samples on most patients. In the studies reported here, serial serum samples were collected, extracted, and tested for the ability to stimulate mouse and human marrow cells to form granulocytic colonies *in vitro*. When studied in this manner, CSF levels appeared to make "biologic sense." All infected leukopenic patients had serum CSF levels, which at least moderately stimulated mouse and human marrow cultures. A fall in CSF levels preceeded or was concurrent with the rise in granulocyte count in almost all patients studied. A striking secondary rise in serum CSF levels occurred in all patients in whom GvHD developed. To date, samples on 10 patients with GvHD have been studied, eight of them serially. In all cases, sera stimulated colony growth markedly at a time when granulocyte counts were in excess of 1000 and most were in excess of 2000. The secondary elevation of CSF was not observed in any of seven patients without GvHD.

Price et al. (16) described a human CSF of low molecular weight (MW 1330) capable of stimulating the formation of granulocytic colonies by human but not murine marrow cells in culture. This low-molecular-weight CSF was chloroform extractable. We tested the ability of CER to stimulate mouse and human marrow growth. No stimulation of mouse marrow was observed in the absence of another source of CSF activity. However, marked enhancement of CFU-c growth was noted when CER was added to a low dose of mouse endotoxin serum. CER stimulated human marrow colony growth in the absence of additional CSF. In three patients studied serially, CER activity appeared to parallel the non-chloroform extractable CSF with a striking rise in activity noted during GvHD.

As indicated earlier (24), in human marrow graft recipients it is often difficult to distinguish between the active graft-versus-host reaction (GvHR) and the resulting disorders of target-organ function (GvHD); hence, both processes have been considered under the heading of GvHD. The elevation of CSF described in this chapter may well be an indicator of lymphocyte-mediated GvHR independent of the subsequent target-organ damage. Furthermore, because elevation of CSF precedes recognition of GvHD, serial measurement of CSF may provide a means of early recognition of GvHR. CSF measurements may also provide an index for the effectiveness of immunosuppressive therapy directed against the GvHR.

SUMMARY

We studied serial serum levels of CSF in patients undergoing bone marrow transplantation to determine if high levels of CSF are present in association with early graft-versus-host disease (GvHD). Daily morning serum samples were collected, chloroform extracted to remove inhibitors, and samples were tested for the ability to stimulate mouse and human CFU-C growth. In our system, cultures without human serum produced no colonies. Cultures with normal human serum produced fewer than five colonies. CSF levels on 14 bone marrow transplant recipients were inversely correlated with the absolute granulocyte count in the early postgrafting perid. CSF fell to control levels shortly before the rise in absolute granulocyte count. The mean CSF value in patients with granulocyte counts greater than 1500 without GvHD was 4 ± 1 (SEM) colonies. In eight patients in whom histologically proved GvHD developed, a marked secondary rise in CSF levels occurred when granulocyte counts were rising or were already above 1500/mm³. The secondary elevation in CSF level coincided with the appearance of skin rashes in patients who acquired GvHD skin involvement. The mean peak level of the secondary CSF rise in patients with GvHD was 34 ± 4 (SEM) colonies. This rise in CSF was independent of infection and usually preceded clinical diagnosis of GvHD. Syngeneic transplant recipients and allogeneic recipients without GvHD had no secondary elevation in CSF levels.

ACKNOWLEDGMENTS

This research was supported in part by designated funds of the Veterans Administration and by grant Number CA 18029 awarded by the National Cancer Institute, DHEW. Dr. Thomas is the recipient of a research career award (AI 02425) from the National Institute of Allergy and Infectious Disease, National Institutes of Health, Bethesda, Maryland.

APPENDIX: CASE HISTORIES

Patient 571: The patient was a 6-year-old girl whose aplastic anemia was diagnosed 1 month prior to transplantation. The patient was preconditioned with Cytoxan 50 mg/kg × 3 and 1000 R total body radiation (TBI). She received 9.7 × 10⁸ marrow cells/kg from an HL-A identical sibling and had early engraftment. A rash was first noted on day 18, which was diagnosed histologically as grade II GvHD. Her GvHD spontaneously resolved and she was discharged on day 60. The patient died of an acute cytomegalovirus pneumonia on day 88.

Patient 544: The patient was a 16-year-old boy whose aplastic anemia was diagnosed 8 weeks prior to transplantation. The patient was prepared with Cytoxan, 50 mg/kg × 3 followed by TBI. He received 2.83 × 10⁸ marrow cells/kg from his 16-year-old sister. A bone marrow on day 14 showed engraftment, and by day 19 his neutrophil count was over 500. On day 26 he developed diarrhea and a macular rash involving his entire body. A skin biopsy done 2 days later was non-diagnositic, but a rectal biopsy done on day 31 demonstrated grade I GvHD and a liver biopsy, grade II to III GvHD. The patient was treated with six doses of goat antithymocyte globulin (ATG), after which he showed some improvement. However, a pneumocystis carinii and CMV inerstital pneumonia developed. The patient died of progressive respiratory failure on day 95.

Patient 532: The patient was a 28-year-old man in whom acute myeloblastic leukemia developed 15 months prior to transplantation. He was treated with combination chemotherapy and went into complete remission, but relapsed after 10 months. The patient's conditioning course consisted of procarbazine 12.5 mg/kg/day every other day for 3 days, with ATG given 774 mg/day i.m., on the alternate days, for three doses. He was then treated with BCNU 720 mg i.v., nitrogen mustard, 1 mg/kg. and TBI. He received 1.4 × 10⁸ cells/kg from an HL-A compatible brother. The patient acquired an early engraftment with evidence of a "take" by day 7. On day 35, an interstitial pulmonary infiltrate was seen on X-ray, and an open lung biopsy was performed. Cytomegalovirus was cultured from lung tissue on day 40. The patient did well and was discharged on day 48. Liver function abnormalties were noted postgrafting with a rise in alkaline phosphatase to 815 IU by day 20. A liver biopsy was done on day 119 and demonstrated GvHD, grade II-III, with abnormalities in a majority of the small bile ducts. The patient relapsed and died on day 183.

Patient 549: The patient was a 19-year-old boy with myelomonocytic leukemia. He had a short remission on combination chemotherapy, but he had a relapse both in his bone marrow and central nervous system. A second short remission was obtained, and he was referred for transplantation 3 months after the initial diagnosis. At the time of admission, his marrow was hypercellular with 90% blasts. He was conditioned with cytosine arabinoside, 600 mg/m², by continuous infusion for 5 days, followed by Cytoxan, 60 mg/kg i.v. daily × 2, followed by 3 days of rest. He then received 1000 R TBI and 2.7 × 10⁸ marrow cells/kg from an HL-A matched sister. Granulocyte transfusions were given postgrafting. No evidence of engraftment was noted by day 21. On day 23, a second infusion of marrow was given from the same donor without additional preparation. Marrow, 7 days after second infusion, showed engraftment of all three cell lines with 25% cellularity. By day 21, postsecond infusion, the leukocyte count rose to a peak of 4500 with 69% neutrophils. A transient skin rash appeared on day 14 postsecond infusion, and a biopsy showed evidence of GvHD, grade II. No therapy was given at that time. By day 32, postsecond infusion the white count had fallen. Despite a trial of corticosteroids the WBC continued to fall. On day 39, postsecond infusion a marked rash appeared on the trunk, and biopsy again showed GvHD, grade II. Two doses of rabbit ATG were given, but was discontinued because of a progressive fall in white count. A third infusion of marrow was given on day 56 postsecond infusion, again without reconditioning. No response to this infusion was noted, and the patient died of bronchopneumonia 96 days after his initial graft.

Patient 567: The patient is a 26-year-old woman with AML who received two courses of combination chemotherapy and entered remission. Preparation for marrow graft included Cytoxan, 60 mg/kg × 2, and intrathecal methotrexate for central nervous system prophylaxis. Following TBI, she received 2.0×10^8 marrow cells/kg from an HL-A-compatible brother. Her granulocyte count was greater than 500 by day 28. On day 40, an idiopathic interstitial pneumonia developed. On day 70, she acquired a skin rash, which was biopsied on day 72 and read as Grade I GvHD. A repeat biopsy on day 84 showed Grade III GvHD. Her alkaline phosphatase began to increase by day 73 and reached a peak of greater than 1000 IU by day 85. She was started on rabbit ATG on day 86 and showed clinical improvement.

Patient 550: The patient was a 28-year-old woman with aplastic anemia diagnosed 6 months prior to transplantation. Preparation for a transplant from an HL-A identical brother included Cytoxan, 50 mg/kg/day × 3 and TBI. Early engraftment occurred. The patient received granulocyte transfusions from the posttransplant days 2–12. A rash was first noted on day 18. Skin and liver biopsy showed grade III GvHD. Treatment with ATG was begun but proved ineffective. The patient died of complications involving GvHD and interstitial pneumonia on day 79.

Patient 576: This patient was an 8½-year-old boy who had acute lymphoblastic leukemia of 5 year's duration. He had an initial remission of 2 years induced by vincristine and prednisone. After three relapses, he eventually proved refrac-

tory to conventional drug therapy. The patient was prepared with Cytoxan, 60 mg/kg × 2 and TBI. He received 6.4×10^8 marrow cells/kg from an HL-A compatible sister. But, rapid engraftment and early GvHD developed. A rectal biopsy and a skin biopsy on day 21 showed grade II GvHD. An idiopathic interstitial pneumonia developed on day 28, and he died of progressive respiratory failure on day 62.

Patient 574: The patient is a 3½-year-old boy who presented 6-months prior to transplantation with juvenile chronic granulocytic leukemia. He entered blast crisis 1 month prior to transplantation. He was Philadelphia chromosome negative, but had a trisomy 8. The patient was treated with 100 mg/kg hydroxyurea and 5 mg/kg dimethylmyceran on day −4. On days −3 and −2, he was given Cytoxan, 60 mg/kg, and on day 0, he received TBI. He received 10.4×10^8 marrow cells/kg from an HL-A compatible brother. By day 22, granulocytes were greater than 500. At no time during the hospital stay did he show evidence for GvHD. He was discharged from the hospital on day 39, doing well.

Patient 568: The patient is a 24-year-old man who had idiopathic aplastic anemia diagnosed 3 months before transplantation. The patient was treated with Cytoxan, 50 mg/kg × 4, and given 4.26×10^8 marrow cells/kg from an HL-A compatible sister. He had early engraftment with a rising granulocyte count by day 10. Platelets were self-supporting by day 18, and the marrow on day 42 had 100% cellularity. There was no evidence of GvHD at any time posttransplantation. He was discharged from the hospital on day 50.

Patient 558: The patient was a 3½-year-old girl in whom aplastic anemia developed 6 weeks prior to transplantation. She was considered "sensitized" on the basis of a positive chromium-release test and was prepared with Cytoxan, 50 mg/kg/day × 3, and TBI. She received 6.6×10^8 marrow cells/kg from an 8-year-old brother. A bone marrow on day 7 showed engraftment, and by day 28 her peripheral white blood cell count was 2700. No evidence for GvHD was noted at any time posttransplantation. She died of a pneumocystis carinii pneumonia 3 months after transplantation.

Patient 564: The patient is a 20-year-old man who had idiopathic aplastic anemia, diagnosed 4 months prior to transplantation. His preparation for transplantation was Cytoxan, 50/kg × 4. He received 4.3×10^8 cells/kg from his HL-A identical sister. He received prophylactic granulocyte support until day 7. Engraftment occurred early, and he was discharged from the hospital on day 24. He has continued to do well on an outpatient basis.

Patient 545: This patient was a 19-year-old woman who had acute myelomonocytic leukemia diagnosed 2 years prior to transplantation. Following two relapses, she was prepared for transplantation with Cytoxan, 60 mg/kg/day × 3, followed by TBI. The patient received additional, unirradiated buffy coat from her twin for 7 days after marrow transplantation. Her white blood cell count nadir occurred on day 4, and by day 15 she had greater than 500 granulocytes. Bone marrow biopsy obtained on day 40 showed 100% normal cellularity with no evidence of leukemia. The patient died of progressive cardiorespiratory, renal, and hepatic failure day 90. At autopsy, no evidence of GvHD was present. Other findings included pulmonary interstitial fibrosis with severe carnification of the lungs and a dilated 280-g heart with widespread biventricular myocardial fibril dropout, which was felt to be consistent with Cytoxan cardiomyopathy. The etiology of the pneumonia was found to be cytomegalovirus and pneumocystis carinii.

Patient 572: This patient is a 16-year-old boy with AML in relapse. The patient's preparation for marrow grafting included hydroxyurea in a total dose of 29 g over 4 days; Cytoxan, 60 mg/kg for 2 days; intrathecal methotrexate; and TBI. He received 1.6×10^8 cells/kg from an HL-A identical sister. Granulocytes rose to greater than 500 by day 27. The patient received horse ATG prophylactically for six doses between days 23 and 33 postgrafting. The patient was discharged from the hospital on day 45 totally asymptomatic. On day 59, postgrafting the patient noted malaise, anorexia, nausea, and icterus. A skin biopsy on day 57 was nondiagnostic. On day 63 a liver biopsy was performed, which exhibited Grade III GvHD. He was treated with steroids and showed gradual improvement in his liver-function tests.

REFERENCES

(1) Aye, M. T. *Personal communication.*

(2) Bradley, T., and Metcalf, D. The growth of mouse bone marrow cells *in vitro. Aust. J. Exp. Biol. Med. Sci., 44*:287, 1966.

(3) Summer, M. Stimulation of mouse bone marrow colony growth *in vitro* by conditioned medium. *Aust. J. Exp. Biol. Med. Sci., 46*:607, 1968.

(4) Cantor, H., and Weissman, I. Development and function of subpopulations of thymocytes and T lymphocytes. *Prog. Allergy, 20*:1, 1976.

(5) Chan, S. H., and Metcalf, D. Inhibition of bone marrow colony formation by normal and leukaemic human serum. *Nature (London), 227*:845, 1970.

(6) Chervenick, P., and Lobuglio, A. Human Blood monocytes: Stimulators of granulocyte and mononuclear colony formation *in vitro. Science, 178*:164, 1972.

(7) Cline, M., and Goldie, D. Production of colony-stimulating activity by human lymphocytes. *Nature (London), 248*:703, 1974.

(8) Foster, R., Metcalf, D., Robinson, W., and Bradley, T. Bone marrow colony stimulating activity in human sera. *Br. J. Haematol., 15*:147, 1968.

(9) Granstrom, M. Studies on inhibitors of bone marrow colony formation in normal human sera and during a viral infection. *Exp. Cell. Res., 82*:426, 1973.

(10) Hara, H., Kitamura, Y., Kawata, T., Kanamuru, A., and Nagai, K. Synergism between lymph node and bone marrow cells for production of granulocytes. II. Enhanced colony-stimulating activity of sera of mice with graft-versus-host reaction. *Exp. Hematol. 2*:43, 1974.

(11) Metcalf, D., Chan, S., Gunz, F., Vincent, P., and Ravich, R. Colony stimulating factor and inhibitor levels in acute granulocytic leukemia. *Blood, 38*:143, 1971.

(12) Mintz, U., and Sachs, L. Differences in inducing activity for human bone marrow colonies in normal serum and serum from patients with leukemia. *Blood, 42*:331, 1973.

(13) Parker, J., and Metcalf, D. Production of colony-stimulating factor in mixed leukocyte cultures. *Immunology, 26*:1039, 1974.

(14) Parker, J., and Metcalf, D. Production of colony-stimulating factor in mitogen-stimulated lymphocyte cultures. *J. Immunol., 112*:502, 1974.

(15) Pick, E., and Turk, J. The biological activities of soluble lymphocyte products. *Clin. Exp. Immunol., 10*:1, 1972.

(16) Price, G., McCulloch, E., and Till, J. A new human low molecular weight granulocyte colony stimulating activity. *Blood, 42*:341, 1973.

(17) Prival, J., Paran, M., Gallow, R., and Wu, A. Colony-stimulating factors in cultures of human peripheral blood cells. *J. Natl. Cancer Inst., 53*:1583, 1974.

(18) Robinson, W., and Pike, B. Leukopoietic activity in human urine. *New Engl. J. Med., 282*:1291, 1970.

(19) Robinson, W., Metcalf, D., and Bradley, T. Stimulation by normal and leukemic mouse sera of colony formation *in vitro* by mouse bone marrow cells. *J. Cell. Physiol., 69*:83, 1967.

(20) Ruscetti, F., and Chervenick, P. Regulation of the release of colony-stimulating activity from mitogen-stimulated lymphocytes. *J. Immunol., 114*:1513, 1975.

(21) Sheridan, J., and Stanley, E. Tissue sources of bone marrow colony stimulating factor. *J. Cell. Physiol., 78*:451, 1972.

(22) Stanley, E., Bradley, T. R., and Sumner, M. Properties of the mouse embryo conditioned medium factor(s) stimulating colony formation by mouse bone marrow cells grown *in vitro. J. Cell. Physiol., 78*:301, 1971.

(23) Stanley, E., McNeill, T., and Chan, S. Antibody production to the factor in human urine stimulating colony formation *in vitro* by bone marrow cells. *Br. J. Haematol., 18*:585, 1970.

(24) Thomas, E. D., Storb, R., Clift, R. A., Fefer, A., Johnson, F. L., Neiman, P. E., Lerner, K. G., Glucksberg, H., and Buckner, C. D. Bone marrow transplantation. *New Engl. J. Med., 292*:832, 1975.

(25) Tigelaar, R., and Asofsky, R. Graft-versus-host reactivity of mouse thymocytes: Effect of cortisone pretreatment of donors. *J. Immunol., 110*:567, 1973.

(26) Wewerka, J., and Dale, D. Colony stimulating factor in patients with chronic neutropenia. *Blood, 47*:861, 1976.

(27) Grebe, S., and Streilein, J. W. Graft-versus-host reactions: A review. *Adv. Immunol., 22*:119, 1976.

INTRODUCTION

Many of the cytotoxic drugs used in cancer chemotherapy damage cells throughout the hemopoietic system, which results in neutropenia and thrombocytopenia with the associated problems of infection and hemorrhage. These symptoms are in large part attributable to depletion of the early precursor cells in the bone marrow, the descendents of which ultimately replenish the peripheral blood supply. However, in addition to the numbers of surviving progenitor cells, their proliferation rate is also an important factor, because upon this will depend their rate of recovery and, for many drugs, their response to further doses of chemotherapy. It is therefore important to measure changes in these two variables in order to predict with more confidence the optimum timing for treatment.

We have used an agar diffusion-chamber technique (9) to investigate these effects of cyclophosphamide on human bone marrow colony precursor cells by measuring the toxicity of the drug and following recovery patterns and changes in proliferation rate after treatment.

Changes in Human Bone Marrow Colony-Forming Cells following Chemotherapy Using an Agar Diffusion-Chamber Technique

M. Y. Gordon

26

Information on the initial toxicities of drugs may be obtained by constructing survival curves when cells from a single sample are exposed to graded doses of the cytotoxic agent. This has been done in the past using *in vitro* systems (12,13) in which the cells are incubated with the drug before plating or are cultured in its presence. It is difficult, however, to reproduce *in vitro* the continuous changes in drug activity that occur *in vivo* as the drug is metabolized and degraded. Furthermore, the concentrations effective *in vitro* may not be achieved *in vivo*, and certain drugs, notably cyclophosphamide, require metabolic activation. Exposure of the cells in diffusion chambers, by injecting the drugs into the host mouse, provides conditions where the drug is activated and degraded, thus overcoming at least some of the difficulties encountered with *in vitro* studies (8).

We have also looked at recovery patterns in the bone marrow of patients following a single 5-g dose of cyclophosphamide, as the ability of the marrow to recover is important when considering repeated doses. Changes in proliferation rate of the colony-forming cells during the recovery period were detected by measuring changes in their sensitivity to the S-phase specific drug, cytosine arabinoside (ara-a) (10).

METHODS

The methods used for this study have been described in detail elsewhere (6,8–10).

The bone marrow cells for culture were suspended in agar medium. This mixture was introduced into the diffusion chambers, which were incubated in the peritoneal cavity of host mice. The hosts for the human bone marrow cells had received 900 rad ^{60}Co γ radiation to ensure colony growth. However, unirradiated host mice gave adequate support to the growth of mouse bone marrow colonies.

The dose-response curves for cyclophosphamide were obtained by injecting groups of chamber bearing mice with graded doses of the drug and transferring the chambers to secondary groups of host mice for incubation 18 hr later. At this time, the survival of the colony forming cells in the femurs of the primary drug-treated hosts was assayed using the agar diffusion chamber technique. The procedure for measuring the sensitivity of the cells to cytosine arabinoside was similar, but in this case the cells were exposed to the drug for 2 hr.

FIGURE 1. Survival of colony-forming cells from (●) human and (■) mouse bone marrow-treated in diffusion chambers compared with (▲) mouse femoral cells treated *in situ* with different doses of cyclophosphamide (8). Courtesy of Cancer Research.

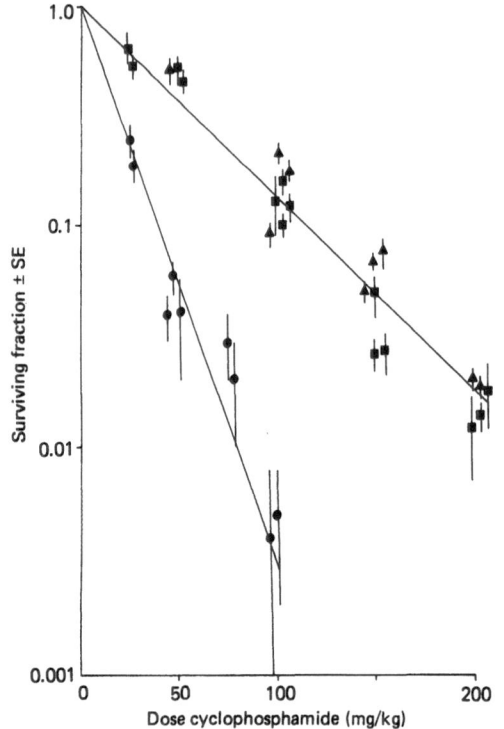

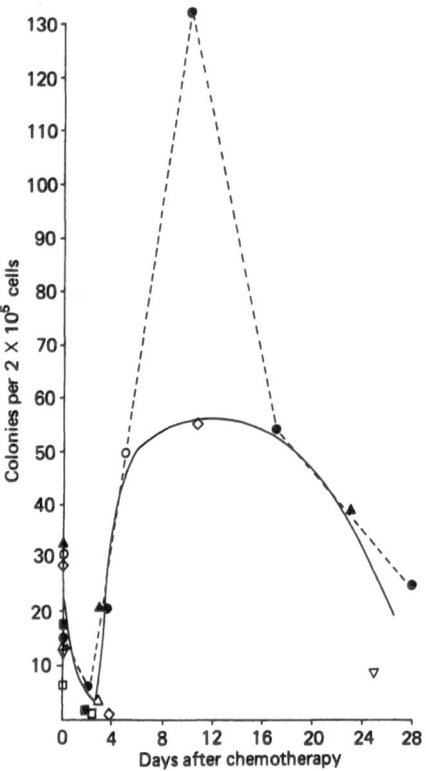

FIGURE 2. Changes in the colony-forming ability of patient's marrow following treatment with 5 g cyclophosphamide. The different symbols refer to data from different individuals.

RESULTS

Figure 1 compares the sensitivity of human bone marrow colony-forming cells with that of mouse marrow exposed to the drug under the same conditions as well as *in situ* in the femur. Human cells are more sensitive to cyclophosphamide than are the murine cells treated in the diffusion chambers. However, mouse cells are equally sensitive to exposure to cyclophosphamide in diffusion chamber or *in situ* in the femur which indicates that for this drug the culture system provides a good estimation of the true sensitivity of the cells in mouse hemopoietic tissue.

The combined recovery data for patients given 5 g cyclophosphamide are shown in Fig. 2. An initial depletion is followed by an overshoot in colony number, which is particularly marked in one of the cases. The peak, at about 12 days after treatment, is followed by a secondary decline to approximately pretreatment levels by the fourth week.

Figure 3 shows the data from one of the patients in more detail and compares the changes in colony yield with the peripheral granulocyte count. The increase and decline in the incidence of

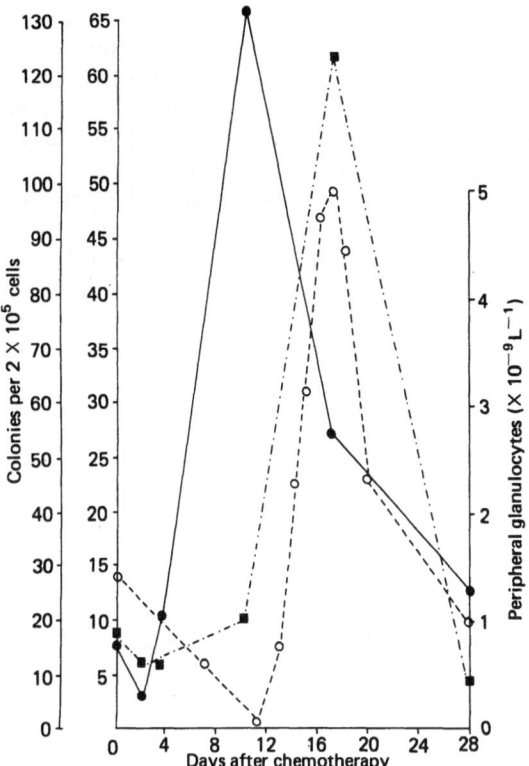

FIGURE 3. Relationship between (●) changes in the incidence of colony-forming cells, (■) the numbers of colony-forming cells per milliliter aspirated marrow, and (○) the peripheral granulocyte count in a single individual following treatment with 5 g cyclophosphamide.

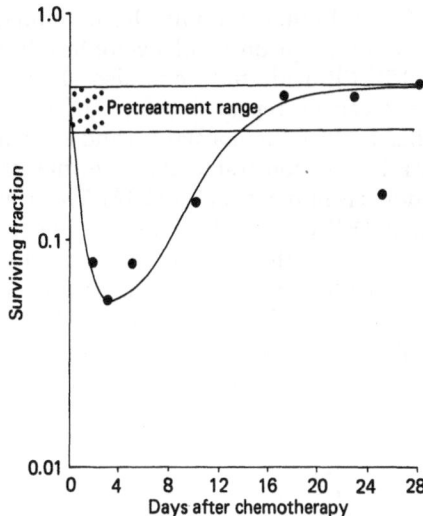

FIGURE 4. Sensitivity to cytosine arabinoside of colony-forming cells before and after chemotherapy. The points are the mean survivals at three doses (100, 200, and 300 mg kg^{-1} mouse body Wt.) on the plateau of the dose-response curve to cytosine arabinoside.

DISCUSSION

The data presented in this chapter demonstrate that the agar diffusion-chamber technique may be used to investigate three aspects of bone marrow damage by the drugs used in chemotherapy. First, some indication of the relative toxicities of different drugs can be obtained by constructing survival curves (8); second, recovery patterns can be determined by following changes in the colony yield after chemotherapy; finally, the changes in the proliferation rate of the cells can be detected by measuring changes in their sensitivity to the S-phase specific drug, cytosine arabinoside (10).

The recovery pattern in humans, following treatment with cyclophosphamide, is similar to that found in mice for spleen colony-forming cells (CFU-s) (4) and for agar diffusion-chamber colonies (7), in that they are characterized by a rapid recovery to supranormal values followed by a secondary decline. The timing of the recovery when the results are expressed as colony incidence (colonies per 2 × 10⁵ cells) differs from that seen when they are expressed as colonies per milliliter aspirated marrow. The incidence of colony-forming cells increases before the marrow cellularity begins to rise, and the results may therefore be interpreted as showing that recovery of the colony precursor cell population precedes recovery of the total marrow pool, which occurs at approximately the same time as recovery of the peripheral blood granulocytes.

colony-forming cells (colonies per 2 × 10⁵ bone marrow cells) is mirrored by the peripheral blood count some 8 days later. When the culture data are expressed as colonies per milliliter aspirate, the separation between the curves is considerably reduced.

The changes in the sensitivity of the cells to the S-phase specific action of cytosine arabinoside, following treatment of the patients with cyclophosphamide, are shown in Fig. 4. Colony survival was measured at three dose levels (100, 200, and 300 mg kg^{-1} mouse body wt.), which fall on the plateau of the dose-response curve to cytosine arabinoside (10). The points on Fig. 4 represent the mean survival at these three dose levels, and the curve shows the changes in the level of the plateau. Therefore, these data provide some indication of the timing of changes in the proliferation rate of the cells. An increase in the sensitivity of the cells is evident during the first few days, after treatment followed by a return to normal sensitivity by the 17th day after chemotherapy.

Several clinical studies have emphasized the value of large doses of cyclophosphamide (2,3,14,15). Clinical study has also shown that treatment can be repeated before the leukocyte count has returned to normal (1) and experimental work has demonstrated that the interval between doses is important (4,5,11,16). The work described by DeWys et al. (4) relates the importance of timed doses to the proliferation dependent action of cyclophosphamide and the changes in the proliferation rate of the cells during recovery.

The present data show that the peripheral blood count does not provide an accurate indication of events in the bone marrow, which occurs several days earlier. There may be recovery in the marrow while the patient is apparently still neutropenic; conversely, the peripheral blood count was highest when the incidence of bone marrow colony-forming cells was falling.

The relationships between the pattern of recovery of the colony-forming cells, their proliferation rate, and the peripheral blood cell count provides a framework for improving schedules of treatment, so that the effectiveness of a drug against the tumor is increased without an unacceptable increase in marrow toxicity.

SUMMARY

Studies have been made of the effects of cyclophosphamide on the early granulocytic precursor cells in the marrow of patients receiving intermittent high doses (5 g) of the drug. This study was designed to investigate the effect of a single cytotoxic agent on normal granulocytic precursor cells; measurements have been made of the initial cytotoxicity, recovery, and changes in proliferation rate.

The growth of granulocytic colonies was assayed in agar diffusion chambers. For this, bone marrow cells were suspended in semisolid agar medium, introduced into diffusion chambers for intraperitoneal incubation in radiated (900 rad ^{60}Co γ) mice. The colonies present in the agar were scored 8–9 days later.

In addition to measuring the effect of treatment on the colony-forming capacity of the patients' marrow, full dose-response curves were obtained by injecting graded doses of cyclophosphamide into chamber-bearing mice—the method allows some account to be taken of the continuous changes in drug activity, which occurs during its degradation in vivo. Comparison of these results with measurements of the sensitivity of mouse femoral cells under the same conditions or exposed to cyclophosphamide in situ in the donor mouse has been used to detect any effect of the culture environment or the response of the cells.

Following treatment with cyclophosphamide, the incidence of colony precursor cells in the patients' marrow has been monitored. Changes in the proliferation rate of these cells during the recovery period have also been estimated by measuring their sensitivity to the S-phase specific drug, cytosine arabinoside.

Information on the timing of changes in the incidence and proliferation rate of granulocytic precursor cells may provide guidelines for improving schedules of treatment in an attempt to reduce the attendant bone marrow toxicity.

ACKNOWLEDGMENTS

The author wishes to thank the National Cancer Institute for their support; Professor L. F. Lamerton, and Drs. T. J. McElwain, N. M. Blackett, and I. D. C. Douglas for their interest and advice. Mrs. B. M. J. Pickering is acknowledged for her excellent technical assistance.

REFERENCES

(1) Bergsagel, D. E., Robertson, G. L., and Gasselbach, R. Effect of cyclophosphamide on advanced lung cancer and the haematologic toxicity of large intermittent intravenous doses. *Can. Med. Assoc. J.*, *98*:532, 1968.

(2) Burkitt, D. Long-term remission following one and

two-dose chemotherapy for African lymphoma. *Cancer, 20*:756, 1967.

(3) Carbone, P. P., Spurr, C., Schneiderman, M., Scotto, J., Holland, J. F., and Schnider, B. Management of patients with malignant lymphoma: A comparative study with cyclophosphamide and vinca alkaloids. *Cancer Res., 28*:811, 1968.

(4) DeWys, W. D., Goldin, A., and Mantel, N. Hematopoietic recovery after large doses of cyclophosphamide: Correlation of proliferative state with sensitivity. *Cancer Res., 30*:1692, 1970.

(5) Eckhardt, S., Humphreys, S. R., and Goldin, A. Effect of antitumour agents on the hematology and transplantability of leukaemia L-1210. *Antimicrob. agents Chemother., 5*:503, 1965.

(6) Gordon, M. Y. Quantitation of haemopoietic cells from normal and leukaemic RFM mice using an *in vivo* colony assay. *Br. J. Cancer, 30*:421, 1974.

(7) Gordon, M. Y., and Blackett, N. M. Stimulation of granulocytic colony formation in agar diffusion chambers implanted in cyclophosphamide pretreated mice. *Br. J. Cancer, 32*:51, 1975.

(8) Gordon, M. Y., and Blackett, N. M. The sensitivities of human and murine haemopoietic precursor cells exposed to cytotoxic drugs in an *in vivo* culture system. *Cancer Res., 36*:2822, 1976.

(9) Gordon, M. Y., Blackett, N. M., and Douglas, I. D. C. Colony formation by human haemopoietic precursor cells cultured in semi-solid agar in diffusion chambers. *Br. J. Haematol., 31*:103, 1975.

(10) Gordon, M. Y., and Douglas, I. D. C. Changes in proliferation rate of human bone marrow colony-forming cells measured by a cytosine arabinoside-diffusion chamber method. *Eur. J. Cancer, 12*:551, 1976.

(11) Lane, M. Preliminary report of animal studies with cytoxen (cyclophosphamide). *Cancer Chemother. Rep., 3*:1, 1959.

(12) Ogawa, M., Bergsagel, D. E., and McCulloch, E. A. Sensitivity of human and murine hemopoietic precursor cells to chemotherapeutic agents assessed in cell culture. *Blood, 42*:851, 1973.

(13) Ogawa, M., Gale, G. R., and Keirs, S. S. Effects of cisdiammine-dichloroplatinum on murine and human haemopoietic precursor cells. *Cancer Res., 34*:1398, 1975.

(14) Samuels, M. L., and Howe, C. D. Cyclophosphamide in the management of Ewing's Sarcoma. *Cancer, 20*:961, 1967.

(15) Sutow, W. W. Cyclophosphamide (NSC-26271) in Wilms' Tumour and Rhabdomyosarcoma. *Cancer Chemother. Rep., 51*:407, 1967.

(16) Venditti, J. M., Goldin, A., and Kline, I. The influence of treatment schedule on the chemotherapy of advanced leukaemia L-1210 in mice. *Cancer Chemother. Rep., 6*:55, 1960.

INTRODUCTION

During the past several years, evidence has accumulated indicating that functionally separate subpopulations of T-lymphocytes are present in various mammalian species; these include suppressor (16), helper (10), and effector (11) T-cells. We recently reported experimental evidence indicating the possibility that at least two different subpopulations of effector cells may exist in the mouse—one capable of causing graft-versus-host (GvH) reactions and the other primarily responsible for antitumor reactions (3,5). Similarly, Fernandes et al. (15) and Kedar and Bonavida (18) reported that separate subpopulations of effector T-cells were responsible for antihost and antitumor reactions. Several investigators have found that *in vivo* or *in vitro* immunizations across histocompatibility barriers could result in a large increase in cell-mediated cytotoxicity against a tumor target with little or no increase in GvH reactivity (9, 23, 24). Mage and McHugh (20) were able to separate lymphocytes with antitumor activities from those with antihost activities by incubating the effector cells on monolayers of allogeneic fibroblasts.

27

Graft-versus-Leukemia, Donor Selection for Adoptive Immunotherapy in Mice

William P. LeFeber,
Robert L. Truitt, William C. Rose,
and Mortimer M. Bortin

Separate (and possibly separable) subpopulations of effector T-cells have important implications for clinical application of adoptive immunotherapy for cancer. In adoptive immunotherapy, immunocompetent cells from a normal donor are transplanted into the tumor-bearing host. We employ the term graft-versus-leukemia (GvL) reaction to signify the adoptive immunotherapeutic effect of transplanted cells against leukemia cells in an immunosuppressed host. Unfortunately, GvH disease has been a major complication of adoptive immunotherapy (21). We have reported the value of adoptive immunotherapy both in long-passage (2) and spontaneous (6,8,25) acute lymphoblastic leukemias in AKR mice while circumventing lethal GvH disease by using a transient GvL reaction, as originally described by Cleton et al. (12) in humans and by Boranić (1) in the mouse.

Ideally, donors of cells for adoptive immunotherapy would show great antitumor reactivity and no antihost reactivity (high GvL and no GvH reactions). In the experiments reported here, we investigated the GvL and GvH reactivity of immunocompetent cells from a panel of 11 potential donor strains in order to select the one most closely approximating the ideal. The potential donors consisisted of (1) unprimed mice mismatched with AKR at the major (H-2) histocom-

239

patibility complex and at minor histocompatibility loci; (2) unprimed congenic mice mismatched with AKR only at a portion of the H-2 complex; (3) unprimed H-2 matched mice; and (4) H-2 matched mice primed with multiple injections of irradiated spleen cells from leukemic AKR mice. As tested in these experiments, mice of the H-2 mismatched SJL strain appeared to be most suitable for their high GvL reactivity and minimal acute GvH reactivity. All other tested strains were less desirable because of lower GvL reactivity higher GvH reactivity, or both.

MATERIALS AND METHODS
MICE

Inbred C57BL/10J (H-2^b), BALB/cJ (H-2^d), DBA/2J (H-2^d), AKR/J (H-2^k), B10.BR/J (H-2^k), CBA/J (H-2^k), RF/J (H-2^k), ST/bJ (H-2^k), AKR.M/Sn (H-2^m), DBA/1J (H-2^q), and SJL/J (H-2^s) mice were obtained from The Jackson Laboratory (Bar Harbor, Maine). The AKR.M/Sn mice were provided by Dr. M. Cherry (Bar Harbor). The mice were used at 8–14 weeks of age, housed in Isocages with filter lids, and given autoclaved mouse chow and acidified, chlorinated water *ad libitum*.

DRUGS AND IRRADIATION

Cyclophosphamide (CY) was provided by the Drug Development Branch, Division of Cancer Treatment, National Cancer Institute (Bethesda, Md.), and was administered i.p. Total body irradiation (TBI) was administered as X-rays employing a Picker Vanguard X-Ray therapy unit at a dose rate of 43 R/min or as γ-rays using twin ^{137}Cs sources in a Gammacell 40 Irradiator (Atomic Energy of Canada, Ltd., Quebec) at a dose rate of 123 R/min. CY was given approximately 6 hr and TBI approximately 4 hr prior to transplantation of cells.

CELL SUSPENSIONS

Bone marrow and mesenteric lymph-node cells were collected, processed into single-cell suspensions for i.v. administration, and tested for viability using methods previously described (7).

LEUKEMIA

A long-passage lymphoblastic leukemia (BW5147) carried in AKR mice (14) was obtained from The Jackson Laboratory, and for the past 5 years has been maintained in our laboratory as an acute lymphoblastic leukemia by weekly i.v. passage. Leukemia cells were obtained from the spleens of AKR mice that had received 2.5×10^4 BW-5147 blast (large) spleen cells 1 week previously. Administration of a known number of viable blast cells results in a highly reproducible mortality pattern (4). The proportion of splenic blast cells in these experiments ranged from 60% to 80%.

IMMUNIZATION PROCEDURES

Mice of the B10.BR, CBA, and RF strains were given five to six i.p. injections of 3000 R γ-radiated BW5147 leukemic blast cells. The mice received an initial injection of 10^4 cells followed by four or five injections of 10^6 cells at approximately weekly intervals. The immunized mice were used as donors of bone marrow and lymph-node cells approximately 1 week after the last immunization injection. To test the efficacy of the immunization procedure, groups of immunized RF and CBA mice were challenged by i.v. inoculation of 10^5 viable BW5147 blast cells. A proportion of unprimed RF and CBA mice were susceptible and died of leukemia following injection of this dose of AKR leukemia cells, whereas B10.BR mice were resistant (3).

EXPERIMENTAL DESIGN

To evaluate GvH reactivity, 2×10^7 bone marrow and 10^7 lymph-node cells from the panel of 11 normal or immunized donor strains were administered to nonleukemic AKR mice that had been immunosuppressed with 185 mg/kg CY and 400 R TBI. The mice were observed daily for survival and for clinical signs of GvH disease. Representative mice were autopsied, and tissue specimens were examined for histologic evidence of GvH disease. Analysis of variance of each replicate experiment employing median survival times (MST) and proportion of AKR host mice surviving 30 to 100 days was used to compare differences between strains in the severity of GvH disease produced by bone marrow and lymph-node cells from the panel of donors. For the purpose of this study, all deaths of nonleukemic AKR mice given cells from allogeneic donors were attributed to acute GvH disease (deaths within 30 days) or delayed GvH disease (deaths beyond 30 days).

A standard bioassay (4) was used to evaluate the GvL reactivity of bone marrow and lymph-node cells from the same panel of normal and immune donors. On day 0, AKR mice were given 800 R TBI followed by i.v. inoculation of 10^4 viable BW5147 blast cells. The TBI was given to vitiate any reaction of the AKR primary hosts against the leukemia and as immunosuppression to prevent rejection of the forthcoming transplants of bone marrow and lymph-node cells. There was no antileukemic effect from the TBI because it was administered prior to inoculation of leukemia cells. On day 1, leukemic AKR primary hosts were given 2×10^7 bone marrow and 10^7 lymph-node cells i.v. from the panel of donors; this transplant of immunocompetent cells was the only antileukemic treatment given. On day 7, the spleen of each AKR primary host was removed and all cells were injected i.p. into an individual AKR secondary recipient. The secondary recipients were observed daily for 60 days. It was assumed that the transfer of at least one clonogenic leukemia cell could cause death of the secondary recipient as a result of leukemia. The theoretical and experimental bases for this assumption and the rationale for the use of the spleen as the most sensitive bioassay organ have been described (4). Survival of the secondary recipients would indicate that the leukemia had been eliminated from (at least the spleens of) the primary recipients. Analysis of variance of each replicate experiment using percent survival of the secondary recipients at 60 days was used to evaluate the relative GvL reactivity of bone marrow and lymph-node cells from the panel of donors.

RESULTS

GvL ASSAYS USING UNPRIMED DONORS

Summarized in Table 1 are the results of the GvL assays. Control groups received no cells (group 1) or cells from syngeneic donors (group 2). Allogeneic donors were matched with AKR at the *H-2* locus, but mismatched at minor histocompatibility loci (group 3), congenic (i.e., matched except at the *H-2* locus, group 4), or mismatched at *H-2* and at minor histocompatibility loci (groups 5-9). Administration of no cells or syngeneic cells to immunosuppressed, leukemic AKR primary recipients had no detectable antileukemic effect; viable leukemic cells were transferred in the spleens of the primary recipients to the secondary recipients, and 99% (89/90) died of leukemia (groups 1 and 2, Table 1). Mice of the ST/b strain are phenotypically identical with AKR at *H-2*, and their cells had no demonstrable antileukemic effect (group 3). Cells from the congenic AKR.M strain provided suggestive evidence of mild GvL reactivity because 13% (4/30) of the secondary recipients did not die (group 4), but this was not significantly different from groups 1 to 3. Significant GvL reactivity was found among

TABLE 1 Survival Data from Assays of GVL and GVH Reactivity in Leukemic and Nonleukemic AKR Mice[a]

DONOR			GVL REACTIVITY[b]		GVH REACTIVITY[c]			
							Survival (%) at day:	
Donor No.	Strain	*H-2*	No. of Mice	Survival (%) at day 60[d]	No. of Mice	MST (days)	30	100
1	No cells	—	33	3	31	48	84	39
2	AKR	*k*	57[e]	0[e]	29	> 100	100	93
3	ST/b	*k*	29	0	30	50	90	30
4	AKR.M	*m*	30	13	30	51	97	23
5	DBA/1	*q*	25	100	30	7	0	0
6	BALB/c	*d*	30	87	30	36	67	0
7	DBA/2	*d*	30	70	30	53	87	3
8	C57BL/10	*b*	30	87	30	41	83	0
9	SJL	*s*	30	100	30	49	83	3

[a]Gvl reactivity of transplanted immunocompetent cells was evaluated in immunosuppressed leukemic AKR mice by means of a bioassay. Survival studies were used to evaluate GVH reactivity of the cells in immunosuppressed nonleukemic AKR mice.

[b]Measured by bioassay of primary recipient's spleen; values are for secondary recipients.

[c]Measured in primary recipients.

[d]No secondary recipients died after day 44.

[e]Data from ref. 4 (using a different bioassay model) have been included for comparison.

the five remaining strains (groups 5–9) in comparison with groups 1–4 ($p < 0.01$). Cells from DBA/2 donors (group 7) had moderate GvL reactivity, as evidenced by 70% (21/30) survival of the secondary recipients, but this was significantly less ($p < 0.05$) than that exhibited by mice of the DBA/1 and SJL strains (groups 5 and 9). Immunocompetent cells from DBA/1 and SJL donors eliminated all viable leukemic cells from the spleens of the primary recipients with resultant survival of all secondary recipients.

GvH ASSAYS USING UNPRIMED DONORS

Also summarized in Table 1 are the results of the GvH assays. The doses of 400 R TBI and 185 mg/kg Cytoxan administered to nonleukemic 8- to 10-week-old AKR mice as immunosuppressive conditioning for the GvH assays were toxic and resulted in an MST of 48 days with 39% 100-day survival (group 1, Table 1). The chemoradiotherapy toxicity could be overcome ($p < 0.01$) by transplanting syngeneic bone marrow and lymph node cells (cf. groups 1 and 2). Cells from *H-2* identical ST/b (group 3) and congenic AKR.M (group 4) mice caused minimal acute GvH disease [10% (3/30) and 3% (1/30) deaths before day 30] and moderately severe delayed GvH disease. Cells from DBA/1 donors (group 5) caused the most severe GvH disease of all donor strains ($p < 0.01$) with a 7-day MST and death of all recipients by day 12. Moderately severe acute GvH disease resulted in 33% (10/30) mortality by day 30 when cells from BALB/c mice were transplanted (group 6). Mortality from delayed GvH disease was severe following transplants of BALB/c, DBA/2, C57BL/10, and SJL cells. Among the strains that showed significant GvL reactivity (groups 5–9),

cells from SJL and DBA/2 donors caused least severe GvH disease: SJL MST significantly longer than BALB/c ($p < 0.01$) and C57BL/10 ($p < 0.05$); DBA/2 MST significantly longer than BALB/c ($p < 0.01$) and C57BL/10 ($p < 0.01$).

CHALLENGE OF PRIMED *H-2* COMPATIBLE MICE WITH AKR LONG-PASSAGE, ACUTE LYMPHOBLASTIC LEUKEMIA CELLS

AKR, RF, and CBA mice were challenged with 10^5 viable BW5147 blast cells following repeated injections of γ-irradiated spleen cells obtained from AKR mice bearing far-advanced BW5147 leukemia. The survival data are summarized in Table 2. With the techniques employed it was not possible to immunize AKR mice successfully against BW5147; the difference in MST and percent survival at 100 days for unprimed and primed hosts were not significantly different (cf. groups 1 and 2, Table 2). Immunization significantly decreased ($p < 0.01$), but did not abrogate, susceptibility of RF mice to this AKR long-passage leukemia (cf. groups 3 and 4). A small proportion of unprimed CBA mice died of leukemia following administration of 10^5 BW5147 blast cells, but this susceptibility disappeared following immunization (cf. groups 5 and 6). From these experiments, it was not possible to state with certainty whether the decreased susceptibility to BW5147 produced in the RF and CBA mice was attributable to immunization against minor histocompatibility or other cell surface antigens, or both, on the normal AKR spleen cells that were present in the irradiated inoculum along with the BW5147 blast cells, or to antigens unique to the malignant cells.

TABLE 2 Survival Data Following i.v. Administration of 10^5 BW5147 AKR Leukemic Blast Cells to Primed and Unprimed, *H-2* Matched Recipients[a]

GROUP NO.	RECIPIENT STRAIN	*H-2*	NO. OF MICE	MST (DAYS)	100-DAY SURVIVAL (%)
1	AKR[b]	k	30	9	0
2	Immunized AKR	k	10	8	0
3	RF[b]	k	30	39	23
4	Immunized RF	k	30	> 100	73
5	CBA[b]	k	30	> 100	83
6	Immunized CBA	k	30	> 100	100

[a]Prior to challenge, immunized AKR, RF, and CBA mice received five or six weekly injections of γ-irradiated BW5147 cells. All deaths were attributable to leukemia.

[b]Data from ref. 4 have been included for comparison.

TABLE 3 Survival Data from Assays of GVL and GVH Reactivity in Leukemic and Nonleukemic AKR Mice[a]

DONOR			GVL REACTIVITY[b]		GVH REACTIVITY[c]			
							Survival (%) at day	
Group No.	Strain	H-2	No. of Mice	Survival (%) at day 60[d]	No. of mice	MST (days)	30	100
1	B10.BR[e]	k	58	5	58	45	76	36
2	Immunized B10.BR	k	15	40	30	16	30	0
3	CBA[f]	k	23	0	59	> 100	85	61
4	Immunized CBA	k	20	20	30	58	93	20
5	RF[e]	k	44	0	60	> 100	98	83
6	Immunized RF	k	25	8	30	59	97	20

[a]GvL reactivity of transplanted immunocompetent cells from primed and unprimed H-2 matched donors was evaluated in immunosuppressed leukemic mice by means of a bioassay. Survival studies were used to evaluate GvH reactivity of the cells in immunosuppressed, nonleukemic AKR mice.

[b]Measured by bioassay of primary recipients' spleen, values are for secondary recipients.

[c]Measured in primary recipients.

[d]No secondary recipients died after day 44.

[e]Data from ref. 4 (using a different bioassay model) have been included for comparison.

[f]Data from ref. 16 have been included for comparison.

GvL AND GvH ASSAYS USING PRIMED H-2 COMPATIBLE DONORS

Summarized in Table 3 are the survival data following GvL and GvH assays using primed B10.BR, CBA, and RF mice as donors of immunocompetent cells. GvL and GvH reactivity of immunocompetent cells from unprimed donors of these strains were assayed previously (4,16): these results also are presented in Table 3. None of the unprimed H-2 matched donors exhibited any appreciable GvL reactivity and caused only mild to moderate GvH disease (groups 1, 3, and 5). Direct comparisons of GvL reactivity were precluded for the unprimed and primed donors because different bioassay systems were used. Nevertheless, cells from primed B10.BR and CBA donors (Groups 2 and 4, Table 3) exhibited significant GvL reactivity ($p < 0.01$), as reflected in the survival data (cf. groups 1 and 2, 3, and 4).

DISCUSSION

LACK OF CORRELATION BETWEEN SEVERITY OF GvH DISEASE AND EFFICACY OF GvL REACTION

Cells capable of causing GvH diseases in nonleukemic AKR mice were not necessarily effective against BW5147 leukemia. As measured by MST, cells from ST/b, AKR.M, DBA/2, and SJL donors (groups 3, 4, 7, and 9, Table 1) caused GvH disease of approximately equal intensity. Yet, cells from ST/b, and AKR.M mice had no or little GvL reactivity, whereas DBA/2 and SJL cells exhibited moderate to high GvL reactivity. Similarly, cells from B10.BR donors that had been primed with AKR leukemic spleen cells (group 2, Table 3) had significantly greater GvH reactivity than did cells from SJL donors (group 9, Table 1), as measured by MST and percent survival at 30 days ($p < 0.01$), yet GvL reactivity was significantly less than SJL ($p < 0.01$). Also, in previous studies we found that cells from C57BR/cd (H-2^k) donors caused acute GvH disease in nonleukemic AKR mice (MST = 30 days and 14% survival at 100 days), yet their cells had no detectable antileukemic effect in a bioassay (3).

Thus, cells from some donor strains caused acute GvH disease but were without detectable GvL reactivity; cells from other strains caused both GvH and GvL reactions. GvH and GvL reactions did not parallel one another, and these experimental results suggest that a donor for adoptive immunotherapy cannot be selected simply on the basis of known GvH reactivity.

If GvH reactive cells were the same population of cells as the GvL reactive cells, a correlation between GvH and GvL reactions would have been observed. Certainly, one of the more plausible explanations for the observed lack of correlation is that there are different subpopulations of effector cells, one reactive against antigens present on normal cells and another reactive against antigens present only on tumor cells. Such an explanation is consistent with the reports of others (15,18). However, our data do not exclude other interpretations. For example, the lack of correlation between GvL and GvH reactivity may have been caused by a differential ability of each strain to recognize normal or leukemic cells, i.e., a difference in the recognitive rather than the effector phase of the cell-mediated immune response.

LACK OF CORRELATION BETWEEN MAJOR HISTOCOMPATIBILITY DIFFERENCES AND EFFICACY OF GvL REACTION

Mice of the AKR.M (H-2^m) and AKR (H-2^k) strains are known to differ only at the D-end of the H-2 region (19); they are identical at the K, IA, IB, IC, S, and G subregions. Like AKR, the congenic strain develops spontaneous leukemia-lymphoma and shows no differences in tumor incidence, tumor types, or tissues and organs involved (22). In the experiments reported here, administration of immunocompetent cells from AKR.M donors to immunosuppressed, leukemic AKR hosts (group 4, Table 1) resulted in GvL reactivity that was not significantly different from that observed when no cells (group 1, Table 1), cells from syngeneic donors (group 2, Table 1) or cells from H-2 matched donors (group 3, Table 1, and ref. 3) were tested. Thus, the mere presence of known disparity at the major histocompatibility complex, or at least at the D-end of the H-2 locus, was insufficient to guarantee significant GvL reactivity against this long-passage leukemia.

UNSUITABILITY OF UNPRIMED OR PRIMED H-2 MATCHED MICE AS DONORS FOR ADOPTIVE IMMUNOTHERAPY AGAINST BW5147

In previous experiments, we found that six unprimed donor strains that were phenotypically identical with AKR at H-2 had no detectable GvL reactivity against BW5147 AKR leukemia (3). This observation held true in the present experiments; cells from ST/b (H-2^k) donors had no detectable GvL reactivity (group 3, Table 1). Other workers (13,17) have shown considerable adoptive immunotherapeutic activity of immunocompetent cells from primed histocompatible donors in different experimental tumor systems. Therefore, three H-2 identical donor strains (RF, CBA, and B10.BR) were primed with γ-irradiated BW5147

cells. In each instance, immunocompetent cells from the primed donors exhibited some level of GvL reactivity (Groups 2, 4, 6, Table 3). However, in each instance, cells from primed donors significantly increased the severity of GvH disease when compared with cells from their respective unprimed donors ($p < 0.01$). In the case of primed B10.BR donors (group 2, Table 3), the severity of GvH disease was significantly greater ($p < 0.01$) than that produced, for example, by SJL (group 9, Table 1), but the GvL reactivity was significantly less ($p < 0.01$). The increased severity of GvH disease produced by the immunization procedure clearly overshadowed the increase in GvL reactivity. Thus, primed histocompatible donors were less desirable as donors of cells for adoptive immunotherapy of BW5147 than were several unprimed mismatched donor strains. Immunization of histocompatible donors with purified tumor-specific antigens might significantly improve GvL reactivity without increasing GvH reactivity.

SELECTION OF "BEST" DONOR FOR ADOPTIVE IMMUNOTHERAPY OF BW5147

Certain strains from the panel of potential donors were manifestly unsatisfactory for adoptive immunotherapy of BW5147: Syngeneic (group 2, Table 1), because of no GvL reactivity; unprimed H-2 matched (group 3, Table 1, and groups 1, 3, and 5, Table 3), because of no GvL reactivity; primed H-2 matched (groups 2, 4, and 6, Table 3), because of high GvH and low GvL reactivity; and H-2 mismatched congeneic (group 4, Table 1), because of no significant GvL reactivity. Among the remaining H-2 mismatched donors, DBA/1 (group 5, Table 1) had high GvL reactivity, but in the cell doses employed caused hyperacute GvH disease. It is possible that a smaller number of DBA/1 cells would still provide high GvL reactivity with acceptably low GvH activity. This seems unlikely to us because of the intensity of the GvH reaction; we believe that mice of the DBA-1 strain would be a poor donor choice for adoptive immunotherapy of AKR leukemia. Cells from BALB/c and C57BL/10 donors (groups 6 and 8, Table 1) had high GvL reactivity, but they caused GvH disease that was significantly greater than that caused by cells from DBA/2 or SJL donors (groups 7 and 9, Table 1). Among the histoincompatible donor strains, cells from DBA/2 and SJL donors caused the least intense GvH disease as measured by MST; however, SJL had significantly greater GvL reactivity than did DBA/2 ($p < 0.05$). Therefore, it appears that among the donor strains tested and within the constraints of the experimental procedures used that mice of the SJL strain most closely approximated the ideal donors for adoptive immunotherapy.

SUMMARY

The optimal donor for adoptive immunotherapy would exhibit great antitumor reactivity and no antihost reactivity. Immunocompetent cells from 11 strains of mice were tested *in vivo* for their reactivity against a long-passage AKR acute lymphoblastic leukemia and against immunosuppressed nonleukemic AKR mice. Donor mice were syngeneic, unprimed *H-2* compatible, primed *H-2* compatible, congenic, or *H-2* incompatible with AKR. Bioassays were used to evaluate the relative graft-vs-leukemia (GvL) reactivity and the relative graft-vs-host (GvH) reactivity of transplanted bone marrow and lymph-node cells from the panel of donors.

No significant GvL reactivity was found when cells from syngeneic, unprimed *H-2* compatible, or congenic donors were tested. *H-2* compatible donors that were immunized with γ-irradiated AKR leukemic spleen cells showed modest GvL reactivity, but associated with the immunization was a disproportionate increase in acute and delayed GvH mortality. Among the *H-2* mismatched donors, mice of the SJL strain appeared to most closely approach the ideal because of least intense GvH reactivity and maximal GvL reactivity.

As measured in these experiments (1) there was no correlation between the severity of GvH disease and the efficacy of the GvL reaction; (2) GvL reactivity in unprimed donors was always associated with *H-2* incompatibility; (3) disparity between donor and recipient at *H-2* did not guarantee an effective GvL reaction; and (4) the increase in GvL reactivity obtained by immunizing *H-2* compatible donors was overshadowed by the increase in GvH disease.

ACKNOWLEDGMENTS

This research was supported by The Leukemia Research Foundation, Inc., the Patrick and Anna M. Cudahy Fund, the Briggs and Stratton Corporation Foundation, Inc., the Margaret and Fred Loock Foundation, and the Board of Trustees, Mount Sinai Medical Center, Milwaukee, Wisconsin. We gratefully acknowledge the technical assistance of Gail Abendroth, Susan Goelzer, and Evangeline Reynolds.

REFERENCES

(1) Boranić, M. Transient graft versus host reaction in the treatment of leukemia in mice. *J. Natl. Cancer Inst.*, *4*:421, 1968.

(2) Bortin, M. M., Rimm, A. A., Rodey, G. E., Giller, R. H., and Saltzstein, E. C. Prolonged survival in long-passage AKR leukemia using chemotherapy, radiotherapy and adoptive immunotherapy. *Cancer Res.*, *34*:1851, 1974.

(3) Bortin, M. M., Rimm, A. A., Rose, W. C., and Saltzstein, E. C. Graft versus leukemia. V. Absence of antileukemic effect using allogeneic H-2 identical immunocompetent cells. *Transplantation*, *18*:280, 1974.

(4) Bortin, M. M., Rimm, A. A., and Saltzstein, E. C. Graft versus leukemia: Quantification of adoptive immunotherapy in murine leukemia. *Science*, *179*:811, 1973.

(5) Bortin, M. M., Rimm, A. A., Saltzstein, E. C., and Rodey, G. E. Graft versus leukemia. III. Apparent independent antihost and antileukemic activity of transplanted immunocompetent cells. *Transplantation*, 16:182, 1973.

(6) Bortin, M. M., Rose, W. C., Truitt, R. L., Rimm, A. A., Saltzstein, E. C., and Rodey, G. E. Graft versus leukemia. VI. Adoptive immunotherapy in combination with chemoradiotherapy for spontaneous leukemia-lymphoma in AKR mice. *J. Natl. Cancer Inst.*, 55:1227, 1975.

(7) Bortin, M. M., and Saltzstein, E. C. A modified technique for collection and processing of mouse fetal hematopoietic tissue for production of radiation chimeras. *Exp. Hematol.*, *10*:27, 1966.

(8) Bortin, M. M., Truitt, R. L., Rose, W. C., Rimm, A. A.,

and Saltzstein, E. C. Adoptive immunotherapy of spontaneous leukemia-lymphoma in AKR mice. In *Proceedings of the 7th International Congress of the Reticuloendothelial Society. (In press.)*

(9) Brunner, K. T., Mariel, J., Rudolf, H., and Chapuis, B. Studies of allograft immunity in mice. I. Induction, development and in vitro assay of cellular immunity. *Immunology, 18*:501, 1970.

(10) Cantor, H., and Asofsky, R. Synergy among lymphoid cells mediating the graft-versus-host response. III. Evidence for interaction between two types of thymus derived cells. *J. Exp. Med., 135*:764, 1972.

(11) Cerrotini, J. C., and Brunner, K. T. Cell-mediated cytotoxicity, allograft rejection and tumor immunity. *Adv. Immunol., 18*:67, 1974.

(12) Cleton, F. J., Tan, B. L., Meindersma, T. E., van Rood, J. J., Thomas, P., Mellink, J. H., de Vries, M. J., van Putten, L. M., Kenis, Y., and Tagnon, H. Bone marrow transplantation after total-body irradiation: A case history. *Exp. Hematol., 14*:44, 1967.

(13) Fefer, A. Immunotherapy of primary Moloney sarcoma-virus-induced tumors. *Int. J. Cancer, 5*:327, 1970.

(14) Fekete, E., and Kent, E. Transplantable mouse tumors. *Transplant. Bull., 2*:61, 1955.

(15) Fernandes, G., Yunis, E. J., and Good, R. Depression of cytotoxic T cell subpopulation in mice by hydrocortisone treatment. *Clin. Immunol. Immunopathol., 4*:304, 1975.

(16) Gershon, R. K. A disquisition on suppressor T cells. *Transplant. Rev., 26*:170, 1975.

(17) Glynn, J. P., and Kende, M. Treatment of Moloney virus-induced leukemia with cyclophosphamide and specifically sensitized allogeneic cells. *Cancer Res., 31*:383, 1971.

(18) Kedar, E., and Bonavida, B. Studies on the induction and expression of T cell-mediated immunity. IV. Non-overlapping populations of alloimmune cytotoxic lymphocytes with specificity for tumor-associated antigens and transplantation antigens. *J. Immunol., 115*:1301, 1975.

(19) Klein, J. *Biology of the Mouse Histocompatibility-2 Complex*, p. 211. New York: Springer-Verlag, 1975.

(20) Mage, M. G., and McHugh, L. L. Retention of graft-vs-host activity in nonadherent spleen cells after depletion of cytotoxic activity by incubation on allogeneic target cells. *J. Immunol., 111*:652, 1973.

(21) Mathé, G. Secondary syndrome, a stumbling block in the treatment of leukaemia by whole-body irradiation and transfusion of allogeneic haematopoietic cells. In *Diagnosis and Treatment of Acute Radiation Injury*, page 191. Geneva: World Health Organization, 1961.

(22) Meier, H., Taylor, B. A., Chen, H. W., Heiniger, H. J., Diwan, B. A., and Cherry, M. Host-gene control of murine C-type RNA tumor virus expression and tumorigenesis: Genotypic programming for single and multiple primary tumors. In Severi, L., ed., *Multiple Primary Malignant Tumors*, pp. 569–680. Perugia, Monteluce, Italy: Division of Cancer Research, 1975.

(23) Rouse, B. T., and Wagner, H. The in vivo activity of *in vitro* immunized mouse thymocytes. II. Rejection of skin allografts and graft-vs-host actiity. *J. Immunol., 109*:1282, 1972.

(24) Simonsen, M. Graft versus host reactions, their natural history and applicability as tools of research. *Prog. Allergy, 6*:349, 1962.

(25) Truitt, R. L., Rimm, A. A., Saltzstein, E. C., Rose, W. C., and Bortin, M. M. Graft-versus-leukemia for AKR spontaneous leukemia-lymphoma. *Transplant Proc., 8*:569, 1976.

Index